The Aral Sea Encyclopedia

Igor S. Zonn · Michael H. Glantz ·
Andrey G. Kostianoy · Aleksey N. Kosarev

The Aral Sea Encyclopedia

Springer

Prof. Igor S. Zonn
Engineering Scientific Production Center
for Water Economy
Reclamation & Ecology
Baumanskaya Str. 43/1
Moscow
Russia 105005
igorzonn@mtu-net.ru

Prof. Michael H. Glantz
Consortium for Capacity Building
University of Colorado
Boulder CO 80309
USA
glantz@mail.ucar.edu

Prof. Andrey G. Kostianoy
Russian Academy of Sciences
P.P. Shirshov Inst.
Oceanology
36, Nakhimovsky Pr.
Moscow
Russia 117997
kostianoy@mail.mipt.ru

Prof. Aleksey N. Kosarev
Lomonosov Moscow State University
Faculty of Geography Dept. Oceanology
Vorobievy Gory
Moscow
Russia 119992
akosarev@mail.ru

ISBN 978-3-540-85086-1 ISBN 978-3-540-85088-5 (eBook)

DOI 10.1007/978-3-540-85088-5

Library of Congress Control Number: 2008936627

Cover design: deblik, Berlin

Printed on acid-free paper

9 8 7 6 5 4 3 2 1

springer.com

The Aral crisis is the most dramatic example of the environmental problems with serious socioeconomic consequences facing, directly or indirectly, all states of Central Asia. The crisis related to the drying of the Aral Sea emerged as a result of the agrarian orientation of economics based on development of irrigated farming and growing volumes of consumptive water use for irrigation.

Fourth Conference of Ministers
"Environment for Europe"
Central Asian States:
Environmental Assessment,
Denmark, Orkus, June 1998

Once upon a time the sea was here
Near the steep slope.
The Aral fishermen enjoyed themselves
Just on its scope.
They were catching fish by fishing tackle,
Lived in peace and concert,
Spent the nights near campfires,
Sang the songs, and never thought
That the sea would disappear here,
And no place would be for them
In the sea expanse.
The Aral went away,
We'll never meet again,
The only thing which left
It is its name...

Olga Krestovskaya
Pupil of the 6th class
Aralsk, 1998

It takes all our strength and resolution not to leave things that will make our future generations feel shameful.

Saigo Takamori,
last samurai of Japan

Contents

Introduction

The "Aral Sea Encyclopedia" is the first one in the new series of encyclopedias about the seas of the former Soviet Union.

Preparing it we faced certain difficulties. The thing is that this encyclopedia is a *monument to the sea* that is disappearing during our lifetime.

The world community considers the situation with the Aral Sea and all changes that occurred in its whereabouts in the recent decades as one of the most serious, if not disastrous anthropogenic environmental crises of the 20th century.

Before 1960, this was a water-abundant sea-lake that was fourth among world lakes after the Caspian Sea (USSR, Iran), the Great Lakes (USA, Canada) and Victoria Lake (Africa). This was a real "pearl" among the sands of the largest deserts, the Karakums and the Kyzylkums. Navigation between the sea ports Muinak and Aralsk and fisheries famous for the Aral breams, barbells, sturgeons, shemaya, and others were developed here. One could find beautiful recreational zones and beaches here. The deltas of the Amudarya, the major river of Central Asia, and the Syrdarya bringing their waters into the Aral Sea were famous for their biodiversity, fishery, muskrat rearing, reed production. The local population found occupations related to the water infrastructure.

However, the development of wide-scale irrigated farming in an attempt to create cotton independence for the former Soviet Union demanded regulation of the Amudarya and Syrdarya flows and construction of water intake structures there. With the expansion of irrigated lands, the water inflow into the Aral Sea diminished and the process of its drying and salinization was set in motion. This led to a practically complete degradation of the historically established ecosystem and, as a result, to the socioeconomic crisis in the whole Circum-Aral area.

By the mid-1980s, the Aral crisis was acknowledged by the whole world and became one of the most significant environmental protection issues. The Aral problem is not global, but nevertheless it stirs global interest. For many years, it was used by various interested parties to stress how quickly human activities may cause degradation of vast expanses on our planet.

I.S. Zonn et al., *The Aral Sea Encyclopedia*, DOI 10.1007/978-3-540-85088-5_1,

The Caucasus and Central Asia

Fig. 1 The map of Asia (http://www.lib.utexas.edu/maps/asia.html)

Former US Vice President and 2007 Nobel Prizewinner Al Gore, who visited the Aral Sea during the period of its drying, wrote that more often many people define their nationality using ecological rather political terms. Thus, the Aral Sea region was populated by the people from some former Soviet republics affected by the regional environmental disaster of the Aral Sea (Al Gore, Earth in the Balance, Ecology and the Human Spirit, 1992).

Today much of the geographical and hydrographic "infrastructure" of the Aral has been lost, and, unfortunately, we have to write about this in the past time. This loss includes islands, bays, capes, arms, and straits. Of course, their contours are changing, and now they are not found among the waves of a blue sea, but in the "sea" of the stiffened, sandy waves of the world's youngest desert – Aralkums. And today, the Aral really turns into a "glass of water" as A.I. Butakov, who studied this sea, wrote in the mid-19th century (although in Butakov's time this "glass of water" was rather full).

The Aralkum "sea" is a museum in the open air. Its main exhibits include remnants of ships that not long ago sailed over the real sea but have now turned into rusty metal hulks, replaced by the live "ships of the desert" – camels.

As is known, today the Aral Sea is shared by two independent states, the Republic of Uzbekistan and the Republic of Kazakhstan. They share the suffering of all of the consequences of the Aral Sea drying. But the same

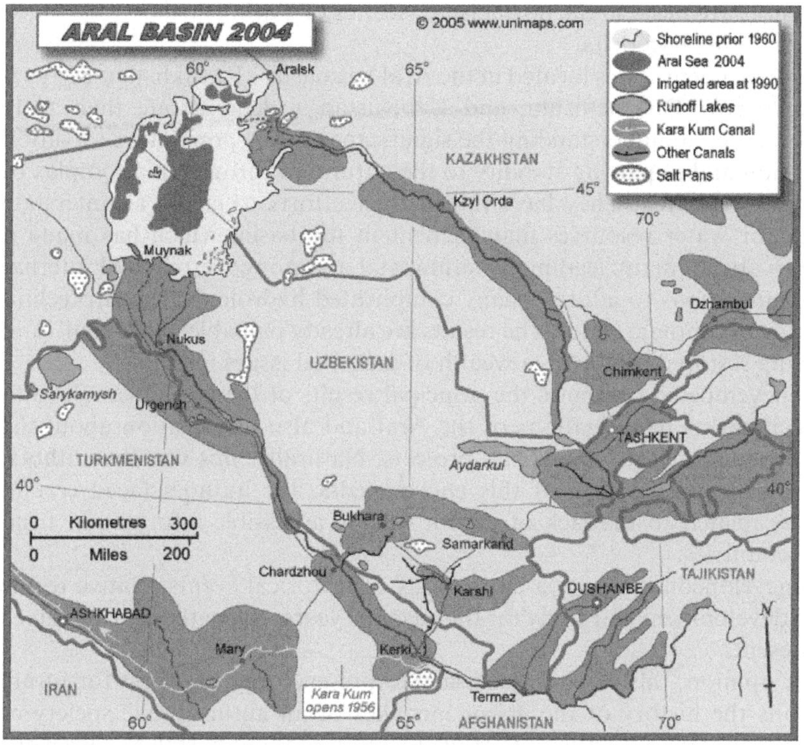

Fig. 2 The map of the Aral Sea Basin (http://unimaps.com/aral-sea/aral-pic.gif)

Fig. 3 The remnants of ships in the Aralkums desert

consequences are faced in the northern territories of Turkmenistan, too, which border the Amudarya delta.

Five independent states located in the Aral Sea basin – Kazakhstan, Kyrgyzstan, Tajikistan, Turkmenistan, and Uzbekistan – demonstrate their truly Oriental wisdom in understanding the significance of the population salvation in this region and imparting stability to the natural-anthropogenic complex of the Circum-Aral area. They have rallied their efforts to create an interstate authority for water resources management in the basin, which has made it possible to attract many leading governmental and nongovernmental international organizations to address many complicated hydrological, hydrotechnical, and socioeconomic issues. The results are already palpable – the Small Aral Sea is being restored; however, a wealth of unsettled issues remain.

This encyclopedia combines the principal results of the fundamental, so to say "benchmark," investigations of the Aral and also information about the leading international programs and projects. Naturally, this was the authors' choice. During preparation of this encyclopedia, the authors faced certain difficulties related to the lack of or not readily accessible information from the Aral countries.

The encyclopedia includes a chronology of historical events relative to the Aral Sea development and study for the past 300 years – from the time of Peter I to the present.

In our opinion, this work is necessary to preserve and highlight for future generations the history of the major mistakes of an authoritative society of "nature conquerors" and attempts at rectification of those mistakes. This work does not claim to be exhaustive in elucidation of the Aral problem. This publication is intended for a wide public – from decision-makers to school pupils and for all those who are interested in the problems of this region – its geography, history, ethnography, economics, and ecology.

We would like to thank Springer-Verlag for the steady interest to the Aral Sea problem, which was initiated by the book by Letolle R. and Mainguet M. "Aral" published in 1993. In 1996 the Proceedings of the NATO Advanced Research Workshop on the Aral Sea Basin, that was held in 1994 in Tashkent, Uzbekistan, were published and till present are cited very often in the scientific publications. The same year Springer published in German the book by Letolle R. and Mainguet M. Der Aralsee (1996). Another interesting book "Sustainable Land Use in Deserts" edited by S.-W. Breckle, V. Veste, W. Wucherer was published by Springer in 2000. In 2005 Springer in association with Praxis Publishing issues "Physical Oceanography of the Dying Aral Sea" by P.O. Zavialov. The present book "The Aral Sea Encyclopedia" continues this very interesting Aral Sea series and starts the new one – "Encyclopedia of the Seas", that will be continued by the following volumes –"The Caspian Sea Encyclopedia" and "The Black Sea Encyclopedia" in 2009. And finally, in 2009 Springer will publish "The Aral Sea Environment" edited by A.G. Kostianoy and A.N. Kosarev.

We acknowledge with many thanks the assistance of Ubbiniyaz Ashirbek Ashirbekov, Director of the Nukus Branch of the Executive Committee of the

International Foundation for Aral Salvation, who supported the idea of this encyclopedia.

We are thankful to Dr. G.S. Kust; Dr. D.Ya. Faschuk (Institute of Geography, Russian Academy of Sciences), Dr. A.N. Urazbaev (Karakalpak Division of the Uzbek Academy of Sciences), Dr. A.G. Tarasov for consultations and graciously provided materials that were helpful for more comprehensive description of research programs and projects. All these people were kind enough to render their assistance and without their participation this publication would not be so comprehensive and accurate.

We are also thankful to our good assistants in Moscow – Tatyana Abakumova for her preparation of the manuscript and Elena Kostianaya (Scientific-Coordination Oceanological Center at the P.P. Shirshov Institute of Oceanology, Russian Academy of Sciences) for her painstaking work on research and selection of information materials, and in Boulder, Colorado, USA – D. Jan Stewart for ceaseless support of all our initiatives, and her goodwill and cooperation.

We sincerely appreciate the excellent copy editing task undertaken by research assistant Gregory Pierce, who works with NCAR's Center for Capacity Building.

We are thankful to Dr. S.M. Shapovalov, head of the Scientific-Coordination Oceanological Center at the P.P. Shirshov Institute of Oceanology, Russian Academy of Sciences, for the support of our activities in preparation and publication of encyclopedias of the Russian seas.

We are thankful to Pavel Kosenko (Russia), who carried out an expedition to the southwestern shores of the Aral Sea in November 2007 and made very professional photos of the Aral Sea environment, which he kindly provided us for illustration of the text (http://pavel-kosenko.livejournal.com).

We are also thankful to Dmitry Soloviev (Marine Hydrophysical Institute, Sevastopol, Ukraine) for his photos of the Aral Sea which he made during the expedition in June 2008 and for a set of satellite images he processed specially for our book.

We are especially thankful to G.V. Ivanov, Manager of "ScanTransRail" (Finland), for support and understanding of the Aral problem's significance for humankind, which enabled publication of this encyclopedia.

Prof. Igor S. Zonn
Prof. Michael H. Glantz
Prof. Andrey G. Kostianoy
Prof. Aleksey N. Kosarev
Moscow, 24 April 2008

A

Actions on radical improvement of the environmental and sanitary situation in the Aral Sea region, more effective management and protection of water and land resources in its basin – Resolution of the CPSU Central Committee and the USSR Council of Ministers No. 1110 of September 19, 1988. It was elaborated on the basis of the report prepared by the Governmental Commission on Environment Condition in the A.S. Basin under the guidance of Yu.A. Izrael. The Resolution cited serious shortcomings in water and land resources management, including the cultivation of newly irrigated lands without appropriate consideration of environmental and social consequences, and provided a brief description of the condition of the natural environment and economy in the region and proposed a complex of actions to improve the environmental equilibrium in the Circum-Aral Area and growth of production forces. It envisaged a growing river inflow into the deltas of the Amudarya and Syrdarya as well as to A.S. in the following amounts: in 1990 – no less than 8.4 cu. km, in 1995 – 11 cu. km, in 2000 – 15-17 cu. km, and by 2005 – up to 20–21 cu. km (in regard to drainage waters). In 1988–2000 projects rehabilitating the irrigation systems on an area of 3.2–3.3 mln ha, constructing and refurbishing the collecting-drainage network on an area of 1.7–1.8 mln ha, and reducing the specific water consumption for irrigation in the A.S. basin by 15% by late 1995 and by 25% in 2000 were proposed. At the same time, projects were proposed to reduce the scale of new cultivation of irrigated lands and from 1991 to suspend the construction of large irrigation systems in the A.S. basin. The urgent construction of water supply projects and improvement of the medical servicing of the population were envisaged. The Resolution also confirmed the need for carrying out research and feasibility studies, including prevention of salt and dust drift from the dried Aral seabed, regulation of the level and water regime of shallow areas of A.S., verification of integrated programs on development of production forces in the Central Asian republics and Kazakhstan, among others. This Resolution played an important role in addressing the Aral problem.

I.S. Zonn et al., *The Aral Sea Encyclopedia*, DOI 10.1007/978-3-540-85088-5_2,

Adjibai Bay* (formerly Ken-Kamysh) – found to the west of the *Rybatsky Bay* (see) and sandwiched between *Muinak Island* (see) and the high, steep mainland bank in the southwest of the Aral Sea. The northern border runs along the line connecting the eastern inlet *Cape Tigrovy khvost* (Tiger's tail) (see) with the Kustau Cape in the west. In the west of the bay between Muinak Island and the mainland southward of the island is *Muinak Bay* (see). The depth in the bay's central part was 6 to 8 m and tends to gradually decrease towards the banks. In the south of A.B. the Urginsky path rambles through the reed thickets. In 1985, the bay dried out completely and an artificially regulated water body was created in its place. It is fed by Amudarya waters via the *Mezhdurechensky reservoir* (see) and the *Kazakhdarya flow duct* (see).

Agreement among the Kazakhstan Republic, Kyrgyzstan Republic, Uzbekistan Republic, Tajikistan Republic, and Turkmenistan on cooperation in joint management of utilization and protection of interstate water resources – historic agreement signed by the heads of water management organizations duly authorized for entering into negotiations on behalf of the governments of the 5 new Central Asian states, on February 18, 1992 in Almaty, Kazakhstan. The Agreement comprises a preamble and 15 articles. Among other things the Preamble states:

> … guided by the need for coordinated and organized settlement of issues of joint management of interstate water resources and with a view of further pursuance of the coordinated policies in the interests of economic development and improvement of the life standard of the population;
>
> proceeding from the historical unity of the peoples living on the territories of these states, their equal rights and responsibility for ensuring the rational management and protection of water resources;
>
> acknowledging the uninterrupted dependence and interrelation of the interests of all states in addressing the issues of joint management of water resources on the common for the whole region principles and just regulation of their consumption;
>
> showing respect of the established structure and principles of distribution and based on the acting regulatory documents on distribution of water resources of interstate sources the aforementioned Agreement was signed.

Pursuant to Article 7 of this Agreement, the parties decided to create on a parity basis the Interstate Coordination Water Management Commission (ICWC) for regulation and rational management and protection of interstate water resources and include into its membership the heads of water management organizations who envisioned convening quarterly and, if necessary, on the initiative of the parties by turn in each country.

In December 1992 in Tashkent the regulations of the ICWC were signed.

The adopted Agreement was ratified by the Government of Kazakhstan on February 29, of Uzbekistan on March 4, of Tajikistan on March 12, of Kyrgyzstan on April 2, and of Turkmenistan on April 20, 1992.

---

* Because many of geographical objects, flora and fauna have changed or disappeared in the Aral Sea, hereinafter asterix means that the description of the term is given for the state in the early 1960s

Agreement on joint actions to settle the Aral and circum-Aral area problems, improve environmental conditions, and ensure the socioeconomic development of the Aral region – Agreement signed on March 26, 1993 in Kyzyl-Orda by the Presidents of five Central Asian states. It confirmed the resolution of five states to further cooperation in management of water resources in the basin. Within the framework of this agreement, several regional organizations responsible for integrated management of water resources were established: the Interstate Council for the Aral Sea Problem (ICAS), the highest ranking body in charge of elaboration of recommendations to the five states on behalf of the basin in general; the Executive Committee and Secretariat of the ICAS; and the International Fund for saving the Aral Sea (IFAS), the highest ranking body in charge of financial support of ICAS activities.

Agurme Peninsula* – located in the middle of the eastern coast of the Aral Sea. It extends from the north to the south for nearly 16.5 km to the left of the entrance into the *Bozkol Bay* (see). The peninsula is low-lying, sandy, and its shallow rugged banks are overgrown with reeds.

Aiderly Cape* – the eastern inlet cape in *Shevchenko Bay* (see), it protrudes far into the sea to the south. It consists of flat, elevated terrain that dips steeply into the bay. The eastern coast of the cape is gently sloping, while the western coast is steep.

Aijarym Island* – located in the eastern part of A.S. Together with *Tasty Island* (see) it lies to the south of *Bozkol Bay* (see). The island was low-lying and was surrounded by shallow waters with depths of less than 1 m.

Aitek-Aral Island* – occupies the northern part of the Ushkol Bay in the north of A.S. It divides the entrance into two straits: eastern and western. The eastern strait is shallow, while the western strait has depths from 1.2 to 1.4 m. The island does not have very high cliffs and is practically devoid of vegetation. A sand bar runs from the northeastern tip to the northwest obstructing the entrance into the Ushkol Bay.

Akbasat Bay* – protrudes into the A.S. eastern coast and makes up the eastern shallow part of the *Kashkynsu Bay* (see). It has several low-lying sandy islands; its depths are not more than 1 m. This bay is the easternmost part of A.S.

Akbasty Island* – located in the east of A.S., 18 km to the south-southeast of *Kaskakulan Island* (see). The island is low, covered with thin shrub vegetation; its shallow waters are overgrown in places with reeds.

Akbidaik Bay* – protrudes into the western coast of *Butakova Bay* (see); the bays are linked with a strait about 1.3 km wide. The strait is very shallow – about 2 m. The commercial fishing industry once located in Akespe on the northern cape restricted the entrance into the bay.

Akchadarya delta of the Amudarya river* – located to the east of its modern delta on its right bank. In the 9th to 2nd centuries B.C., this area was covered by vast tugai wetlands that received 5 to 10 km^3 of water a year. In the 2nd century B.C., artificial irrigation was practiced in the delta. At this time, the delta's marshes were lost, and as a result the Amudarya river rushed to the Aral Sea and drained the valley.

The delta was formed in the Late Pleistocene and Holocene and consists of two parts: the southern, located to the south of Sultanuizdag ridge, and the northern that stretches southward and eastward of Beltau. Its surface gradually merges with the modern Amudarya delta in the west and *Zhanadarya* (see) in the northeast. The southern and northern deltas are linked via the Akchadarya corridor separating the Western and Central Kyzylkum. The southern A.D. is broken up by large and small river channels up to 5–10 m deep. Some of them were used in the past and are used at present as irrigation canals, while their greater part is deflated and filled with sand. These sands cover large areas here. All these river channels join together in the Akchadarya corridor, the width of which is no more than 2–4 km. To the north of this corridor the river channels become fan-like, irrigating a vast territory of the northern delta. Flat takyr surfaces with relic uplands composed of parent rocks and separate sand massifs prevail here. Similar relic highlands are found in the southern delta, too, and are located mostly to the north and east of the delta at a height of 40–80 m. The southern delta was used for irrigated farming, but as the Amudarya waters have stopped flowing it has dried out for the most part.

Akchadarya Lowland – located to the east and northeast of the Sultanuizdag mountain ridge. In ancient times only one of the Amudarya arms, the Akchadarya, flowed into this area. It ran around Sultanuizdag on the southeast, gradually forming a vast delta. Later on the Akchadarya delta deposits were diminished to a great extent and the delta became a sandy desert.

Akdarya – one of the Amudarya branches. In the early 1980s, waters from the Amudarya flowed via it into A.S near the Uoredobay settlement.

Akhmeta Island* – located to the north of the A.S. (Small) in the northeastern part of the *Greater Sarychaganak Bay* (see) directly before *Aralsk City* (see). It covers the *Aralsk Bay* (see) to the southeast.

Akkala, Cape* – located in the south of A.S., separates the Djiltyrbas and *Adjibai* (see) bays.

Akkol, Bay* – located 20 km to the south-southeast of the Syrdarya mouth and encroaches the land eastward for 10 km. The entrance into the Bay is bound on the north by the sandy *Kosaral Peninsula* (see), the northern part of which accommodates the settlement of Karateren where fishermen live. An underwater bar overgrown with reed and rush runs from this peninsula for 2.6 km in the south-southeast direction. To the south, the entrance into the Bay is bound by the flat sandy Karashokat Cape. Nearly the whole water area is overgrown. The depth here is approximately 3 m.

Akpetkinsky (Karabailinsky) Archipelago* – located in the southeastern part of A.S. It comprises about 230 islands, though other sources claim up to 300.

It takes its name from nearby shoal Ak-Petki and Rusengir Island. Its area is over 2,000 sq. km. The Archipelago was formed in the early 20th century when the Aral Sea, after the level rise, intruded into the Kyzylkum sands for 40–50 km near the ancient delta of *Djanadarya* (see). As a result, many low islands and islets with numerous bays, including *kultuks* (see) and *uzyaks* (see), were formed here, while on the continental coast, many shallow lakes were connected with the sea via uzyaks. The Archipelago coastline configuration is variable because it depends on the sea level. The Archipelago and the whole eastern coast as far as the Kuilyus Bay are covered with reed and rush. As the Aral waters dried out, they were gradually replaced with *collection-drainage waters* (see) which formed an intricate system of lakes on the exposed seabed.

Aksaga Bay* – located in the north of the *Akpetkinsky Archipelago* (see) 9 km to the south-east of the *Kendyrli Island* (see) and extending for about 28 km southwards. On the east it is confined by the *Seleuli Island* (see) and a meandering continental coastline. The bay depth is up to 10 m. The western edge of the bay is sheltered by the *Greater Chushka* (see), Little Chushka, and Kamyshovy islands.

Aktumsuk, Ak-Tumsuk Cape* – located on the western coast of A.S. 40.5 km southwards of the Djidelibulak Cape. The coast near the Cape is high and steep. On the south-east the Cape has two escarpments formed by blocks with a complicated configuration. The Cape has a steep underwater slope. At a distance of 350 m from it the depth reaches 10 m, while at a distance of 4 km, the depth is 40 m (1960). From 1948–1964, a sea observation station was located here, while at present there is a meteorological station.

Fig. 4 Aktumsuk meteorological station. Photo by Dmitry Soloviev, June 2008

Aktykendy Cape* – located 22 km to the south-southwest of the *Baigubekmuryn Cape* (see). It is high and steep, and its far end descends smoothly to the sea. The Cape has a steep underwater slope. At 350 m from it reaches a depth of 10 m, while at a distance of 3.5 km to the east of the Cape a bottom trough 69 m deep runs parallel to the coast.

Aktyubinsk region (Kazakh – *Aktobe oblysy*) – formed on March 10, 1932, it belongs to the Republic of Kazakhstan, located in its western part. Its area is about 300,000 sq. km (about 10% of the Kazakhstan territory). Population: 682,000 (1999). The region includes 12 administrative districts, 7 cities (Alga, Zhem, Kandyagash, Temir, Khromtau, Shalkar, Emba), and 4 urban-type settlements. Center: Aktobe (former Aktyubinsk); population: 278,000. The greater part of A.R. is a flat terrain broken by river valleys. Prevailing altitudes: 100–200 m. The central part of the region is covered with the Mugodjary Mountains (the highest is Greater Baktybai, 656 m). The western part of A.R. is occupied by the Poduralsky Plateau passing in the southwest into the Circum-Caspian Lowland. The Turgai table area is in the northeastern part of A.R. The southern part represents massifs of hummocky sands: *Circum-Aral Karakums* (see), *Greater and Lesser Barsuki* (see) and others. Here the region goes out to the Aral Sea. The following deposits are found here: chromites, iron pyrite, nonferrous metals, phosphate rocks, bauxites, oil, black and brown coal, potassium salts. The climate in A.R. is sharply continental and dry. The average temperature in July in the northwest is +22.5°C and in the southeast +25°C; the average temperature in January is –16°C and –15.5°C, respectively. The precipitation in the north in the center of the region is about 300 mm a year, reducing sharply southwards. The vegetation period varies from 175 days in the northwest to 190 in the southeast. All rivers in A.R. run to the drainless basins of the Caspian Sea and other small lakes. The largest river here is Emba. Among other rivers are tributaries of the Ural – Or and Ilek as well as Irgiz, Uil, Turgai and Sagiz. The rivers are mostly shallow and in summer they dry out. More than 150 lakes are found in A.R. The northwestern part of the region is covered with cereal-wormwood steppes composed of dark chestnut soils. The valleys of the rivers are overgrown with thickets of shrubs; the asp, birch and poplar groves are also found here. The middle and northeastern parts are covered with wormwood-cereal vegetation growing on light-chestnut slightly alkaline soils. In the south wormwood-saltwort deserts composed of brown solonetz soils spread. The main industries developed here are mining, chemical, machine-building, and meat production. Power generation is based here on Karaganda coals. The industry is mostly concentrated in Aktobe. The leading branch of agriculture is rainfed grain farming. In late 1950 over 2 mln ha were cultivated here. The northwestern part of A.R., with its well-developed farming and animal husbandry, specializes in rearing large-horned cattle (meat/milk breeds) and pigs, while in the south mutton-fat and mutton-wool sheep are tended. The Orenburg–Tashkent railroad crosses the region from northwest to southeast, while the Atyrau–Orsk rail line crosses from the southwest to the northeast.

Akushpa Lake – located in the southern part of A.S., its area is 308 sq. km. It makes up part of the wetlands of *Sudochie Lake* (see), covering 70% of their area. The maximum lake length is 20 km, width is 6.5 km, and depth is no more than 1.5 m; the coastline runs for 62 km. In 2000–2001 during a disastrous low-water period, the lake nearly lost its flow and dried out completely. In 2003 after intensive filling the water level in the lake reached 52.5 m abs. elev.

Altai Island (former Uzun-Kair)* – located in the eastern part of the Aral Sea to the north-east of the *Uyaly Island* (see). Viewed somewhat as Uyaly's extension. The island is low and sandy and merges with the mainland line. The banks are covered with dense thickets of reeds.

Altynkol Bay* – lies in the eastern part of the Aral Sea 3.5 km eastwards of the *Karatma Bay* (see). It stretches meridionally for 27 km. The prevailing depths in the bay are 2–4 m, the maximum depths (6–7 m) being found in its central part.

Amudarya, Amu-Darya (*Oxus* (Lat.), *Ox* (ancient Greek name, a changed local name "Vakhshu"), Oke or Okey, Araks (Antique); Djeikhun (Arab) translated as "Wild", "Amudario" (Uzbek) – by watershed and water flow, the largest river in Central Asia. "Amu" – from the city, Amul (Amue, Amu, former Charjou), located on the river; and "Darya" from the Persian, "great full-water river." The Amudarya is mentioned in the "History of Northern Courts" (5th century) and in later publications under the name of "Uhu" and the ancient Persian name, "Veh-rud." Beginning in the 14th–15th centuries, the name Amudarya came into local use. It flows over the territories of Tajikistan, Turkmenistan, and Uzbekistan, though its watershed basin also includes Kyrgyzstan. It originates in Afghanistan at the Vrevsky glacier (altitude: 4900 m) where it is called "Vakhadjir," then it begins flowing as "Vakhandarya." After confluence with the Pamir River, it becomes the "Pyandj." Below the confluence of the Pyandj with the Vakhsh, it is called "Amudarya." The length from the confluence of the Pyandj and Vakhsh Rivers is 1450 km; the total length of the river from the origin of Pyandj is 2574 km. It flows into the Aral Sea from a total watershed area of 465 thou sq. km, of which only the mountainous area (227.8 thou sq. km) generates runoff. The main tributaries are found in the mountainour area of the basin: Gunt, Bartang, Yazgulem, Vanch, Kyzylsu, Kafirnigan, and Surkhandarya. The tributaries join the river in its first 180 km stretch: at the 12th km from the left the Kunduz (Surkhab) River (Afghanistan); at the 38th km from the right, the Kafirnigan River; at the 137th km, the Surkhandarya; and at the 180th km, the Sherabad River. Downstream from the Surkhandarya mouth, the Amudarya runs out to the *Turansky Lowland* (see) and receives water from no other tributaries before reaching the sea (1200 km). In the plains, the Amudarya. flows over the Karakum and Kyzylkum Deserts. In the lower reaches, it forms *the delta* (see) with an area of approximately 9,000 sq. km. Its average many-year flow is evaluated at about 70 cu. km, a number subject to significant variations depending on water abundance in a year. Out of the total flow, 19 cu. km or 24% comes from the territory of Afghanistan. Currently, the flow is almost completely regulated and is withdrawn mostly for irrigation purposes, the main cause of the drying up of the Aral Sea (the water level in the sea dropped from 53 m abs. elev. in 1960 to 29.6 m in 2006).

In its lower reaches, the river enters from the *Tuyamuyun narrow* (see) and then runs for about 260 km over the valley as far as the *Takhiatash settlement* (see). The total length of the river from Tuyamuyun to the Aral Sea is 452 km. The width of the valley reaches several dozen kilometers. Its slopes smoothly merge with the surrounding terrain. The river floodplain widens to 6–10 km and abounds through lakes and meandering arms. During high-water periods this area floods. The riverbed is highly furcated, though in areas of parent rock outcrops (Djumurtau, Kipchak, Takhiatash) it narrows to 0.3–0.4 km. The lower reaches are heavily affected by bank erosion (*deigish*, see).

The basin is sharply broken into a mountainous area, where runoff forms, and a flat area, where runoff spreads. In the past, about 2600 lakes were found in the lower reaches, though at present nearly all of them are dried out due to insufficient flow coming to the delta and a dropping of its base level of erosion. The Amudarya has glacier-rainfall recharge and its water regime is characterized by a high flow in summer and a low flow in winter. The greatest water flows are observed in July–August and the lowest are in January–February. It freezes only in its lower reaches, mostly within the delta area (near Nukus for approximately 4 months).

Water flow begins increasing in March-April, with the March-May period characterized by non-stop rainfall augmenting the general rise of water level, causing sharply pronounced peaks of small duration. In June, July, and August, the water flow is the highest due to glacier and snowfield melting.

Before regulation, the spring flood coincided with the beginning of snow melting and rainfall, while the summer flood was fed with thawing waters from glaciers and permanent snow. The water in A.R. is very turbid, taking first place in Central Asia and among the top rivers in the world by this parameter (its sediment flow is twice as large as that of the Nile). In the peak of the summer flood (June), the river resembles a mud flow running at a speed of 15 km/hour (4.17 m/s).

In the past, the lower reaches of the river were navigable; however, large flow velocities (over 1 m/s) and a great number of shallows, a breakdown of the riverbed into arms up to 1 m deep, and a great quantity of suspended sediments cause difficulties for shipping. To maintain normal conditions, extensive channel-improving and bottom dredging efforts are needed along with releases of about 250 cu.m/s of water to the river mouth (downstream of the last water intakes). In recent years, the water intake from A. in springtime (March–April) for irrigation purposes has grown, so the lower reaches do not receive the water flows necessary for normal shipping.

In the A. basin there were 88 hydraulic structures, of which 36 are water intakes, 341 km of which are canals of interstate significance, and more than 100 of which are hydrological stations, among others.

The Karakum Canal is intensively used for irrigation water supply (at present called the Karakum River or Turkmendarya). Via the Amu-Bukhara and Karshi canals, A. is linked with the piedmont drainless areas of the Zarafshan (378 km long) and Kashkadarya (877 km long). One of the spawning rivers for *bastard sturgeons* (see) is found 1800–2600 km from the *fattening area in the Aral Sea* (see). Such cities as *Urgench, Nukus* (see), and Termez are also found on A.

According to Moslem myths from the late Middle Ages, four of the world's largest rivers have their origin in Edem, flowing from under a crystal dome into the world. These rivers are the Nile, the Tigris, the Euphrates and the Djeikhun (Amudarya).

The Amudarya, in the basin of which such ancient Central Asian states as Khorezm (in the river mouth), Sogdiana, and Bactria (in the middle and upper reaches) were found, has been known from Ancient Times. In the Neogene a powerful predecessor of A. – Pra-Amudarya – flowed through the central part of the Karakum Desert and further westwards to the Caspian Sea. About 70 thousand years ago it turned to the north and, having cut a deep narrow near Tyuya-Muyun, reached the Khorezm trough where a large lake was formed. The enormous quantity of sediments brought here with water gradually deposited in the lake turned it into a flat plain. About 10 thousand years ago, A. flowed westwards and reached the *Sarykamysh Depression* (see) having turned it into a lake. Fresh waters that filled the Sarykamysh partially flowed from it along the *Uzboy* (see) to the Caspian Sea. Deposits were gradually built-up in the river delta and soon its flow to the Sarykamysh became obstructed. About 4 thousand years ago or, according to other sources, about 10 thousand years ago, A. turned to the north and flowed into the huge *Aral Depression* (see) that later on became a lake-sea.

Arab geographers Ibn-Khordabek (about 847), Ibn-Rust (between 903 and 913), Masudi (died in 956), Istakhry (about 951), Ibn-Khaukal (976) evidenced that A. (Djeilhun) flowed into the Aral Sea.

Fig. 5 Amudarya River near Khiva (http://cache.eb.com/eb/image? id = 69539&rendType Id = 4)

Amudarya basin museum – opened in 1993. It is located in a specially constructed building near the Takhiatash dam on the right bank of the Amudarya River. The exposition included materials showing the history of irrigation and hydraulic construction development in the Amudarya basin.

Amudarya navy fleet – a unit of the military department of tsarist Russia. It was created in mid-1888 on the Amudarya River with a view to support transportation during construction of the Trans-Caspian railroad and to safeguard waterways. The base in Chardjui (Charjou, presently Turkmenabad) was under command of the Turkestan military area. In 1897, the flat-bottomed ship "Great Duke" built in Abo (presently Turku) in Finland was brought to Charjou. In 1901, the fleet comprised 6 paddle steamers, 2 steam cutters, and 13 barges. In 1917, it took the side of the Soviet power.

Analogs of the Aral Sea problems – on the globe level, there are some lakes that face problems similar to those of the Aral Sea, in particular drying out due to excessive withdrawal of flow of the rivers feeding them. Among such analogs are lakes Mono, Pyramid, and Salton in the USA, Lobnor Lake in China, Lake Victoria in Africa, Murray-Darling River basins in Australia (analog of the Amudarya-Syrdarya river basins). These analogs are presented in a small booklet, "Brothers in Misery: Analog Problems of the Aral Sea Basin" published in 1997 by *NIC MKVK* (see). It should be noted that in 2003 in Liege (Belgium) at a special *NATO Meeting "Dying and Dead Seas"* (see) it was stressed that the situation observed in the Aral Sea is similar to the problems of the Dead Sea, Balkhash Lake, Kingai-Hu and Ebi-Nur lakes (China), Eyre and Korangamite lakes (Australia), Chad, Quota, Rudolf, Tanganyika, Nyasa lakes (Africa). Of course, the causes, both natural and man-made, of water level fluctuations in these lakes differ, but the consequences are similar to a great extent.

"Anthropogenic degradation of lands in the Aral Sea basin" map – a map at scale 1:2500000 prepared in 1993 by the Institute of Desert Studies of the Turkmen Academy of Sciences. Applying a system of symbols, this map provides characteristics of the following processes: vegetation degradation, deflation, water erosion, irrigated land salinization, land salinization due to a water level drop of the Aral Sea, man-made desertification, pasture waterlogging in the zone of irrigation canals, air pollution in urban areas. All criteria are grouped into three classes: (1) by degree of geosystem degradation – from undisturbed to slightly disturbed; fairly disturbed; from heavily disturbed to complete loss of biological productivity; and (2) by class of geosystem degradation – slight, moderate, and strong; and (3) lands that are practically not used in the economy, such as natural drifting sands, solonchaks, outcrops of parent rocks and territories close by their utilization regime to nature preserves (a strip along the frontier of the former USSR). As there are no data about the "background level"' of geosystems, the first class includes geosystems undisturbed or slightly disturbed. Here the authors proceeded from the fact that practically the whole territory of the Aral Sea basin was affected by anthropogenic activities. These lands are not anthropogenically affected and that is why their degradation was not assessed.

Antipolder – a system of small water bodies created in the dried part of the Amudarya delta. The development of such system was elaborated by *SANIIRI* (see).

Arabian-Aral water transportation route (AAWTR) – a project developed by the Water Problems Institute of the Uzbek Academy of Sciences about joint interstate utilization of a part of the flood flow of the Indus and Ganges rivers for the purposes of the socioeconomic development and environmental improvement of the habitat in arid territories of a group of states of Southern, Western, and Central Asia. Some specific features of AAWTR are as follows: an interstate regulation of the flood flow by large reservoirs, pump water lifts, and high power consumption. AAWTR construction may ensure lower damage from floods of donor-rivers; navigation links among states of the region with access of the Central Asian republics to the sea routes; irrigation and development of arid zones in Beludjistan, Systan, Gerat, Khorasan; employment possibilities for the growing population; harnessing of hydropower resources; environmental enhancement in the Circum-Aral region; and better water quality in the Amudarya River basin. Water sharing between the states of the Iranian Plateau and Central Asia is nearly equal. According to preliminary estimates, the cost of AAWTR construction will be 20–30 bill US Dollars depending on the headwater intake volume. Annual costs (conventionally 10 states) may be as high as 120 mln US dollars. Feasibility studies and feasibility report preparation are to be carried out jointly by all interested states.

Aral – a fundamental monograph prepared by French scientists René Létolle and Monique Mainguet published in 1993 by the French "Springer-Verlag-France" Publishers. The preface to this book was written by Professor N.F. *Glazovsky (see)*. The book includes 357 pages, 120 black–white, and 47 color photographs. Both French scientists visited the Aral more than once as international experts participating in investigations and addressing Aral problems. The book has 8 chapters: Chapter 1 "Introduction", Chap. 2 "Between Europe and Asia: Geography and Geology of the Aral Sea Basin", Chap. 3 "History of the Aral Region: Crossroads of Civilizations", Chap. 4 "Live Nature, Soils, and Vegetation of Turan: Agriculture, Animal Husbandry and Fishery Development", Chap. 5 "Aral Region Management: Giantism and Fragility", Chap. 6 "Aral Tragedy: Complex of Problems", Chap. 7 "Que faire?", Chap. 8 "Conclusion".

Aral-88 – the first All-Union Integrated Scientific-Publishing Expedition organized on the initiative of editorial boards of "Pamir" and "New World" journals. The expedition was led by writer and journalist G.I. Reznichenko. The expedition conducted investigations in the Amudarya, Syrdarya, and Aral basins. One of the results of this expedition was the book *"Aral Disaster"* (see) by G.I. Reznichenko published in 1992.

Aral and its problems – a special heading that appeared in 1999 in the Journal, *"Problems of Desert Development"* (see) (Ashgabat, Turkmenistan), that publishes articles devoted to investigations of the Aral Sea.

Aral Asp (*Aspius aspius ibboides*)* – anadromous fish of the carp family (*Cyprinidae*). The body length is up to 80 cm and weight up to 5 kg. Most often found in the Aral Sea basin. In spring (February–March) and autumn (October–December) the fish migrates into rivers. It reaches maturity in its 4th–5th year. Spawning: early spring. A predatory fish, it feeds on sand smelt, roach, sichel, and white-eye. Commercial fishing tools used are nets and shore seines.

Fig. 6 Letolle R., Mainguet M., 1993. Aral. France: Springer-Verlag

Aral Barbell (***Barbus brachycephalus***)* – the migratory fish of the carp family (*Cyprinidae*). Its length is up to 1 m and weight up to 20 kg. It is fattening in the open sea. The maturity is reached in the 5th–6th year of age. It migrates to the Amydarya and Syrdarya 10–12 months before spawning. It usually selects the grounds for spawning behind sand bars with solid ground where the water current is not high and a depth is 1 to 2 m (the spawning peaks during the highest water level in a river). The eggs are pelagic, reaching 4.6–6.8 mm in diameter, and fecundity is high (193–540 thousand eggs). After laying eggs, the fish return back into the sea and feed mostly on mollusks there. The greater part stay in the river for no more than one year where they eat profusely and then migrate to the sea. Most often found in the Aral Sea basin and in the Chu River. Valuable commercial fish. Prior to 1960, up to 20 thou tons were produced here. In the sea it is caught with shore seines and in the river it is caught with drift nets. The monograph "Aral Barbell," written by L.P. Pavlovskaya, was published in 1975.

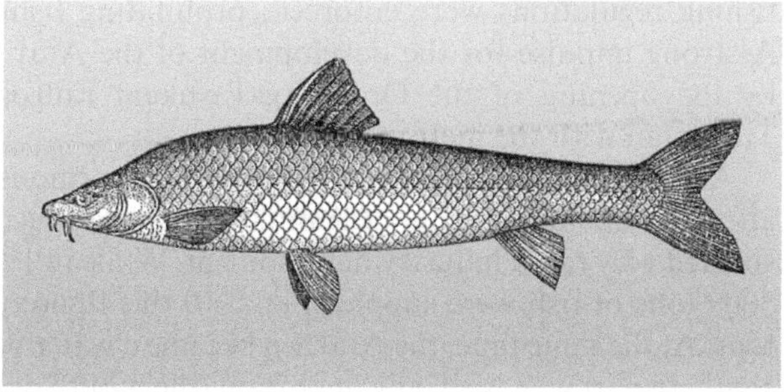

Fig. 7 Aral barbell (*Barbus brachycephalus*)

Aral bastard sturgeon (***Acipenser nudiventris***)* – the only representative of the sturgeon family (*Acipenseridae*) in the Aral Sea. Apart from the Aral it is found also in the Black, Azov, and Caspian seas as well as in the Balkhash Lake where it resettled from the Aral in 1933. Young bastard sturgeons that were hatched in the Balkhash basin are already found in the Ily River. The A.B.S. are mostly 12 to 21 years in age. They lay eggs in rivers with stony beds in April at water temperature 10–15°C. The sturgeon may hibernate in a river till next spawning. They are referred to as early-run fish. Most of the eggs are eaten by the barbells and other fish. From 1936, a high death rate was recorded among the sturgeons due to suffocation caused by attacks of parasite *Nitschia sturionis* that, most probably, was brought here from the Aral together with the starred sturgeon. The catches of the bastard sturgeon dropped sharply from 3497 quintals in 1933 and 6209 quintals in 1936 to 417 quintals in 1937.

Aral beautiful fish – fishing of sturgeon in the A.S. was started by the Urals Cossacks who in 1875 were expelled from the Ural Cossack Troops and deported to the Turkestan Territory for opposing the new code that stopped the election of chieftains. Most of 2500 Cossacks with their families were included into the Kazalinsky military workers battalion. At first, fishing was strictly local and fish was caught for eating. Later, fishing of the Aral *ship* (bastard sturgeon) (see) reached commercial scales and here the fishing of Asian *shovel-nosed sturgeons* (see) was widespread locally. Because of their "long" tails, the aboriginal population called them mouse- and snake-tailed fish, wizard fish, devil, witch and mirage. It was not only a bad omen to eat such fish, but also to catch or even cast a glance at it. In their regulations, the Shi'ah Moslems prohibit eating sturgeon after its descaling, while other theological schools refer to this fish as unclean food. That is how food restrictions of the Islam played a key role in protection of the sturgeons in the Aral basin before Central Asia was included into the sphere of the Russian interests. According to incomplete data, in 1885 the catch of bastard sturgeons was 38500 specimens. The fish was delivered via caravan routes to Orenburg, Tashkent, and Merv, while fishery producers – the brothers Vanyushins – brought it to the Caspian Gulf Mertvyi Kultuk. After establishment in 1892 of the Turkestan Farming and State Property Department at the Ministry of Farming and State Property, fishing regulations were enforced, prohibiting fishing locations and periods. A strong impulse for the development of the Aral fishery industry was given by the opening of the Orenburg-Tashkent railroad. In 1905, it connected Turkestan with the metropolis.

A possibility to supply Aral fish to the enormous state market immediately attracted the investments of Astrakhan fish producers (for the most part) and spurred very rapid industry development. While in 1905 a total of approximately 3000 tons of fish were supplied, in 1910 this figure increased 11-fold to 33,400 tons. At the same time, the Aral Sea became a water body of state significance. Its southern part was especially important (the mouth of the *Amydarya* (see) because it provided up to 65% of the whole fish output.

The Aral bastard sturgeon migrated to the Syrdarya from mid-April to late August. The greatest catches were in summer, from June through mid-September. The Aral fishermen used such fishing tools as drag seines; stationary, drift and racing nets; akhans; fuke hoops; trotlines; baiting lines; karmak; and trandada. The aboriginal population used such tools as "*kazy*", "*syuzeke*" (landing net) and "*chanishke*" (fish spear). The specific feature of the Aral catches that distinguishes them from the Caspian catches was the dominance of bone fish. At the record high catch in 1908 (1275 tons), the share of bastard sturgeon in the total fish supply was approximately 8%, but in the 1920s–1930s was steadily about 1% of the total catch. The fish producers applied twice for permission to introduce here the starred sturgeon and twice received negative responses from the fishery industry department of the Russian Empire. In 1925, the Soviet Government decided to introduce the commercial fish. In 1934, 300 tons of bastard sturgeon caught here were transported for the first time from Astrakhan. In 1936, a mass death among bastard sturgeons started in the Aral affected by the Caspian *Nitzschia sturionis*, a sturgeon-specific parasite. After 1940, the catches of bastard sturgeon became much less and did not exceed 10 t/year. In 1971, Aral sturgeon fishing ceased to exist completely.

Aral canal* – an artificially deepened waterway connecting the *Aralsk Bay* (see) with the *Greater Sarychaganak Bay* (see). The depths of 3.5–4 m are maintained by periodic dredging. The canal length is 1.6 km, and its width ranges from 60 to 100 m.

Aral-Caspian region, Aral-Caspian basin, Aral-Caspian closed area of internal flow – by its size (4900 thou sq. km), one of the most significant regions in Eurasia. Termed a drainless area because its runoff does not reach the World Ocean but lingers in inland bodies of water, it is second only to the eastern region of Northern Africa in this regard. It is usually divided into two large parts – the Caspian Sea and *the Aral Sea* (see). Natural scientist A. Humboldt, taking into account the common physico-geographical features of the Aral and Caspian seas, linked together both sea-lakes and gave them one common name, the "Aral-Caspian basin." Apart from its drainless nature, the other important feature is the obvious aridity of the region because it is located in desert and semidesert zones. The region covers the following physiographical areas:

circum-Caspian lowland, plains and plateaus of Western and Interior Kazakhstan, Betpak-Dala and Mainykum plateaus, Turan lowland, the Ustyurt and Mangyshlak plateaus, offspurs of the North-Afghanistan Paropamiz (Karabil and Badkhyz low mountains), Messcrian plain, Gorgan and South-Caspian lowlands, Lenkoran lowland, Kura-Araks lowland, Terek-Kuma lowland, and Kuma-Manych depression.

Administratively, the Aral-Caspian region completely covers the territories of the Uzbekistan and Turkmenistan Republics and includes parts of the territories of the Russian Federation, Kazakhstan, Kyrgyzstan, Tajikistan, Azerbaijan as well as small parts of the territories of both the Islamic Republic of Iran and the Republic of Afghanistan.

"Aral" consortium – a Union-Republican consortium formed in December 1990. Its founders are the governments of Uzbekistan, Kazakhstan, Kyrgyzstan,Turkmenistan and Karakalpak Republics, State Concern "Vodstroy," Khorezm, Kyzyl-Orda, and Tashauz Regional Executive Committees. The consortium received financial support from the USSR, republican, and local budgets. Its activity was coordinated and controlled by the State Commission of the USSR Council of Ministers on Emergency Situations. The Statute of the consortium, pursuing a single scientific-technical and investment policy, identified as its key targets the elaboration and implementation of the program on the improvement of the situation and living conditions for the population in the circum-Aral region and also the revival of the Aral Sea.

Aral crisis – a monograph by *N.F. Glazovsky* (see), Doctor of Geography and Corresponding Member of the Russian Academy of Sciences. It was published in 1990. The monograph reviews the history of the Aral problem, provides an overview of the natural environmental state and the economic condition in this region, identifies the causes of such crises, defines likely ways for addressing them, and discusses basic tasks of scientific research.

Aral crisis (historical and geographic prospects) – a collection of articles prepared by the *Research Coordination Center "Aral"* (see) and the N.N. Miklukho-Maklai Institute of Ethnology and Anthropology of the Russian Academy of Sciences. It was published in Moscow in 1991. The materials of this collection mostly deal with the history of water and land resources development in the Aral Sea basin.

Aral depression* – the center of the Circum-Aral area located north of the Turan plain on elevations below 60 m (abs.). It has a complicated morphological structure. In the east and south it joins the Aral Karakum lowland (90–100 m abs.) and Aral Kyzylkum lowland (130–160 m abs.). On the west, it is rimmed by the relatively elevated structural-denudation, the *Ustyurt Plateau* (see), that was formed as a result of recent uplifts of the bottom of the sea that existed here in the Miocene. There is an opinion that the A.D. originated mostly due to exogenous processes such as erosion, salt carryover, and deflation (wind destruction of mountain rocks). About 2 million years ago these factors contributed to shaping its basic relief forms and those of the surrounding

Aral-Sarykamysh depression. Originally, A.D. was much larger than today and represented a system of deep (150–200 m) deflation basins having appeared in the Upper Pliocene. Wind erosion was the main factor in the A.D. formation. The volume of wind drift was 15 thou cu. km. The north of this depression borders on the plains of denudation composed of marine sediments that in some places were partially blown from the surface. In such places Aeolian sand areas were formed. To the northeast, east, and south of the A.D., the differently aged alluvial-delta lowlands of Syrdarya and Amudarya, including the Near-Sarykamysh (50–100 m), Akchadarya (60–100 m), Circum-Aral (60–80 m) lowlands as well as the Sarykamysh (minus 20–55 m) and Assake-Audan (30–100 m) depressions, extend. They have formed from the Late Pliocene until the present. They slope toward the A.D. and gradually disappear under the sea forming its most shallow areas. In the meridional direction, between the Muinak and Kulandy Islands, A.D. is crossed by a tectonic rampart. The Vozrozhdeniya, Lazareva, Komsomolsky, and other islands form its highest parts. Such locations of the basic genetic types of the relief on the terrain around the Aral may be attributed to the strictly asymmetric structure of the depression. The eastern part of the depression is characterized largely by a flat topography with prevailing slightly inclined, slightly terraced shelf plains with dominating slopes making fractions of a degree. Such smoothly sloping plains surround the flat bottoms of the basins. The largest of them covers the central part of the Large Sea, while little ones are found in the north within the Small Sea. The structure of the western part of the Aral Depression is somewhat different. Here, the long and narrow Pre-Ustyurt trough runs between the Muinak-Kulunda tectonic rampart and the eastern chinks of Ustyurt. It is associated with the greatest sea depths. The elevations of the trough bottom in its deepest parts reach 19 to 13 m below sea level. The trough has rather steep sides (over 3–5 degrees) with the steepest slopes becoming cliffs in some places and coming together under the Ustyurt chinks, around the Barsakelmes Island, and nearby some abrasion coasts of the Small Sea made up of the parent rocks.

All circumstances of the A.D.'s appearance – its filling with water and formation of the modern water basin – have been studied, but insufficiently, although the basic stages of their geological history are known. The Aral evolution is connected with geomorphological (formation of the sea basin) and hydrological (its filling with the river flow) processes. Their interactions are complicated. An important role in Aral paleohistory was played by numerous migrations of the Amudarya riverbed and related periodical flooding and drying of the Sarykamysh Depression and the ancient riverbed of Uzboy.

Aral disaster – the problem of the Aral Sea drying out with all its consequences; often mentioned in scientific and popular publications.

Aral disaster – a diary of the expedition "Aral-88" written by well-known journalist G.I. Reznichenko that was published in 1992 in Moscow. The book discusses the issues related to the death of the Aral Sea and deterioration of the environmental situation in the Aral region. The author was the chief of the

Fig. 8 Reznichenko G.I. Aral Disaster (1992)

scientific and journalistic expedition that was carried out in 1988. He tells the readers about the critical situation in this region and calls to take urgent actions to remedy it.

Aral environmental crisis, Aral crisis, Aral disaster, degradation of the circum-Aral natural complex – a serious environmental problem that arose in the 1960s and that is connected with the Aral Sea's drying out. The sea water level in 2007 became 23.7 m lower than 1960 at an elevation of 29.9 m (January 2007). Sea water salinity reached 90–150 g/l (2006). As a result, about 4.5 mln ha of the sea bed became exposed and turned into a vast solonchak desert – *Aralkum* (see). The area of solonchaks has increased from 85 thou ha to 273 thou ha. The groundwater level now drops to 8 m depending on the distance from the sea shore. The ingress into the river beds is up to 10 m. The *desertification processes* (see) are going on; there is a real danger of a merging of the *Ustyurt* (see), *Karakums* (see) and *Kyzylkums* (see). In the Circum-Aral area within a strip 150–200 km wide, climate change and a growing amplitude of annual temperature variations have been recorded. The summers have became hotter and the winter colder. The soil cover has also changed: the area of hydromorphic soils has shrunk from 630 to 80 thou ha. In the lower reaches of the Amudarya and Syrdarya a hundred thousand hectares of pasturelands that once might have boasted great species diversity have dried out. The area of *riparian forests* (see) has been reduced from 1300 to 50 thou ha, along with a 20-fold reduction in reed thickets, from 600 to 30 thou ha. Many valuable plant and animal species (in particular birds and fish) have disappeared. Many bays and lakes around the sea have dried out quickly. The surface area of lakes in the Amudarya delta was reduced from 400 thou ha in the 1960s to 26 thou ha in 2001. Progressing intensively are the deflation of bottom sediments – sand, dust, salt, and their drifts – during strong windstorms at a rate of 0.1 to 2.0 t/ha. The annual volume of their drift is 500 km or so and is as high as 75 mln tons. The width of the area affected by dust-salt storms is 40 km and the length is 400 km. The dust and salt spread with winds and atmospheric precipitation deposition in the *Circum-Aral area* (see) where 3.5 million people now live.

Aral Karakums (Aralmany Karakums) – ancient lowland with absolute elevations 90–100 m and channel-like depressions up to 70–80 m. It extends over the north–east of the Circum-Aral Area to the north of the Syrdarya lower reaches. Its relief was formed on the alluvial-deltaic, mostly sandy deposits of the Late Pliocene with a thickness up to 14–20 m. It was strongly affected by eolian processes. At present, a sandy desert topography is developed here: the sands are characterized by the hillocky and cellular and, to a less degree, the ridge-barkhan relief (the Terentikum, Yesentkum, Buyalykum, Kesekum sands, and others). The depth of its dissection varied from 1.5–3.0 m to 5–10 m. In the low-lying desert there were found some isolated structural-denudation plateaus (Yakhshiklych with elevations up to 150 m and others). A gradual transition of the lowland into the Aral Depression was observed, with altitudes dropping to 68–72 m abs. in the north-east.

Aral Kyzylkums – confined by the Zhanadarya deltaic plain on the north-east, east and south. Its western part located mostly within the Akkyr-Kumkialinsky tectonic saddle separating the East-Aral Depression from the Syrdarya Depression. The plain was formed in the Late Pleistocene-Holocene by the Syrdarya waters. This flat, mostly clay (takyr) plain dropped from elevations 130–135 m in the east nearby the Syrdarya to the level of terraces of the Aral Depression. It was composed of sands and loams up to 20 m thick. The plain was dissected by recent channels of Syrdarya arms, such as Inkardarya, Zhanadarya, Kuvandarya and others traced from the Syrdarya as far as the Aral. Not long ago (in the 1970s) some of them (Zhanadarya and Kuvandarya) were used for periodic releases of waters from the Syrdarya. The massifs of eolian sands and remnants composed of clays with wind-formed sandy ridges were met in some places on the plain. The sandy massifs on the plain were represented either by the remnants separated by deltaic arms from the Kyzylkums, the dissection depth being 7–10 m, or by the eolian Late Pleistocene and Holocene alluvial deposits with dissection of 3–5 m.

A.K. came out to A.S. on the east and was submerged by its waters. This was a sandy low plain of accumulation located in the East-Aral tectonic depression. Its surface sloped smoothly forward with Aral lowering from absolute elevations of 100–130 m to the level of the ancient Aral terrace with absolute altitudes of 58–59 m. It is composed of thick Early and Middle Pleistocene alluvial sands with loam and clay interbeds with total thickness of 20–30 m on average. Affected by wind erosion, the surface sands formed ridges that ran in the direction close to the meridional. Nearly in the same direction the plain was dissected by several takyr and solonchak lowlands up to 1–2 km wide with the channels of the ancient arms of the Syrdarya that once flowed into the Aral. The slopes of sandy ridges were generally fixed with vegetation, while their crests were often bare and subject to eolian processes.

Aral – my sorrow – a book by the well-known Karakalpak writer Orazbay Abdirakhmanov published in 1992 in Nukus. The book recounts in documentary form about the tragedy of the Aral Sea. The author analyzes the causes and consequences of the environmental disaster of the circum-Aral region. The author was awarded the UNESCO Medal for this book.

Aral navy fleet – creation of the Russian fleet. (1) The Aral Navy was created on the initiative of *A. Butakov* (see). The Russian fleet sailed over the Aral Sea and in the Syrdarya and Amudarya rivers from 1852–53 to 1883; the fleet was based in the port Raim (Aralsk) and later in Kazalinsk. The A.N. included two steam ships ordered and constructed in Sweden in the Mutalsky docks. The names of these ships were chosen personally by Russian Emperor Nicholas I in honor of two General-Governors of the Orenburg Area, “Perovsky” and “Obruchev.” Otherwise, the fleet also included transportation ships (the schooners “Konstantin” and “Nicholas,” 2 iron barges, 2 launches, 5 boats, 1 flat-bottomed boat, and 4 yawls) – a total of 16 floating units. In 1880, the fleet had 6 armed ships, the “Aral” and the “Syrdarya” among them, and 9 barges. It ensured security of shipping, conducted hydrographic inventories, and supported the transport of troops to Bukhara (1868) and Khiva (1873) during military campaigns of the Tzar’s troops. It was disbanded in 1883. Part of its property was handed over to the *Amudarya Navy Fleet* (see).

In 1868–1871, Pavel Pavlovich (1834–1871), the grandson of famous seafarer I.F. Kruzenshtern, served in the A.N.

(2) In Soviet times, the A.N. was created again in January 1920 but existed only until 1921. It comprised two motor schooners, two steam schooners, two steamers, and other vessels. During the Civil War (1917–1923), the fleet supported military actions in the Turkestan battlefield against British troops, the White Guard, and *basmaches* (members of the anti-Soviet movement in Central Asia). The fleet carried troops, cargo, and also conducted hydrographic inventories.

Aral postage stamps – postage stamps devoted to the drying of the A.S. and the loss of its biodiversity. The first stamp, depicting vessels among sand dunes on the dried Aral seabed, was issued by the USSR Postal Service in 1991. It had a nominal value 20 kopecks. The top of the stamp read: "Aral – zone of environmental disaster." In 1996, the countries of Central Asia, such as Kazakhstan, Uzbekistan, and Tajikistan, issued a block of five stamps, "Save the Aral Sea." On these stamps, against a map of the drying Aral Sea, depictions of the Amudarya bastard sturgeon (*Pseudoscaphirhynchus kaufmanni*), caracal (*Felis caracal*), hyena (*Hyaena hyaena*), Aral salmon (*Salmo trutta aralensis*), pike-like asp (*Aspiolucius esocinus*) were placed.

Fig. 9 The Aral Sea stamps (http://www.stamp.elcat.kg/images/big/115-119.jpg)

Aral problem and landscapes in the Amudarya delta (Structural-dynamic state of landscapes in the Southern Circum-Aral Area in connection with the Aral problem) – monograph of V.A. Popov (Department of Geography, Uzbek Academy of Sciences) published in 1990. For the first time, the geographical investigations of Central Asia provided a forecast of the landscape dynamics based on the theory of Marcovian chains. This enabled a comprehensive analysis of the dynamics of the landscape borders in the Amudarya delta at the present stage of desertification. Balance 3D models of landscape dynamics and a probabilistic model of landscape transition in the Amudarya delta were constructed. These models were helpful in producing a quantitative (in units of area) forecast of the probability of mutual or directed transition of landscapes that differed by wetness.

Aral region – most often understood as the Aral Sea proper with its surrounding territories, although while considering the environmental problems this notion may also include the whole *Circum-Aral area* (see). Some treat the A.R. as the Aral Sea Watershed Area.

Aral roach (*Rutilus rutilus*)* – commercial fish of the carp family (*Cyprinidae*). It has two varieties: anadromous and dwarf or reed. Its body length is 30 cm. Spawning at a temperature of 1–5°C, both in fresh water and in water with salinity to 11‰, in places with previous-year vegetation. Each female is capable of laying 54–142 thou eggs. In the coastal zone, the adult roach feeds on mollusks and vegetation, mollusks and freshwater shrimps offshore, and vegetation in rivers. In winter, they do not stop feeding. In spring and autumn, they make vertical migrations to catch shrimps that rise to the surface at night, while in the daytime they descend to the bottom. Most widespread in the Aral Sea and in the backwater pools and lakes of the Amudarya and Syrdarya.

Aral salmon (*Salmo trutta aralensis*)* – migratory fish of the salmon (*Salmonidae*) family featuring an oblong head and jaws. The fish is large, reaching up to 1 m in length and weighing 13–14 kg. Its way of life is not sufficiently studied. It is found mostly in rivers of the eastern and southern parts of the Aral Sea, in particular in the Amudarya (up to Turtkul). It has no commercial significance.

"Aral," Scientific-Research Coordinating Center (SRCC) of the State Committee for Science and Technology and the USSR Academy of Sciences – established in 1989 by the Resolution of the USSR Council of Ministers. It was headed by Academician V.M. Kotlyakov, Director of the Institute of Geography of the USSR AS. The main objectives of the Center were the outlining of the basic lines of research on the Aral problem, coordination of research efforts, and information activities. The Center supported the publication *"Aral Crisis"* by N.F. *Glazovsky* (see) and the *collection of articles "Aral Crisis"* (see), among other works. The Center ceased to exist in 1993.

Aral Sea* (Uzbek –"*Orol Dengizi*", Kazakh – "*Aral Tenizi*" meaning "Island sea") – an inland, drainless saline lake (called a sea) located in the sandy deserts of Central Asia and Kazakhstan at an elevation of several dozens meters above the sea level. Its specific geographic location, including complete isolation from the world's oceans, makes the Aral quite a unique natural feature. The Aral combines the properties of a sea and a lake. According to A.I. Butakov, the Aral Sea reminds one of "a spilled glass of water".

A.S. is located on the territory of two states – Uzbekistan and Kazakhstan. The interstate border between them crosses the western coast along parallel 45°24′N and the eastern coast crosses along 44°30′N, dividing the sea into two parts. The northeastern part with the Kug-Aral and Barsakelmes Islands belong to Kazakhstan, while the southwestern part with the Muinak and Vozrozhdeniya Islands belongs to Uzbekistan (Republic of Karakalpakstan).

Before 1960, A.S. was one of the world's greatest lakes. Its surface area was less only than the Caspian Sea, Lake Superior in North America, and Lake Victoria in Africa.

The Aral appeared only 10–20 millenia ago, thanks solely to river flow, the reduction of which is now causing its demise. Inflow of river waters over several thousand years into a desert erosional (Aral) depression, then their subsequent

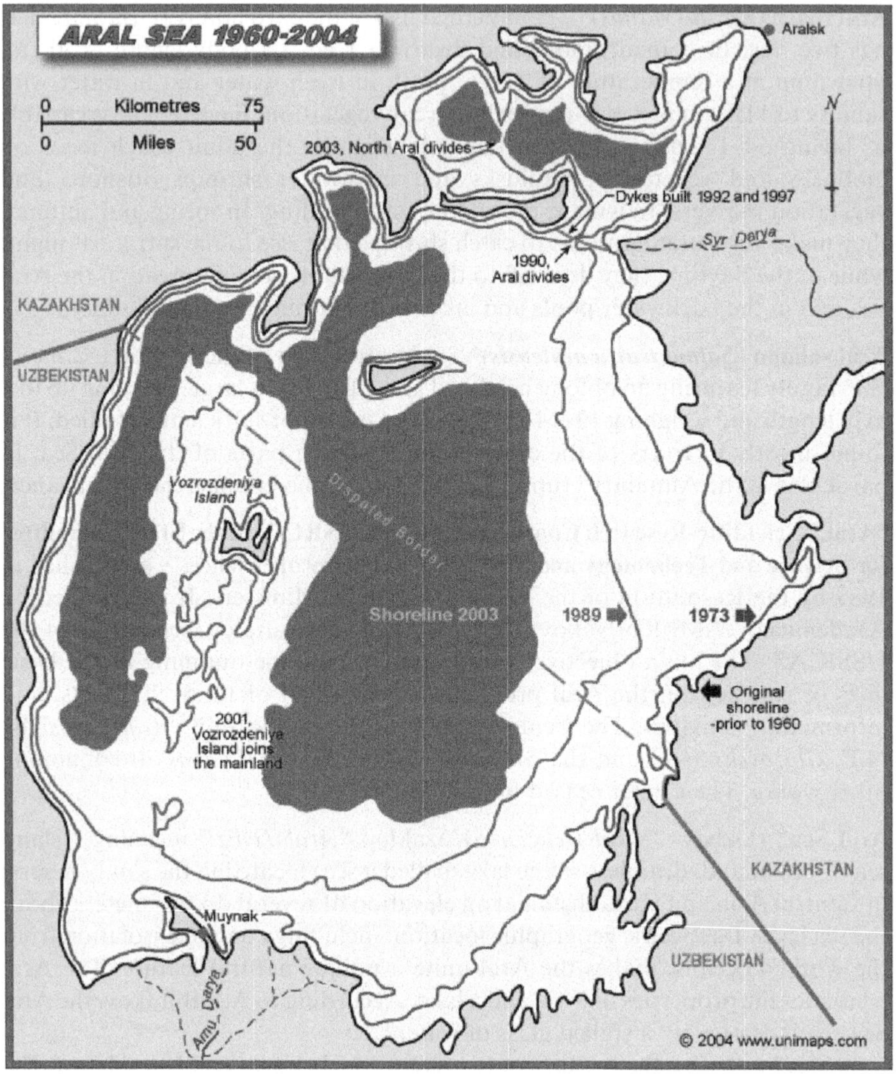

Fig. 10 The Aral Sea desiccation in 1960–2004 (http://unimaps.com/aral-sea/aralmap.gif)

metamorphosis in the course of a water-salt exchange with the atmosphere and groundwater led to a unique formation of the lake-sea water body.

The sea has rather high and steep shores in the west and northwest and low and flat shores in the east and south.

The bottom relief of the shallow A.S. is rather rugged. In its western part, an underwater ridge runs in the meridional direction. In some places, it rises out of the water, forming the *Lazarev* (see), *Vozrozhdeniya* (see), *Komsomolsky* (see) and other islands. Westwards of this ridge along the sea shore was a narrow and deep trough (50–55 m) where the maximum depth of the Aral (69 m) was located.

Fig. 11 The Landsat satellite view on the Aral Sea in 1973 (http://earth.google.com/outreach/images/case_study/unep_fig1a_lg.jpg)

The fact that this rather small sea is located in a zone of extratropical deserts is responsible for a sharply continental climate there. The summer is dry and hot, while the winter is cold with varying weather. The whole vast Circum-Aral area and the sea itself are in the region influenced by the winter Siberian anticyclone and summer South-Asian low. They determine the barometric situation in this region along with other multiple synoptic situations. In autumn and winter, the northwestern winds prevail. The air temperatures in November are negative and in January they average –12 to –14°C, though over the sea the air is warmer.

During warm seasons, the offspur of the Siberian anticyclone dissipates, leaving the Aral and its surrounding territories open to the effect of the South-Asian low. Eeastern winds dominate. The air temperature in spring rises quickly from 5–10°C in March to 20°C in April. In July, for example, the average monthly temperatures are 27–28°C, while the average daily temperature may be as high as 35°C. Over the sea, the air temperature is 3–5 degrees lower than in the nearby land areas.

The Aral water is warmed during spring and summer and cools in autumn and winter. The surface water temperature before strong dessication was rather uniform over the whole sea area and in summer it reached 23–25°C. Due to intense winter cooling, the temperature of the surface water layers becomes lower, while underlying water layers became lower. At the maximal depths it was 2-4°C.

The average annual salinity of the Aral waters at a relatively stable water level (before 1960) varied from 9.6 to 10.3‰. The annual dynamics of water salinity in the sea was rather weak. In winter, it was somewhat higher due to ice formation. In summer, especially in August, when the river flow into the sea was maximum, the waters in the southwestern part of the sea became less saline and there the lowest salt content was recorded – 9.3 to 9.4‰. High salinity (10.4–10.5‰) has been observed in the eastern shallow areas due to high evaporation. Today (2008) salinity in the Large Aral Sea reaches 90–150‰.

One of the specific features of the A.S. water chemistry was a permanently high level of dissolved oxygen. High saturation of the bottom layers with oxygen provided favorable conditions for bottom vegetation development. As a result of photosynthesis, the near-bottom waters became oversaturated with oxygen by 140–150%.

An autumn-winter convection is well discernible in the Aral Sea. It started with the beginning of the surface water cooling, when the average daily air temperature was steadily less than the average daily water temperature, which was observed in the first half of August in the Western Aral and in the second half of August in the Eastern Aral.

Unlike all other seas in the Northern Hemisphere, the Aral Sea, prior to its water level lowering, was characterized by its anticyclonic movement of waters.

Prevailing weak winds in combination with shallow depths generate usually short (up to 1 m) but steep waves. When winds are blowing along the greater axis of the sea, seiches are sometimes generated that usually do not decay for long.

The input part of the sea water balance is made up largely by the inflow of waters from the Amudarya and Syrdarya and also by a small quantity of precipitations falling over the sea surface. The output portion of the balance is composed of high water evaporation from the sea surface.

From the onset of the 20th century and until the 1960s, the water balance of A.S. was constant. The annual inflow of river waters (50–52 cu. km) and precipitation falling on the sea (8–10 cu. km) compensated for the water losses to evaporation (60–62 cu. km/year). But still there were slight fluctuations of the sea level around elevation 53 m over the mean ocean level that was the average multi-year value. The sea lived its natural life that had been formed and existed for centuries, had its contours and, in general, rather stable natural conditions.

Before the 1960s, the A.S. stretched out from the southwest to northeast, from the Barja-Kultuk (*Adjibay Bay* (see)) Bay to the top of the *Little Sarychaganak Bay* (see) – approximately for 430 km. The *Kokaral Island* (see) divided it into the northeastern part called the Small Sea, and the rest was called the Large Sea. Both parts (Large and Small seas) were linked via the Berg Strait which was about 20 km wide and 10–11 m deep. The greatest width of the A.S., from the Tamak cliff (in the northwestern corner of the sea) to the top of the *Sulu Bay* (see), was 292 km. More than 1100 islands were found in the sea, mostly concentrated in the southeast, forming the *Akpetkinsky Archipelago* (see). The largest islands were Kokaral, *Barsakelmes* (see) and Vozrozhdeniya; in 1950, the total area of all islands in the sea was 2235 sq. km. The coastline of the Aral was 4920 km long. The area of its water surface reached 66 thou sq. km, the volume was 1060 cu. km, the average depth was approximately 16 m, and the greatest depth was 69 m.

The water level fluctuations were strictly seasonal: the maximum water rise in summer resulted from the flood flow of the Amudarya and Syrdarya; the drawdown in autumn was associated with evaporation from the water surface that reached its maximum after the river flooding. The minimum water level in winter was due to the very small quantity of water brought by rivers into the sea.

Although A.S. lies in the southern latitudes, every year it is covered with ice. The ice formation usually started in the northern and northeastern coastal regions approximately in the second week of November. The ice cover was greatest in mid-February. The ice cover breakup started in late February and early March.

The level of biogenous substances, especially phosphates, in A.S. is low. The main source of biogenous matter for the sea was the Amudarya and Syrdarya flow because they were fed mostly from high-mountain glaciers and snow. In spite of the favorable conditions, the specific feature about A.S. is the scarcity of its flora and fauna, which is due to the isolation of this water body and peculiarity of its regime. In the Aral, there are not many groups of animals whose populations are found in other inland seas.

Before dessication the phytoplankton of A.S. comprised, by various sources, only 40–70 species, among which the diatomic algae and flagellates were dominant. Zooplankton contained 25 species, but over 70% of its mass was crawfish diaptomus. This restricted utilization of the food resources of this water body. Zoobenthos numbered 48 species, out of which 20 are mostly bivalve mollusks. Biomass of zoobenthos was significant – about 20 g/sq. m on average.

The warm Aral Sea with its low saline waters was habitat to 20 species of fish of fresh-water variety, out of which 12 carp species and 3 perch species were of commercial significance. There were other valuable fish species, such as Aral bastard sturgeon, barbell, pike perch, and sea roach. Muskrats lived in the Amudarya and Syrdarya deltas and nearby territories. The coastal growths – tugai – abounded in waterfowl. The Turanian tiger was also found here. At that time it was like an oasis in a desert.

Many projects were undertaken to improve the fishing activity of the Aral by introducing species of zooplankton, zoobenthos, and fish. In 1940–1950, fries of the Caspian starred sturgeon, bullheads, and Baltic herring were released into the sea and after this zooplankton consumption increased and its standing stock was reduced.

In the Circum-Aral area under a quasistationary regime a specific structure of the economy was established. It was connected with the sea (fishery, muskrat breeding, sea transport). The sea produced a mitigating effect on the climate of nearby territories. Its very existence was favorable for the environmental and socioeconomic situation in the region.

A.S. played a significance role in the economics of the Central Asian republics in Soviet times. It was used as a transport waterway and as a fishery. About 80% of the population living near A.S. was connected with fishing, fish processing, and fish delivery to port *Aralsk* (see) from where it was transported by railroad to all parts of the Soviet Union.

Every year the fishing provided 400–500 thou quintals of high-quality fish. Muskrat rearing gave about 1 mln pelts a year. Sea transport carried 200–250 thou tons of cargo every year.

A.S. was a key factor in formation of the environment in the Central Asian region. Its effect ensured higher air humidity, decreased the recurrence and intensity of dry winds, and led to relatively higher levels of ground water. In addition, the moisture evaporated from the sea surface percolated down into soils and once "irrigated" the desert here. The environmental impact of the sea on the coastal zone was enormous.

Irrigated farming had been practiced and developed for a long time in the basins of the Amudarya and Syrdarya. The irrigated areas were constantly expanding and by 1950 they reached 2.9 mln ha and continued growing. The volumes of water taken from the Amudarya and Syrdarya were also growing. From the onset of the 20th century and upon until 1960, the water withdrawal from them nearly doubled, though the natural regime of the Aral Sea had not changed. This phenomenon may be explained by the fact that the increase in the consumptive water use with the expansion of irrigated lands until 1960 outbalanced the higher water flow in the Amudarya and Syrdarya simultaneous to a decrease in water losses to transpiration of wild vegetation and an increase in irrigation water runoff after construction of a collecting and drainage network.

Stability of the natural regime in the Aral for nearly five decades caused some experts to assert that irrigation development would not decrease inflow into the sea. Beginning in 1961, however, water flow into the sea experienced a rapid level drop along with a reduction of sea surface area and water volume, which in turn affected the salt balance in the sea.

The sea began drying out and shrinking. Its contours changed significantly, particularly in the shallow eastern, southeastern, and southern areas. Thus, such large shallow bays as *Djiltyrbas* (see) on the southern coast, *Bozkon* (see), *Akkol* (see), and others on the eastern shore disappeared completely. All islands of the Akpetkinsky Archipelago turned into mainland, and new islands appeared in place of former shoals. The relatively large Kokaral Island became a peninsula.

By early 1980, the large bays of A.S. – Muinak, Sarybas, Adjibai and others – had dried out. They received water from the Amudarya and due to surge waves they were recharged with sea waters. In the past, these bays were desalinizable water bodies of the Aral. Steady drying out of the Aral created a crisis situation in this vast region, and the negative consequences of the process became felt in different fields. Water recess from the original coast of 15 km and in some places up to 120 km resulted in the complete immobilization of such ports as Aralsk, Muinak, Urga, Uchsai, Taili, Uyaly, and others. All port equipment and facilities were dismounted. The climate in the Circum-Aral area within a zone of some 500 km from the sea became worse. The annual difference of air temperatures had increased by 1.5–2.0°C on average: summers became still hotter and winters colder. The non-frost period became shorter, air humidity became lower, dry winds and dust storms became more frequent and stronger. The sea no longer produced a mitigating effect on the climate of the nearby territories. The lake-marshy and solonchak complexes appeared on the vast expanses of the dried out sea bed. Dust storms originating in the dried out sea bed were recorded for the first time from space. Sulfates and chlorides, once contained in sea water and then deposited on the sea bed, together with the dust, rise high into the atmosphere and are carried away for hundreds of kilometers. As a result, the salinity of fertile soils in the circum-Aral area became higher, their productivity dropped, and the quality of agricultural products declined.

The increased water salinity in the drying sea led to a nearly complete disappearance of once unique fish: barbell, bastard sturgeon, shovel-nosed sturgeon as well as more widespread species, such as asp, catfish, bream, sea roach, and pike perch, among others. The sea became practically lifeless.

The retreat of the sea was accompanied by the disappearance of the coastal vegetation. Rare animals became really "rare." Rare birds such as white herons and flamingo-like birds are nearly all gone.

By the late 1980s, the irrigated lands in the sea basin had nearly doubled compared to 1950. Land development was carried out by extensive methods, first of all, by means of expansion of irrigated lands to the areas remote from river beds and located on higher elevations, which is why the natural increase of water flow in rivers could no longer compensate for water withdrawals. As a result, less and less water reached the Aral. At the same time, consumptive water losses were growing. The water intake by the Karakum Canal that takes 1/4 of the Amudarya flow every year also increased. Quite frequently, considerable amounts of water from the Syrdarya (in water abundant 1969–20 cu. km) were released into the Arnasai depression, and water from the Amudarya were released into the *Sarykamysh depression* (see), turning them into large lakes. The canals along which irrigation water was supplied had no concrete lining, so there were great seepage losses and, as a result, enormous water losses in general. The irrigation practices were poor. The strategy of production facilities located in the whole region was not appropriate in many respects. In particular, it was oriented to monoculture such as cotton growing, to water-intensive industrial enterprises, and the like.

Such a sorrowful future for the Aral Sea was not a surprise. It was foreseen in the early 1960s, but all assessments of the situation were made largely on the basis economic indices. For example, some specialists noted that the sea, if unused and set in the center of a desert, was no more than a giant evaporator. Allegedly, it was more beneficial in economic terms that the waters of the Amudarya and Syrdarya evaporated not from the surface of the Aral, but from irrigated fields. Estimates were made to demonstrate that the losses related to the sea disappearance (loss of fishery, muskrat rearing, transportation) would be paid off hundred times by development of irrigated farming with full utilization of the Amudarya and Syrdarya waters; however, the economic criteria alone were and could not be sufficient basis for basing a decision on the fate of the Aral Sea.

Restoration of the Aral Sea as it had been with its natural regime seems impossible in the foreseeable future. But the sea level drop can be stabilized at the present-day elevation. This, in fact, has been done with the Small Aral in Kazakhstan after construction of a through-flow embankment. This implies, first of all, that an economic use for the water exists if the large quantities lost now to seepage and evaporation from canals can be saved. Considerable water saving may also be attained by utilization of advanced water application practices: sprinkler irrigation, drip irrigation, construction of subsurface irrigation networks, etc. as well as introduction of water recycling and other water saving technologies in industry. It is advisable to end sending water to the local runoff basins, such as the Sarykamysh, Arnasai and Ainarsky depressions where collector-drainage waters are discharged after irrigation.

Further salinization of the Aral Sea may be prevented by ending the disposal of drainage waters from the Amudarya basin into the shallow Aral.

With a view to stabilize the level of A.S. and decrease its salinization effect on the surrounding territories, it has been proposed to make it a water body with a regulated regime. For this purpose, about 10 years ago it was proposed to separate the western deep-water part of the Large Aral from the shallow eastern part with two dams with a regulated spillway. Thus, the whole surface runoff will go to the eastern part of the Large Sea. With accumulation here of some excessive amounts of water (due to the excess of inflow over evaporation) there will be a discharge via spillways into the western part of the Large and Small seas. In this case, A.S. will turn into several regulated water bodies.

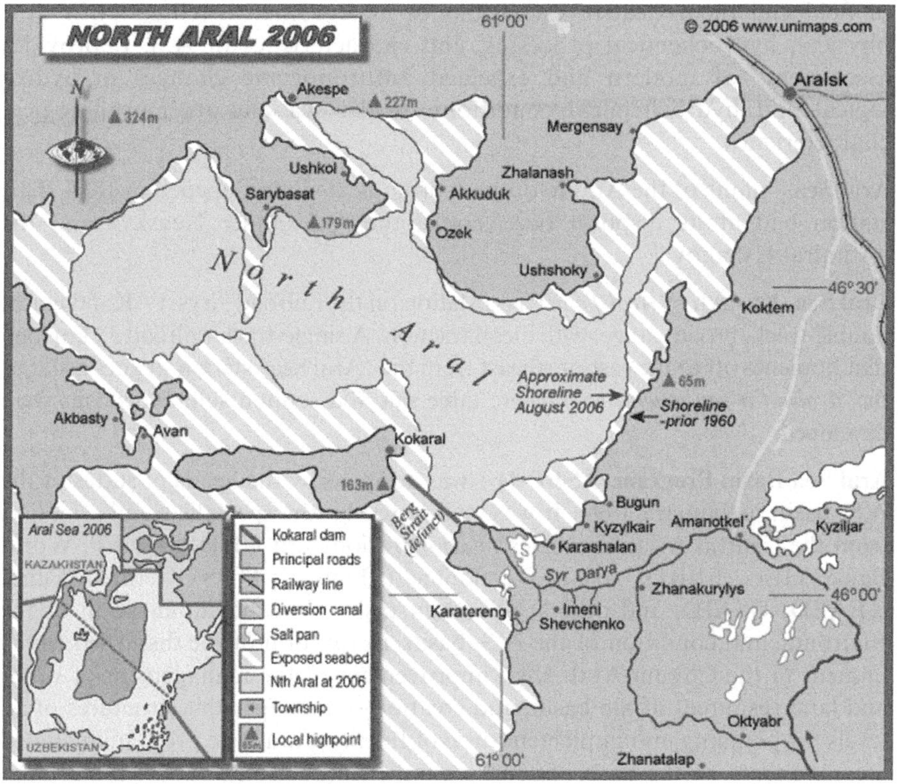

Fig. 12 The North Aral and a new Kokaral dam constructed in the Berg Strait in August 2005 (http://unimaps.com/aral-north/northaral.gif)

At present concrete actions aimed at improvement of the environmental situation in the A.S. region are being developed. They include, in particular, wide-scale works on rehabilitation of irrigation and drainage systems. The wide-scale expansion of irrigated lands has ended. To ensure guaranteed inflow of river waters into the Amudarya and Syrdarya deltas and A.S., limits will have to be set on water utilization for irrigation and other purposes, to improve water treatment, etc.

Scientists from the region countries together with experts from international organizations are involved in addressing the Aral problem and developing an international program for restoration of the environmental equilibrium in this region.

Aral Sea – monograph prepared within the framework of the project "USSR Seas," in the series "Hydrometeorology and Hydrochemistry of the USSR Seas," volume VII, by the State Oceanographic Institute edited by V.N. Bortnik and S.P. Chistyaeva and published in 1990. It presents the results of field and theoretical investigations of meteorological (climatic), hydrophysical, hydrochemical processes, and characteristics of A.S. It provides assessments of modern and expected anthropogenic changes in hydrological and hydrochemical conditions of the sea and its morphological characteristics.

Aral Sea – name of the Aral hydrometeorological station opened in 1884. This station had at its disposal two vessels, the "Alexander Nevsky" and the "Admiral Ushakov."

Aral Sea (Kazakh – "*Aral Tenizi*") – a station on the railroad Arys-1 – Kandagach, double-track, broad-gauge, with diesel traction. A single-track railroad 32 km long that branches off to the east-northeast from the "Aral Sea" station. It goes as far as the *Aralsulfat settlement* (see) where table salt and sodium sulfate deposits were developed.

Aral Sea Basin Program (ASBP-1) – was adopted by the heads of states of the A.S. basin in January 1994. It was elaborated by the specialists and scientists from the Central Asian countries in cooperation with UNDP, UNEP, World Bank, GEF, EBRD, Asian Development Bank, TACIS, German Fund, KFAED, USAID, and others. The main tasks were the stabilization of the environmental condition of the A.S. basin, restoration of the disturbed environment in the Circum-Aral Area, improvement of the management of water and land resources of the basin, and creation of management structures of all levels for planning and implementation of the program. The program included 8 components on the basis of concepts foir projects, some of which had been already well developed:

1.1. Regional strategy of water resource management;
1.2. Ensuring stability of dams and reservoirs;
2.1. Hydrometeorological services;
2.2. Regional system of environmental information;
3.1. Water quality management;
4.1. Restoration of wetlands;
4.2. Restoration of the northern part of the Aral Sea;
4.3. Environmental studies in the Aral Sea basin;
4.4. Project on the Syrdarya River flow regulation and delta development;
5.1. Pure water, sanitation, health – Uzbekistan;
5.2. Pure water, sanitation, health – Turkmenistan;
5.3. Pure water, sanitation, health – Kazakhstan;
5.4. Medium-tern water supply;
6. Integrated project on water and land resources management in the upstream watershed;
7. Operative management of water resources;
8. Development of potential.

The work under ASBP went in two stages: the preparatory stage of 3 to 5 years and approximately US$ 30 million; and the second implementation stage of 10 to 15 years with implementation of works evaluated at approximately US$ 500–750 million.

Apart from budget allocations, the international organizations provided grants for 47.7 mln US dollars and various loans for 278 mln US dollars. The funds allocated by donor organizations were spent mostly on scientific research. The international community of donors agreed to finance ASBP in 1994. After analysis conducted by the World Bank in 1996 the second stages of the Program (ASBP-2) was prepared.

The Aral Sea: diagnostic study for development of a plan of action for conservation of the Aral Sea – a report prepared in 1992 by an international working group of experts (USA, France, Germany, Poland, Australia, Czechoslovakia) and representatives of WMO, UNEP and ILEC in response to the dramatic situation in the Aral Sea region. The purpose of this report was to provide UNEP with a diagnostic document of the environmental, demographic, and economic problems underlying the A.S. situation. For preparation of this report, the Working Group met four times: in Nukus (Uzbekistan) in 1990, Moscow (Russia) in 1991, Almaty (Kazakhstan) in 1991, and Geneva (Switzerland) in 1992. The report was prepared using numerous publications of the Soviet Union and in consultation with the academic, scientific, and research organizations of the Central Asian republics. The studies were supported by government leaders of Karakalpakstan, the Khorezm Region in Uzbekistan, the Tashauz Region in Turkmenistan, and the Kyzylorda Region in Kazakhstan. The Aral report comprised 10 chapters covering the following issues: nature and history of development; manpower, natural resources, and economic potential; basic changes in land ecology; changes in the water-salt balance; changes in water ecology; environmental pollution; the state of public healthcare; the basic reasons of the Aral crisis; and conclusions and a summary.

Aral Sea: experience of a physiographical monograph – the first scientific classical fundamental work of *L.S. Berg* (see), summing up his investigation results during expeditions to the Aral Sea (1902–1903). The author speaks as a natural scientist studying the sea holistically, in its interrelation with the whole Aral basin. Published in 1908, this work (580 pages) is divided into 10 chapters in which quite vividly and in a scientifically substantiated form there are presented, against the background of past investigations, the geography, topography, hydrography, shore morphology and climatology, and hydrogeology together with the analysis of the Aral level, flora and fauna, geology, and the general historical overview of A.S. Berg convincingly disproved the then reigning views that the Aral should become shallower and its level should drop due to "progressing desertification." He proved the existence of alternating wet and dry periods. For this work, Berg was awarded the academic degree "Doctor of Geography," the Golden Medal of the Emperor's Russian Geographical

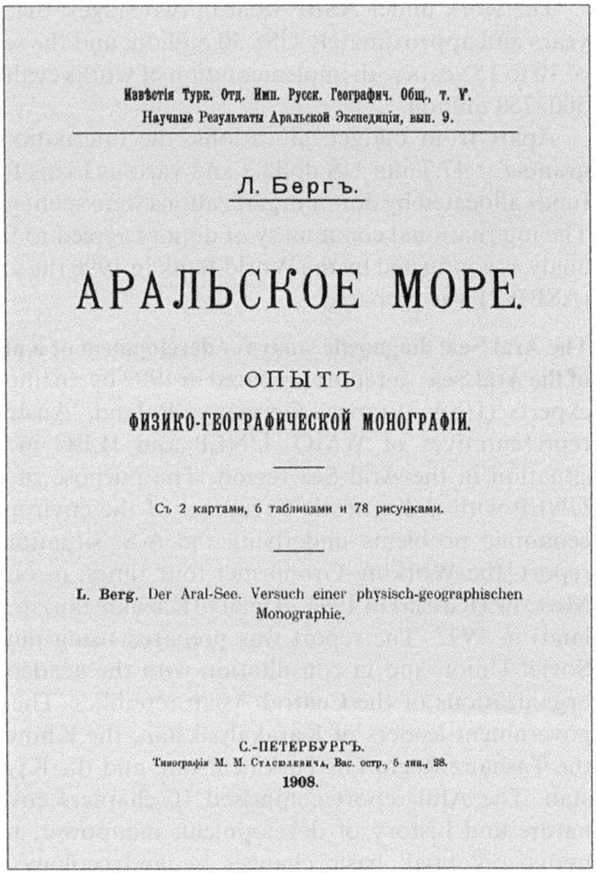

Извѣстія Турк. Отд. Имп. Русск. Географич. Общ., т. V.
Научные Результаты Аральской Экспедиціи, вып. 9.

Л. Бергъ.

АРАЛЬСКОЕ МОРЕ.

ОПЫТЪ
ФИЗИКО-ГЕОГРАФИЧЕСКОЙ МОНОГРАФІИ.

Съ 2 картами, 6 таблицами и 78 рисунками.

L. Berg. Der Aral-See. Versuch einer physisch-geographischen Monographie.

С.-ПЕТЕРБУРГЪ.
Типографія М. М. Стасюлевича, Вас. остр., 5 лин., 28.
1908.

Fig. 13 Berg L.S., 1908. The Aral Sea. An Experience of Physico-Geographical Monograph. Izv. Turkestanskogo Otd. Russk. Geogr. Obshchestva, St.-Petersburg, V.5, N 9, 580 pp. (in Russian)

Society named after P.P. Semenov-Tyanshansky, and the G.P. Gelmersen Award from the Academy of Sciences.

Aral Sea level – in the geological past, after the final shaping of the water body depression, the water level changes in it were determined by two kinds of fluctuations: deformation and volume. The first were related to temporary redistribution of the water masses due to external impacts, including surge events and *seiches*. The volume fluctuations resulted from the changes in water balance. They were much more significant and long term, and they affected the hydrological regime of the sea and had priority practical significance. The volume variations of the level depended on the elements of water balance: river flow and atmospheric precipitation ensuring inflow into the sea and evaporation from the water surface leading to water losses. In the recent centuries, the water balance was in equilibrium: the river inflow (52 cu. km a year) and atmospheric precipitation (9 cu. km a year) were outbalanced by evaporation (61 cu. m a year). As a result, the average water level was close to 53.3 m. Beginning from 1961, however, increased water intake for irrigation and other agricultural and municipal-domestic needs occurred. The water level started steadily lowering and, as a result, the Aral region turned into a zone of large-scale environmental disaster. In other words, the direct cause of the A.S. level drawdown was disturbance of its water balance, excess of water spending over water inflow due to a sharp anthropogenic-related reduction of the Amudarya and Syrdarya flow (80%) (i.e. irreversible withdrawal of water for irrigation) and natural low water availability (20%).

The basic milestones reflecting the negative changes of the Aral natural regime and the subsequent drop of its level were the following:

1961 – beginning of the A.S. level drop;
1970 – water level: 51.4 m. From 1970–1980, the decreased inflow into the Aral was the result of natural events – these were the low-water years;
from 1974 due to nearly complete utilization of the Syrdarya water for economic needs and construction of several dead earth-fill dams on the main river channel within its delta, the inflow into the sea practically stopped;
1980 – water level: 45.8 m. Beginning from 1982 the diversion of the Amudarya waters to A.S. along the main channel was stopped; near Kyzyldjar settlement, a dead earth-fill dam was constructed. The residual flow was directed to irrigation of lands on the left bank of the river and to water supply to the dried water bodies in the delta. A small portion of water ran to the sea along small arms and via a system of lakes. In 1982, 1985, and 1986, there was no inflow into the sea. From 1982 to 1987, flow of the Syrdarya into A.S. ceased completely. Small quantities of water got into the sea only in relatively water abundant years, for example 1988–1994;
1988–1990 – the Aral, having a level of 39.7 m, became two isolated seas – the Large and Small; a dry cutoff wall between them appeared;
1990–38.2 m. In the rather water abundant years that followed (1990–1994, 1997–1998), a part of the Amudarya and Syrdarya waters again reached the Aral;
1999 – a dam dividing the Large and Small Aral seas was broken by the big storm on April 22; by September 1999 the Small Aral sea level decreased by 2.5 m;
2000 – 33.3 m;
2001 – Vozrozhdeniya Island joined mainland to the south;
2005 – a new dam was constructed in the *Berg Strait* (see) for separation of the Large and Small Aral;
2008 – 29.2 m (April).

Aral Seashore – former recreational zone in *Muinak City* (see). It was created here because of a combination of favorable factors that existed, including the

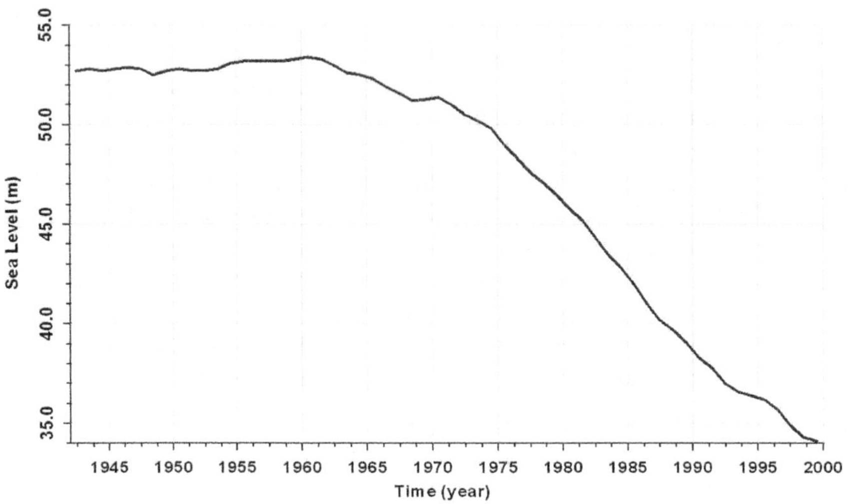

Fig. 14 Sea level variability in the Aral Sea in 1942–2000

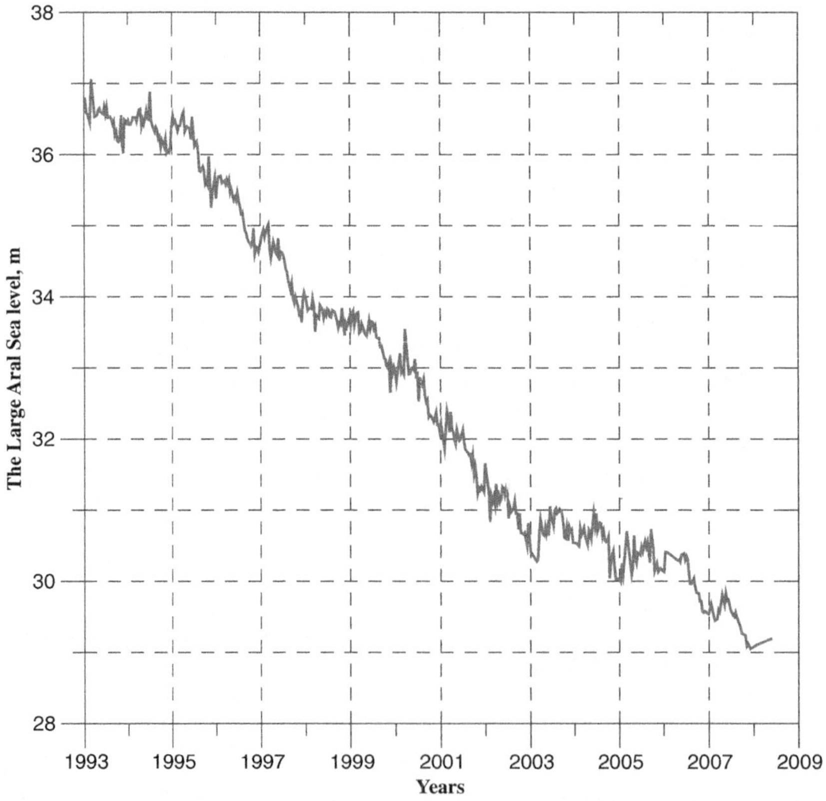

Fig. 15 Sea level variability in the Large Aral Sea in 1993–2008 derived from satellite altimetry data of *TOPEX-Poseidon Jason-1* (raw data source: http://www.pecad.fas.usda.gov/cropexplorer/global_reservoir/#disclaimer)

properties of the Aral water, temperature, mineralization, aeration, and solar radiation.

Aral Shemaya (*Chalcalburnus chalcoides aralensis*)* – fish of the carp family (*Cyprinidae*). The body is narrow and long (up to 30 cm). Lives and lays eggs in the sea. The males reach maturity at the age of 2–5 years at 15 cm long, while the females mature at the age of 6 years, reaching the length of 17 cm. Spawning in the coastal zone, largely on stony and more seldom on the sandy bed. Sometimes eggs get stuck to the roots of floating reeds. After spawning, the fish go to the open sea where they feed intensively on shrimp and the mummies of bloodworms. At night Shemaya stays near the water surface, while in the daytime it descends. Its commercial significance is low.

Aral Sturgeon (*Acipenser nudiventris*) – a common name for the *Aral bastard sturgeon* (see).

Aral-type bay coast* – a type of a shallow coast characterized by rather high roughness of the coastline; formed as a result of sea encroachment on the Aeolian relief. Most widespread on the eastern shore of the Aral Sea with its ridge-cellular structure. Comparing maps of the Aral Sea of 1850 and 1937, over a 100 year period the Aral Sea ingressed 40 km into the Karakums.

Aralkum – the name given to a new desert, the world's youngest sandy-solonchak desert, formed on the dried bottom of the Aral Sea (similar to the Karakum and Kyzylkum). Its area by mid-2000 reached 2 mln ha.

Fig. 16 Aralkum - a new desert formed on the dried bottom of the Aral Sea (November 2007). Photo by Pavel Kosenko, http://pavel-kosenko.livejournal.com

Aral-Kum, project – a project of the European Union (EU) within the framework of the Inco-Copernicus Program implemented in 2000–2003. The main objectives of this project were elaboration and validation of the integrated mathematical instrument for understanding the quantitative assessment of the forecast and control of desertification processes occurring in the Aral Sea region. The proposals included the combining of investigations and expertise of scientists from the EU and CIS in hydrology, limnology, oceanography, climatology, geography, and geology on the basis of digital modeling and database control with a view of obtaining a present-day integrated picture and systems understanding of the climate system affecting the Aral region and a response to the processes of interaction between land-ocean-atmosphere and desertification that result partially from anthropogenic impacts (land use and water resource management) and natural climatic fluctuations in order to identify likely synergetic interrelations among them. The investigations used the accumulated and analyzed (in-situ) data of remote sensing and also data of digital modeling as instruments for integrated environmental management within a watershed basin (regional scale, territories surrounding the Aral Sea, the Amudarya and the Syrdarya) and prevention or alleviation of the effect of ongoing desertification. During this project implementation, a book, *"The Aral Sea: Selected Bibliography"* (see), was prepared.

Aralsk (Kazakh – "*Aral*") – a city (established 1938), port (former Raim), center of the Aral district, and Kyzyl-Orda region in the Republic of Kazakhstan. Located near the northeastern end of the Aral Sea, in the northern part of the *Aralsk Bay* (see), on the coast of the *Sarychaganak Bay* (see) among sands and solonchaks. The population was 32000 people (1999). It was the best equipped port on the Aral Sea. This port was accessible for ships with a draft of 3 m. The city also has the railway station, "Aral Sea." In the Soviet times, a fish-refrigerating plant, dockyards and ship-repair plants, oil plants, glass works, and others were built here. There was shipping traffic to the Amudarya lower reaches (wharf Muinak). In the past, port Aralsk had permanent shipping traffic with *port Ushsai* (see). Cargo-passenger vessels of the *Central-Asian State Shipping Company* (see) made regular voyages. From port Ushsai and from the commercial fishing establishments, fish, cans, and cotton were brought to Aralsk, from where timber, grain, fertilizers, industrial and food products were transported to port Ushsai. Near Aralsk there is salt and sulfate deposit development. It appeared in 1905 due to constructions of the railroad Orenburg – Tashkent. From 1884 to 1983, the sea monitoring station operated here. Originally, the main part of the population here were Urals Cossacks. Today the main population are Kazakhs. This city is officially included into the circum-Aral zone of environmental disaster. After construction in 2005 of the dam nearby the *Kokaral settlement* (see), the level in the Small Aral Sea started rising. The North Aral Sea's surface increased from 2550 sq. km in 2003 to 3300 sq. km in 2008. And the sea's depth increased from 30 m in 2003 to 42 m in 2008. Fish, sea birds and reptiles have begun to repopulate the Small Aral Sea and surrounding area. In the recent past only one species of fish remained in the Small Aral Sea. Today 15 different species have been recorded - bringing back work and income to about 100 local fishermen. The sea is expected to increase further in level and its border will come close to the city, which gives hopes for resolving some urban socioeconomic problems. Fish exports have restarted and the local industry is growing. In the last year, two processing plants and three fish receiving centers have opened, and two more processing plants are scheduled for the nearest future..

Aralsk Bay* – juts into the northern coast of the *Greater Sarychaganak Bay* (see). The bay width in its northern navigable part was no more than 550 m. The bay was confined by the sandy coast of the mainland on the one side and *Akhmeta Island* (see) on the other. The coastal zone is deserted. The *city of Aralsk* (see) is located in its northern part. The bay is shallow, only in its northern part the depth sometimes reached 3-4 m. It is covered with ice for 110–120 days each year. Storm winds prevail in the autumn and winter.

Aralsk port – see Aralsk.

Aralsulfat – an urban type settlement in the Aral district of the Kyzyl-Orda Region in the Republic of Kazakhstan. It is located 15-20 km from the Zhaksyklysh Lake and 21 km from the *railway station "Aral Sea"* (see). Its population is 5000 people. Sodium sulfate is mined in the Zhaksyklysh Lake.

Aralvodstroy – a special construction association established in 1988 with a view to implement a complex of nature conservation, water saving, and irrigation and drainage works and also to construct artificially regulated water bodies near *Muinak* (see) and in the Amudarya delta. These were designed to reduce salt blowout from the dried-out bottom of the sea. This association fulfilled many volumes of research on the rehabilitation of the main and interregional irrigation and drainage networks, the creation of water bodies in the delta, and housing construction to improve the living conditions in this region.

Arid ecosystems – a scientific journal founded in 1995 by the Caspian Biological Resources Institute of the Daghestan Scientific Center, RAS, following the decision of the General Biology Division Bureau of the Russian Academy of Sciences (RAS). The journal presents fundamental investigations and results of applied research on arid ecosystems and man-made desertification control on regional and global scales. From time to time, the journal publishes articles about recent investigations of the problems of A.S. and the Circum-Aral area. The journal has 4 issues a year.

Arid soils – soils formed in the arid regions, in dry steppes, semideserts, dry desertified savannahs, and deserts. Due to an excess of evaporation over precipitation, arid soils are characterized by a "non-flushing" water regime. Arid soils include chestnut, brown semidesert, reddish-brown soils of dry desertified savannahs, various desert soils, and other types of soils.

Aridity – climate dryness caused by insufficient precipitation for vegetation development. Aridity is the principal feature of the climate in the Circum-Aral area.

Arkhangelsky, Andrei Dmitrievich (1879–1940) – the Soviet geologist who founded the Moscow scientific school of "tectonists." From 1926, he was an academician. From 1913 to 1924, he was a member of the Geological Committee. Professor of the Moscow State University (1920–1932) and Moscow Mining Academy (1924–1932). Director of the Geological Institute of the USSR Academy of Sciences (from 1934). Together with I.M. Gubkin, he studied the Kursk magnetic anomaly. Headed the expeditions to study the geological structure in Kazakhstan (1936–1938) and the European part of the USSR (1939–1940). Laureate of the V.I. Lenin Prize (1928). The works of A. cover various issues of regional geology, stratigraphy, tectonics, petrography of sedimentary rocks, paleogeography, among other topics. His first investigations were devoted to the study of the geology of Povolzhie, stratigraphy of Paleogene, and Cretaceous deposits. His monograph, "Upper cretaceous deposits in the east of European Russia" (1912), initiated the application of the comparative-lithological method in the USSR. In 1927, A. and N.M. Strakhov, aboard the vessel "Pervoye Maya" (First of May), studied the geology and morphology of the Black Sea trough. The results were included in the monograph, "Geological structure and history of Black Sea development" (1938), which analyzed the results of observations over the Black Sea bottom sediments and presented a map of their distribution. A. devoted some of his works to conditions of mineral deposit formation: "Conditions of oil formation in the North Caucasus" (1927), "On the origin of bauxites and prospecting for their deposits" (1937). In his works "On the relationships among gravity anomalies, magnetic anomalies and geological structure in Eastern Europe" (1924), "Geological significance of gravity anomalies in the USSR" (1937, as a co-author) and others, A. introduced geophysical methods in regional geotectonic investigations. In his work, "Geological structure of the USSR. European and Central-Asian parts" (1932, 4th edition – "Geological structure and geological history of the USSR", vols. 1–2, 1947–1948), A. revealed some common regularities in evolution of the Earth's crust and provided a modern interpretation of the theory of geosynclinals and platforms.

Arkhangelsky rampart* – a large narrow tectonic uplift, or underwater ridge, extending from the former Muinak Peninsula (now merged with the mainland) via the Lazareva, Vozrozhdeniya, and Komsomolsky Islands to the Kulanda Peninsula. This rampart, with a flat western slope and steep eastern slope, divided the Aral Sea into two asymmetric and unequal by area basins: the western (little), which was deep-water, and the eastern (greater), which was shallow-water. It was named in honor of the outstanding Russian geologist and academician *A.D. Arkhangelsky* (see). Some geologists call it the Aral-Kyzylkum rampart or Aral-Kyzylkum uplift. By 1990, the rampart had practically reached the surface, having united the above mentioned islands. A deep fault runs along the rampart that is sometimes called "the Arkhangelsky structural line."

Artemia (*Artemia*) – crustaceans of the *Branchiopoda* subclass. Their length is 0.8–1.0 cm. They are usually found in salt lakes and pools in semideserts, deserts, and in marine brakish lagoons with salinity levels of 9–330 g/l. Their mass development stops at salinity 230 to about 260 g/l or at a deficit of Mg, which is needed for synthesis of chlorophyll by algae and chitin by A. They actively proliferate at +10°C to +37°C and feed at a temperature no lower than +5°C. At 15–10°C their feeding activity drops. They usually consume green algae. A. are the ideal initial feed in aquarial and commercial-scale fish farming (fresh-water and marine). *A. parthenogenetica*, a typical inhabitant of hyperhallinic lakes, was first found in the Large Aral Sea in 1998, while by 2000 this species dominated in the A.S. plankton taking 99%. A. is highly valued on the world market and there is competition for fishing zones. In recent times, US and Belgium representatives considered a possibility to use the *fishworks* (see) in Muinak to start fishing for this crawfish. A joint venture "Aral Artemia Company" was established in *Nukus* (see), Karakalpakstan.

Fig. 17 Artemia (http://www.cefe.cnrs.fr/ecogev/siteGB/artemia.jpg)

Aryk – an artificial canal or ditch supplying water from a river or reservoir for irrigation purposes. Quite often it is constructed not over the lower-lying terrain, but along a water divide or slope so that the water may be supplied by gravity to any lower-lying land sites.

Asian Chernobyl – this is a reference by ecologists of the Republic of Karakalpakstan because of the drying out of the Aral and its coast infiltrated by salts, chemicals, and different wastes. In 1991, the journal, "World and I," published an article, "Quiet Chernobyl," by M. Glantz (USA) and I.S. Zonn (Russia), describing the situation in the Aral region and the consequences that, similar to the Chernobyl disaster, should be expected in the future.

Assake-Audan depression (Kazakh – "*audan*" meaning "natural area, district") – a large depression, or basin, located northwest of *Sarykamysh* (see) that cuts deeply into *Ustyurt* (see). It is linked with the Sarykamysh Depression only via a narrow strait at absolute elevations 45–50 m. Originating as a narrow strait northwest of the Sarykamysh Depression, it runs for 90 km westwards. Its width is 20–40 km. It is rimmed with chinks and separated from the main depression by steep, seldom flat, clear-cut slopes up to 8–10 m high composed of the Miocene limestones and marls. The bottom is rugged. The lowest absolute elevations are 30 m. Clearly discernible on the slopes of the depression are the pebble and sandy-pebble shore ramparts and lacustrine terraces with mollusk shells indicative of the fact that they were flooded by the Amudarya waters more than once.

Atalyk Island* – located near the eastern shore of A.S., 27 km southwards of the *Agurme Peninsula* (see). The island is covered with sand dunes 2–5 m high in some places with scant vegetation. The island is surrounded by shoals less than 2 m deep; in some places, small underwater bars branch from it.

Atherina (*Atherina mochon pontica*) – marine, schooling, pelagic, eurihallinic fish of the Atherinidae family. Length is up to 15 cm. Imported to A.S. Reaches maturity at the 2nd year. Spawning is by portions in the coastal zone at salinity 7–42% in the spring. The eggs are laid on the seabed, have thread-like protuberances, and stitch to underwater vegetation. Larvae are kept in the surface water layers. Fertility is 600 eggs. Eggs are large, up to 1.9 mm in diameter. Feed on plankton. Commercial significance is not large.

Fig. 18 Atherina (*Atherina mochon pontica*)

Atlas of desertification/degradation of lands in the southern and eastern circum-Aral area – elaborated in 1998 by a group of authors under guidance of G.S. Koust, Doctor of Biology, Professor of the Soil Science Faculty, Moscow State University. It includes the following independent blocks: (a) color maps with legends; (b) sheets to be superimposed on color maps in the form of contours or hatchwork, some of which have special legends; (c) sheets that make it possible to obtain brief information about a particular point on a map; (d) databases making it possible to obtain full information about the current contour in a special window.

The Atlas includes the following maps: (1) Genetically homogeneous types of territories; (2) Relief; (3) Lithology; (4) Depth of groundwater occurrence; (5) Groundwater salinity; (6) Type of groundwater salinity; (7) Soils; (8) Main causes of desertification; (9–24) Block of binary maps illustrating distribution of desertification having different causes: water intake from rivers; discharge of collector-drainage waters into rivers; secondary changes in the natural systems as a result of water intake; discharge of collector-drainage waters into drainless depressions, seepage of filtration waters from irrigated lands and from a collector-drainage network; supplementary drainage of territories after their irrigation is stopped; secondary initiated changes in the natural systems as a result of watering; changes in soils and vegetation as a result of overgrazing; changes in vegetation at undergrazing; secondary initiated changes in the natural systems at distant-range cattle breeding; direct effects of drilling works on the soil and vegetation cover; direct effects of construction of linear structures on the soils and vegetation; secondary initiated changes in the natural systems due to construction of linear structures; destruction of forests and shrub vegetation; (25) Types and subtypes of desertification; (26–37) Block of binary maps illustrating desertification subtypes: biogenic posthalomorphic takyr development; biogenic posthydromorphic takyr development; biogenic posthalomorphic sand formation; biogenic posthydromorphic sand formation; biogenic post-takyr sand formation; abiogenic posthydromorphic salinization; abiogenic anthropogenic salinization; abiogenic posthydromorphic sand formation; abiogenic post-takyr sand formation; abiogenic posthalomorphic sand formation; abiogenic postautomorphic sand formation; abiogenic disturbance of clay deserts; abiogenic disturbance of crushed-stone deserts; (38) Relief desertification; (39) Soil desertification; (40) Plant desertification; (41) Desertification penetration.

The Atlas provides a possibility to use it with a geoinformational system (GIS) and contains, apart from the noted maps, information in such special databases as: (1) Types of territories; (2) Relief; (3) Lithology; (4) Ground waters; (5) Soils; (6) Vegetation; (7) Main causes of desertification; (8) All causes of desertification; (9) Types and subtypes of desertification; (10) Soil desertification; (11) Vegetation desertification; (12) Desertification penetration.

Atlas of ice of the Aral sea – prepared and published in 1970 in Tashkent by the Uzbek Department of Hydrometeorological Service. It describes conditions of ice formation, borders of ice distribution, and ice regime.

Auzy-Kokaral, Auzykokaral strait – dividing the Small Aral and the Large Aral, it is located between the western shore of the *Kokaral Island* (see) and the *Karatyup Peninsula* (see). The strait is shallow, the depths being no more than 2 m for the most part. The channel is narrow, meandering and complicated. In some places its width reaches 8 m. Dried out in 1968.

Fig. 19 Ice and snow in the Aral Sea Region (28 February 2008, MODIS-Terra). Image courtesy by D.M. Soloviev, Marine Hydrophysical Institute, Sevastopol, Ukraine

Avan* – a settlement on the eastern bank of the *Auzy-Kokaral Strait* (see) in the northwestern part of the *Korakal Island* (see). The settlement and fish plant were the largest on the western coast of the Small Aral.

B

"Badai-Tigai," nature reserve – located on the right bank of the Amudarya in the southwestern part of Karakalpakstan, 85 km south of the Takhiatash dam (see) and 130 km north of the Tuyamuyun dam. Its length is 17.5 km, and its width is 1.5 to 2 km. It was created in 1971 for conservation of typical intra-zonal tugai ecosystems that in the Amudarya lower reaches and delta have been practically lost by now. It was the model ecosystem of the Amudarya lower reaches. In the past, *tugai* (see) covered approximately 70% of the territory in the Amudarya lower reaches, but due to regulation of the Amudarya flow and agricultural development of floodplains, they disappeared nearly completely. At the time of the establishment of the preserve, it contained about 167 varieties of higher plants. By 1985, their quantity had shrunk to 61. In addition, it became a habitat of many species of animals now rare in the lower Amudarya, such as Bukhara deer (re-introduced), fox, jackal, badger, rabbit tolai, pheasant, little owl, long-eared owl, purple heron, peregrine, black kite, little cormorant, and others.

Baigubekmuryn, Cape* – an insignificant southward projection from the western shore of A.S. Here, the steep shore descends smoothly, step-like, until at the extreme end it goes down steeply into the water. The sea near the cape was deep, exceeding 50 m only 3 km from the shore.

Barsakelmes, Barsa-Kelmes* (meaning in Kazakh "*if it goes, it will not return*") – (1) the third largest island in the Aral Sea. Its area is 183 sq. km. It is located in the northern part of the sea, 26 km to the south-east of the *Izendyaral Cape* (see) and 85 km from the Syrdarya mouth. The relief of this island is rather high and undulating, but gradually lowering to the north. The western and southern shores are high and steep. In some places, in particular nearby the *Butakov Cape* (see), its cliffs are about 50 m high, steep, and come up closely to the water edge. Along the low northern shore, a narrow belt of sand beaches stretch. On the northwestern part of the island, a group of small lagoon-type saline lakes is found. In 1949, B. was assigned the status of a nature reserve to be managed by the Kazakh Academy of Sciences. It was organized for protection of the endangered animals: Persian gazelle, saiga, and koulans that were brought here recently. Apart from wild hoofed animals, this island is a habitat for

I.S. Zonn et al., *The Aral Sea Encyclopedia*, DOI 10.1007/978-3-540-85088-5_3,

large-toothed souslik, jerboa, copperhead snakes, and many birds, mostly water fowl. In 1997, the island became connected with the parent shore. (2) Sor (solonchak) in the southeastern part of the *Ustyurt Plateau* (see).

Bartold, Vasily Vladimirovich (1869–1930) – one of the most renowned and outstanding representatives of the Russian Oriental science of the late 19th – early 20th centuries. He studied on the Faculty of Oriental Languages in the Petersburg University. At the age of 32, he became a Professor of Petersburg University, since 1913 – academician. He was a member of nearly all European academies and scientific societies and a wide specialist in Oriental disciplines. The basic lines of his scientific activities are history of the peoples and states of Central and Middle Asia, the interaction of Oriental and Western cultures, and the history of Islam. The works of B. contain an enormous amount of factual material supported by archeological and numismatic data and represent a real encyclopedia of historical and philological knowledge about ancient and medieval Central Asia. B. made his significant contribution in the development of a network of scientific establishments, educational institutions, and libraries in Central Asia and in collections of manuscripts and their study in local archives. In 1902, B. published his work, "Data about the Aral Sea and Amudarya Lower Reaches from the ancient times to the 17th century," and in 1914, he published, "About the history of irrigation in Turkestan." In 1924, in Baku he read a cycle of lectures, "Place of Circum-Caspian areas in the Moslem world history" to the young intellectuals.

His most important works are "Turkestan in the epoch of Mongolian invasion" (in two volumes, 1898–1900); "History of Oriental studies in Europe and Russia" (1911); "Ulugbek and his time" (1918); "Islam" (1918); "Turkestan History" (1922); "Tractates" vols. 1–9, M. (1963–1977); and "Works on historical geography" (2002), to name but a few.

Fig. 20 V.V. Bartold (1869–1930)

Basin of the Aral Sea – see Watershed Basin of the Aral Sea.

Basin Water Management Associations (BVO) – in 1987, pursuant to the resolution of the October (1985) Plenum of CPSU CC, the BVO "Amudarya" and

"Syrdarya" were established within the framework of the USSR Ministry of Water Management. They were in charge of allocation of water resources among republics and operation of water intake structures and waterworks. They were located in Urgench and Tashkent. On their basis were formed the Kurgan-Tyube, Chardjou, Urgench (*UPRADIK*), Nukuss (BVO "Amudarya") as well as Gulistan, Uchkurgan, Chardara, Chirchik (BVO "Syrdarya") territorial production authorities on control of water resources utilization and operation of water intake structures. In 1992, BVOs were turned into executive and inter-departmental control bodies of the Interstate Coordination Water Management Commission (*MKVK*).

BVOs were vested the following powers: allocation of transborder water resources and water supply to water users in the Aral delta and A.S. following the MKVK decisions; control of the operating regime, approved by MKVK, of a cascade of reservoirs on the transborder surface water resources; implementation of nature conservation actions within water protection zones of transborder rivers and reservoirs in compliance with the legislations of the parties upon agreement with the local administration; preparation of basic materials for MKVK meetings in accordance with their agenda on water resource management, water use, improvement of the environmental situation and promotion of management and procurement establishments; and support of relationships with other international organizations.

The hierarchy of water resource management in BVO includes three levels, each one subordinate to a higher-ranking one. The first level is BVO Administration that reports to MKVK and addresses the issues of planning, management, control, and water allocation among states. Informationally, BVO is linked with the Ministries of Water Management of Uzbekistan and Kyrgyzstan, Tajikistan, and Turkmenistan, research centers of MKVK, and the hydrometeorological services of Central Asian states. The second level is represented by four territorial authorities of waterworks that, pursuant to the water intake quotas approved by BVO, ensure water supply to water users. Each authority controls the objects of a water management complex and ensures transportation, formation and utilization of water resources within their command territories. The territorial authorities directly subordinate to the BVO administration. The third level are control and management points that include hydraulic structures and gauging stations which are on a balance of territorial authorities.

Basin Water Management Association (BVO) "Amudarya" – In 1987, pursuant to the resolutions of the October (1985) Plenum of CPSU CC and the USSR Government and order of the USSR Ministry of Water Management, a basin department on inter-republican allocation of water resources in the Amudarya basin was established. Later, it was renamed the Basin Water Management Association (BVO) "Amudarya." Its administrative center was located in Urgench (Uzbekistan). It has 5 divisions. According to its Statute, BVO acts as an interstate organization. In its activities, it is guided by the BVO Statute approved by MKVK, interstate treaties, protocols, and other regulatory acts. It has an independent balance, exercises the rights of a legal entity, has its seal, and special and current accounts with state banks. BVO is headed by a chief appointed by MKVK and is financed by three states – Uzbekistan, Tajikistan, and Turkmenistan.

At present, by agreement of the regional states, the BVO "Amudarya" commands not the whole Amudarya basin, but only the main channels of the Pyanj, Vakhsh, Kafirnigan, and Amudarya from their origin to A.S. BVO "Amudarya" controls all pumping stations on the main channels of these rivers and on interstate canals as well as part of river water intakes not put on the BVO balance. These are the waterworks on the Vakhsh, the Karakum Canal, and the Tuyamuyun waterworks with the reservoirs on the Amudarya. In addition, this Association is in charge of operation of the interstate main canals in the Amudarya lower reaches downstream, the Tuyamuyun waterworks.

For management of transborder water resources on such vast territory 4 authorities were created in BVO "Amudarya" responsible for operation of water intake structures, waterworks, and interstate canals: Upper Darya Waterworks Authority (UGU) with its central office in Kurgan-Tyube (Tajikistan); Middle Darya UGU (Turkmenabad, Turkmenistan); Lower Darya UGU (Takhiatash, Karakalpakstan, Uzbekistan), and UPRADIK (Urgench, Uzbekistan).

The Kurgan-Tyube Waterworks Authority (a new name for the Upper Darya Waterworks Authority) is in charge of operating 8 water intake structures and controls water intakes from the Pyanj, Vakhsh, and Kafirnigan Rivers as well as from the Amudarya stretch of 246 km long as far as the Kelif water gauging station.

The Turkmenabad Waterworks Authority (a new name for the Middle Darya Waterworks Authority) controls water intakes on an Amudarya stretch of 552 km long between gauging stations in Kelif and Darganata. Nine major river water intake structures are overseen by the authority.

The Authority for Amudarya Inter-Republican Canals (UPRADIK) is responsible for the operation of 11 river water intakes, 52 hydraulic structures on main canals, management and operation of 386 km of the main canals, control of the water intakes on the river stretch from the Tuyamuyun waterworks to the Pichpak gauging station (167 km in length). UPRADIK controls 3 major irrigation systems: Tashsakinsky, Klychniyazbaisky, and Kupchak-Bozsu.

The Nukus Waterworks Authority (a new name for the Lower Darya Waterworks Authority) undertakes operation of the Takhiatash waterwork, the head river intakes on the Khan-yab and Djumabaisaka canals, and controls all water intakes on a river stretch from the Kipchak gauging station to A.S. (a length of 283 km).

The Tajikistan, Turkmenistan, and Uzbekistan Republics passed 84 hydraulic structures (including 36 headworks of river water intakes), 169 gauging stations, 386 km of interstate canals and related engineering communications (roads, communication lines, power supply lines, technical facilities, etc.) to the BVO "Amudarya" balance for temporary use.

Basin Water Management Association (BVO) "Syrdarya" – In 1987, pursuant to the resolutions of the October (1985) Plenum of CPSU CC and the USSR Government and orders of the USSR Ministry of Water Management, it was decided to establish a basin department on inter-republican allocation of water resources in the Syrdarya basin. Later, it was renamed into the Basin Water Management Association (BVO) "Syrdarya." Its administrative center is located in Tashknet. It has 4 divisions. Its structure includes: Charvak Reservoir Authority (settlement Charvak, Uzbekistan); Upper-Chirchik UGU (Chirchik City, Uzbekistan), Naryn-Karadarya UGU (Andijan City, Uzbekistan), and Golodnaya Steppe Waterworks and Canal Authority "Dustlik" (Gulistan City, Uzbekistan). BVO "Syrdarya" organizes the supply of Syrdarya waters for the MKVK sovereign member states, undertakes operation of water intakes and waterworks, and attempts to improve the environmental situation and control of the quality of utilized water resources. The Association controls the flow regime of the Naryn, Karadarya, Chirchik, and Syrdarya Rivers as far as the Chardarya reservoir. The association is financed by MKVK member states on the parity and share-related basis in proportion to the volumes of withdrawn and used river water. (In fact, some states fail to provide financial support on a regular basis, which makes the association's functioning difficult.)

The association manages head water intake structures on the Syrdarya and its main tributaries and also canals of interstate significance (e.g., the "Dustlik" canal and the Greater Ferghana Canal); all are owned by the states in which territories they are located. BVO manages 203 hydraulic structures, including 21 on the main riverbeds of Naryn, Syrdarya, Karadarya, and Chirchik. Water discharges of these structures vary from 20 to 250 cu. m/s, while the flows of "Dustlik" and GFC reach 400 cu. m/s.

The main water user in the basin is farming; the total area of irrigated lands here is equal to 3.4 mln ha. About 1.73 mln ha are irrigated directly from the river. Metering of water intake from rivers and commanded canals is done at 445 points comprising 21 head water intakes, 36 permanent pumping stations, and 172 temporary pumping plants. Consumption of surface river waters is controlled mostly by the hydrometeorological services of the republics and on water intake structures like those of BVO "Syrdarya" and other water management bodies of the Central Asian states.

BVO "Syrdarya" has a three-level management structure: first level – central authority in Tashkent, second level – territorial authorities, and the third level – control and management divisions. The central authority accumulates information about water supply to projects; estimates water needs; plans water allocation among four states and A.S., including for each water intake on the Syrdarya and interstate canals; plans operation of the Naryn-Syrdarya cascade of reservoirs; and accumulates information about river water quality. The second level of this hierarchy includes 4 territorial authorities that are responsible for repair of waterworks and directly control the structures in the course of their operation. They control compliance with the quotas on water intake from rivers within the set borders together with the district and regional water management and agricultural bodies of the states in which territory they operate. They also control the quality of river waters and compliance with the environmental requirements. The zones of action of the territorial authorities cover the Syrdarya proper (two authorities), its main tributaries (Chirchik and Karadarya), and also the Charvak reservoir zone. The third level of BVO includes the control and management divisions covering head water intakes, dams, pumping stations, and gauging stations. The tasks of these divisions are to amass information about the condition of the water management complex and to understand the impacts of its control.

Bastard Sturgeon (*Acipenser nudiventris*) – fish of the sturgeon family (*Acipenseridae*). In 1828, G. Meendorf, a member of the embassy in Bukhara and Khiva, brought one specimen of the Aral B.S., and A. Lovetsky, using this specimen, described B.S. as a new species. B.S. was often called the Aral sturgeon. It was up to 2 m long. Anadromous. The males reached fertility at the age of 6–9 years, while females matured at 12–14 years. They spawned in the Syrdarya and Amudarya in the second half of April. Their fertility varied from 280 thou eggs to 1 mln eggs. The eggs were about 3 mm in diameter. Prior to introduction, B.S. fed largely on mollusks. This fish was a typical bentophag. Later on there were proofs that it became predatory fish. In the stomach of the Aral B.S. there were found bullhead, sand pipers, aterina, shrimp. A valuable commercial fish, its reserves were severely depleted.

Bay-type coast – see Aral-type coast.

Begdulla-Aidin Lake – a part of the marshlands of the Sudochie Lake in the south of A.S., it covers 5.5% of its water area. The maximum length of the lake is 4 km, while the width is 2.5 km. The depth does not exceed 1.2–1.5 m, and the length of its coastline is 11 km. During the disastrous low-water period in 2000–2001, it dried out completely.

Bellinsgauzen Island, Shoal Bank* – a former shoal, it appeared in place of a water eroded island to the north of the Lazarev Island. This shoal extends for over 10 km from north-north-east to south-south-west. With the drying of the Aral Sea, this shoal first turned into an island and then became linked with Vozrozhdenia Island.

Beltau, Bel-Tau Upland* – an anticlinal upland running along the A.S. coast in the south-east for 30 km in a sublatitudinal direction. It is composed of Paleogene and Neogene clays and sandstones. Its maximum height is 146 m. It rises up to 85 m over the delta adjoining the upland on the south. The southern and western slopes are steep, in places even steep, with structural terraces; some places are covered by wind-blown sands. The B. northern slope gently descends towards A.S. where it is confined by a sea terrace. For over

20 km, the scoured parent rocks to the west of B. form small uplifts coated with wind-blown sands. The prevailing relief is ridge-honeycomb. Some depressions are occupied with saline lakes, the largest of which is *Karateren* (see).

Bening Shoal* – located to the north of the Komsomolsky Island 39 km to the east-south-east of the *Baigubekmuryn Cape* (see) on the bottom rise extending from the Kordjendy Peninsula to the *Muinak Island* (see). The shoal extended in the meridional direction for about 7.5 km. The surrounding depths were 8 to 10 m.

Berg, Fedor Fedorovich (Frederick William Rembert) (1793–1874) – a military and state figure, traveler, geodesist, general field marshal (1866), aide-de-camp general (1831), Count (1856). He studied at the Derpt University. Took part in the Patriotic War of 1812 and in foreign campaigns of the Russian Army in 1813–1814. Then he traveled over Europe. In 1820–1822, he was on a diplomatic service in Munich and Rome. Prepared a military-statistical description of Turkey. In 1823 and 1825, he headed the expeditions to the Trans-Caspian Area during which he collected the material for the military-topographical description and maps of the territory between the Caspian and Aral Seas. Participated in the 1828–1829 Russian-Turkish War. In 1828–1830, he supervised surveys of the northeastern part of Bulgaria and Balkan Mountains, Romania and on the shores of the Kamchia River. From 1843, he was the General-quartermaster of the Headquarters. B. guided preparation (from 1845) of a 3-*versta* (*versta* = 3.500 feet) military topographical map of Russia, including a military-statistical description of provinces; under his surveillance, the application of photographs during topographical surveys began. In 1845, he became a founding member of the Russian Geographical Society. B. was the Honorary President of the Joint Staff Academy (1861) and an Honorary Member of RGS (1870).

Berg, Lev Semenovich (1876–1950) – a well-known Russian geographer, biologist, and natural scientist. A corresponding Member of the USSR Academy of Sciences from 1928, academician from 1946, RSFSR Honored Worker of Science (1934), and Honorary Member (from 1934) and President (1940–1950) of the USSR Geographical Society. Laureate of the USSR's Stalin Award (State Award) (1951) for the monograph "Fresh-Water Fish in the USSR and Neighboring Countries" (1946, 4th edition, Parts 1–2, 1948–1949). 1899 begins the Central Asian (Turkestan) period in Berg's life. For several years, he studied A.S., the Issyk-Kul and Balkhash Lakes, the Greater Barsuki sands, and climbed glaciers in the Turkestan ridge. He studied the problems of ichthyology and fisheries in the Aral and in the Amudarya and Syrdarya deltas. In 1904, B. was invited to Petersburg to work at the Zoological Museum (now Zoological Institute of the Russian Academy of Sciences). In 1908, his fundamental book named *"Aral Sea: Experience of Physiographical Monograph"* (see) was published. He submitted it as his thesis work for the degree of Master of Geography, but in 1909 was awarded the doctoral degree. In 1913, B. began his pedagogical activities, first as professor of ichthyology and hydrology at the

Moscow Agricultural Institute, then after 1916 at Petrograd (Leningrad) University in the Geographical Institute which, in 1925, was included as a department at the University. In 1915, B. was awarded the Great Gold (Konstantinov) Medal. In the same year, the Moscow Society of Natural Scientists elected him its honorary member. In 1925, B. visited A.S. again. He studied fishery and conducted hydrological investigations.

Berg conducted his investigations in Western Siberia, Central Asia, Povolzhie, in the Caucasus, the East-European Plain, and in other regions of the USSR. His scientific interests cover the theory of geography, teachings on landscapes and landscape zones, regional geography, history of Russian geography, ichthyology, limnology, climatology, palogeography and also geomorphology, lithology, soil science, glaciology, zoogeography, and others. B. developed further the V.V. Dokuchaev's theory on natural zones, developed a theory of landscapes, and proposed the soil theory of loess formation. A prominent place among the scientific issues addressed by Berg is his studies of the Caspian Sea. One of the chapters in his book, "Essays on the History of Russian Geographical Discoveries" (1949), was devoted to the first Russian maps of the Caspian and Aral Seas.

In honor of his outstanding works in the field of geography, biogeography, and ichthyology, the Russian Academy of Sciences instituted in 1995 the award of the L.S. Berg Gold Medal (Departments of Oceanology, Physics & Atmosphere, and Geography).

In 2001, in the non-recognized Circum-Dnestr Moldavian Republic (B. was born in Bendery), B. was honored on a commemorative coin (100 rbls., silver 925).

His principal works are: "Will Central Asia Dry Out?" (1905); "Aral Sea. Experience of Physiographical Monograph" (1908; for this work B. was awarded the academic doctoral degree in geography and the medal of the Russian Geographical Society named after P.P. Semenov-Tyanshansky); "Climate and Life" (1922); "Essays on the History of Russian Geographical Discoveries" (1929); "Relief of Siberia, Turkestan, and the Caucasus" (1936); "Nature in the USSR" (1937); "Essays on History of the Russian Geographical Discoveries" (1946); "One Century of the All-Union Geographical Society. 1845–1945" (1946); "Landscape-Geographical Zones of the USSR" (Part 1, 1931); "Geographical Zones of the Soviet Union" (Part 2, 1952); and "Selected Works" (Vols. 1–2, 1956–1958).

Berg Strait – Named in honor of *L.S. Berg* (see), it was located between the *Kokaral Lake* (see) and the Syrdarya mouth. Its width was 15 km with a maximum depth of 13 m. In the early 1980s, bottom dredging works in the existing waterway were carried out. In 1989–1990, it dried out. Until 1992, instead of this strait there was a small strait with a rather strong current from the Small Sea to the Large Sea (the result of the water level rise in the Small Aral). In 1992, a cofferdam was constructed that in several days was destroyed. Then, a stronger cofferdam was constructed that was destroyed in early 1993 due to the water level rise in the Small Aral. In 2005 a solid dam was constructed here.

Biiktau Island* – located near the southeastern shore of A.S., 6.3 km to the south-east of the *Uyaly Island* (see). Low, sandy, with some small hills. The northern and eastern shores of the island are flat. The southern part of the island gets flooded in some places.

Biodiversity of the Aral Sea Region – In the past the uniqueness of the Aral Sea Region contributed to richness and diversity of the biota, which could be compared with Africa. There were 500 kinds of birds, 200 species of mammals and 100 species of fishes, thousands of insects and invertebrates. Before 1960 the river deltas were home to over 70 kinds of mammals and 320 types of birds. At present a half of them remain. In lowstreams of the Amudarya and Syrdarya rivers thousands hectares of alluvial soils became salt-marsh, and swamp and medow-swamp soil became dry. This resulted in the destruction of several herbs needed for fodder for sheep, horses, camels and goats. Diseases and death of cattle began, musk-rats cultivation stopped, and sheep live-stock decreased sharply. The regional flora was impressive and included 1,200 flowers, 560 types of tugai forests of which 29 are endemic to Central Asia. The change in water balance caused mineralization of the water in the Aral Sea basin, which led to the loss of unique biocenosis and a number of endemic species of animals. The formerly flourishing sea ecosystem included 24 species of fishes, including bream, sazan, aral roach (vobla), carp, perch, sturgeon, salmon, sheat-fish and spike that disappeared completely due to tenfold rise of salinity.

Bioecology Institute, Karakalpak Branch of the Uzbek Academy of Sciences – was established in 1994 on the basis of biological laboratories of the Integrated Institute of Natural Sciences and the Department on Circum-Aral Environmental Problems (former Research Center "Aral" of the USSR Academy of Sciences). This is the only research institution in the Central Asian region dealing with this issue. The main lines of research activities of this institute are: the study of environmental consequences of anthropogenic impact on the natural environment of the Aral basin, addressing the issues of conservation of biological diversity, and the wholeness of the ecosystems in the region and criteria for assessment of their state.

The Institute includes laboratories, research stations, and groups: Ecology of Microorganisms, Fish Ecology, Hydrobiology and Hydrochemistry, Soil Science, Phytocenology, Ecology and Physiology of Plants, Ecology of Land Animals, Parasitology, Ustyurt Desert Station, Muinak International Biological Station, the Group of Herbaria of the Southern Circum-Aral Area, and a Zoological Museum. The Institute also conducts 2 research workshops: Problems of Circum-Aral Ecology and Biodiversity & Specially Protected Natural Territories.

Bioplateau – a method of biological treatment of polluted drainage and waste waters by the artificial planting of higher water plants as biofilters in settlement ponds and collecting-drainage canals. Among the higher aqueous plants, the most practical for water treatment are reeds, narrow-leaved cattail, lacustrine

Fig. 21 Camels at the watering place. Photo by N.V. Aladin.

rush, hyacinth aqueous, hornwort, among others. The best water treatment has been achieved by strip planting of rush and cattail across the direction of the polluted water flow. This method is environmentally safe and less costly compared to other methods.

Biyurgundy Island* – located in the north of A.S., 2.6 km westwards of the *Kendyrli Island* (see). High-elevated and undulating, in general, its northern coast ends with rather high and steep cliffs. In the southwestern direction, the island becomes lower and its southwestern coast is low-lying and flat.

Blaramberg Ivan Fedorovich (1800–1878) – a geographer, traveler, and military topographer of Dutch origin. He finished a law degree at Hessen University, and in 1823 went to Moscow where he studied the Russian language and improved his knowledge in French literature, mathematics, history, and drawing. In 1824, he naturalized in Russia. He graduated from the Institute of the Corps of Railway Engineers and took part in the Caucasus campaigns against highlanders as a General Staff officer. In 1832–1836, he completed a major work on a description of the Caucasus. In 1836, he was awarded the rank of captain. In the same year, B. took part in the G.S. Karelin expedition to the southeastern shores of the Caspian Sea, where they made a detailed description of the eastern shores of the Caspian Sea. From 1841 to 1856, he took part and headed topographical surveys and geographical studies of the Northern and Eastern Circum-Aral area. He selected places for construction of forts (fortifications), including Ural (Irgiz-Kala, presently Irgiz settlement in the Aktyubinsk Region, Kazakhstan), Orenburg (Turgai, presently Torgai of the Kostanai Region, Kazakhstan), and Aral (Raim, presently Raim aul of the Kyzylorda Region, Kazakhstan) fortifications. He also took part in the seizure of the Ak-Mechet fortress (later Perovsk, presently Kyzylorda, Kazakhstan) and was promoted to the rank of general. In 1850, he published the "Topographical and statistical description of the eastern coast of the Caspian Sea from the Astrabad

Bay to the Tyuk-Karagan Cape" (more correctly Tyub-Karagan) in the Proceedings of the Emperor Russian Geographical Society. In 1853, he published in the same proceedings the "Record book made during the expedition for study of the eastern coast of the Caspian Sea in 1836" and a "Statistical review of Persia." At the end of his days, he published "Reminiscences" that in 1978 were translated from German to Russian and published in Moscow.

Fig. 22 I.F. Blaramberg (1800–1878)

Board of the International Fund for saving the Aral Sea (IFAS) – see International Fund for saving the Aral Sea.

Bottom soils of the Aral Sea* – the soft, unconsolidated soils prevailing in the sea. Stony soils are practically absent, except for small areas adjoining the Kulandy Peninsula (*Izendyaral and Uzunkair capes* (see)), *Vozrozhdenia Island* (see)) (eastern shore), small Chagaly Island, and also *Lazarev Island* (see) (eastern shore). The coastal strip is composed of sandy soils. Near the eastern shore overgrown with thickets of cane and reed and abounding in algae some places are made of silts and silty sands. The sandy soils occur at depths to 10 m, and next to them there is a narrow strip of silty sands replaced further on by sandy silt. In many places, shells of *Adacna minima, Dreissena polimorpha* and *Cardium adule* are found. Significant accumulations of these shells occur to the south of the Kulandy Peninsula, near the western shore of the sea, to the east of the Baigubekmurun (Bai-Kubek) cape, to the north-west of the Uyaly Island, and nearby the northern and eastern tips of the Kug-Aral Island. The greater part of the sea bottom is composed of gray silt overlain with a thin layer (2–5 cm) of very soft (semi-fluid) light-gray silt. In the central part of the sea is an area of more compacted gray-colored argillaceous silt overlain with a thin layer of light-brown semi-fluid silt. The underwater upland stretching from the Kulandy Peninsula nearly as far as the Lazarev Island is made up of sandy soils passing southward into silty sands. The soils in the deep-water depression near the western shore beginning from depths 35–40 m were represented by black, fluid, slimy silt smelling of hydrogen sulfur. Small areas composed of black silt may be found in the middle of the northern part of the *Chernyshov Bay* (see) and to the south of the Karatyube Peninsula. Black silt (with a considerable addition of sand) is found not only at great depths but also in some shallow-water bays near the eastern shore.

Fig. 23 Dried bottom of the Aral Sea (November 2007). Photo by Pavel Kosenko, http://pavel-kosenko.livejournal.com

Bozkol Bay* – located on the eastern shore of A.S., stretching for 40 km in the northeastern direction and representing a vast shallow water basin limited on the west by the *Agurme Peninsula* (see), on the north and east by the mainland, and on the south by *Tasty* (see), *Takhtaatau, Eshkeatau, and Aijarym Islands* (see). Except for some fishery areas, the bay was not suitable for navigation of even small vessels because it was overgrown nearly completely with reeds; in many places, it dried out and was filled with water only during large floods on the Syrdarya.

Bugun Bay – located in the south-east of the *Small Sea* (see), 22 km to the south-south-west of the *Kishi-Karatyup Bay* (see). The banks around the bay are low and sandy. In some places, hills up to 10–15 m high are found 100 to 200 m from the water edge. Bugun settlement is located on the southeastern shore of the bay, 2.5 km to the east-south-east of the Kara-Karnau Cape. Earlier, the Bugun fishery base "Goslov" and a fish processing plant were located here.

Bulrush (***Scirpus***) – a grass plant of the sedge family, mainly perennial though sometimes annual grasses. Approximately 200 varieties of B. are known. They grow mostly in areas with moderate climate of the Northern Hemisphere along the banks of rivers and lakes and in wetlands. In Russia and a Central Asia, about 40 varieties are found. They grow at depths up to 3 m and their leaves are up to 2.5 m long with stems 3 to 4 m high. Roots creep under water, while flexible stems rise above it. The most well-known variety is lacustrine bulrush (*S. lacustris*) with cylindrical stems nearly without leaves up to 2.5 m high. B. creates protective zones and contributes to ground silting which creates favorable conditions for algae growing. Their creeping roots facilitate spreading and form a solid, net-like cover on the bottom of shallow overgrown water bodies. B. always grows in water or near water. At depths up to 3–5 m, it forms dense thickets that by a wide belt go into a water body and replace *cane* (see) zones. Together with the cane, B. overgrows lakes. B. is often called cane, but the latter is markedly different from B. and is classified as a cereal. Some varieties of wildrye and selinum growing on sands in Central Asian countries and in the south of Russia are called sandy or dune B. Bulrush was well developed in low-saline waters of the Aral, in particular near the Akpetkinsky Archipelago (1960).

Butakov Alexey Ivanovich (1816–1869) – circumnavigator, rear-admiral (1864), investigator of waterways in Central Asia and A.S., and the "Magellan of the Aral Sea," as the great A. Humboldt called him. He graduated from the Sea Corps (1832), and I.F. Kruzenshtern recommended leaving him in the Officer Class, which he finished in 1836 when he was directed for work on the corvette "Lvitsa" (Lioness), then the frigate "Kastor," and still later the vessel "Bogatyr" (Athlete). In 1840–1842, with a rank of senior officer, he navigated round the world on the military transport "Abo." From 1845, he was a member of the Russian Geographical Society (RGS), and from 1848, he was chief of the expedition to survey and describe the A.S. shores. In spring 1848 in Orenburg, he organized production of parts for and assembly of the scow schooner "Konstantin" near Aralsk. In the summer and spring, sailing on this vessel, he described the northern, southern, and western shores of the sea and revealed the sea level fluctuations. In 1849, he was promoted to the rank of captain-lieutenant. For his hydrographic works, he was awarded the Saint-Vladimir Order of the 4th degree. Later, he headed the *Aral fleet* (see). In the western part of the sea, he discovered the Vozrozhdenia Island and other smaller islands. In 1850, B. went to Sweden to order two iron vessels for the Aral fleet. In 1852, a centerboard vessel "Perovsky" and a steam longboat "Obruchev" were delivered via Orenburg in a dismantled state to Raim and in the next year they were launched. These were the first steam vessels on the A.S. In 1850, information about the expedition was published in a short article, "Recent expedition for study of the Aral Sea" in the "RGS Geographical Proceedings." In Orenburg, he prepared the first map of the sea in the Mercator's projection and included with it albums with pictures of the exiled *T.G. Shevchenko* (see). He was reprimanded by the Emperor for inviting Shevchenko for work.

For investigations of A.S., on the A. Humboldt's initiative, in 1853 he was elected the Honorary Member of the Berlin Geographical Society and was awarded the RGS Demidov Prize.

In 1853, on the vessel "Perovsky", he crossed the Syrdarya and took part in the seizure of the Ak-Mechet fortress in Kokand. For building of the vessel, he was awarded the St. Ann Order of the 2nd degree. In 1854, commanding a unit of 300 Cossacks and 50 Bashkirs, he delivered the parts of seven iron row vessels for the fleet from Orenburg to the Aral fortress. He organized construction of a shipyard, a port for the fleet, and a slipway for pulling a vessel onshore. In 1855, he was promoted to captain, rank II. He prepared an inventory of the Syrdarya from the Kum-Sukhta area to the mouth. In 1857, with three hundred Cossacks, he took part in actions against rebellious Kirghizs near the Aryk-Balyk area on the Syrdarya. In 1858, with his fleet he carried the first hydrographic investigation of the Amudarya River for supporting the Russian diplomatic mission. Among the members of this expedition were O.V. Struve (later on the director of the Pulkovo Astronomical Observatory) and fleet lieutenant A.F. Mozhaisky (future inventor of a plane). In 1859, he helped the Kungrad Bek that separated from the Khiva Khan to raise the siege of Kungrad. He also continued his descriptions of the Amudarya delta.

In 1860, he was promoted to the rank of aide-de-camp and sent to England to order two iron vessels, a floating pontoon drydock, and barges. After the execution of this order, he was sent to North-American United States. In 1861–1863, he supervised the assembly of a floating pontoon drydock and the vessels "Aral" and "Syrdarya," which had been delivered from England, and made an inventory of the Syrdarya from the fortress Perovsky for 807 *verstas* upstream and in the Turkestan and Tashkent estates for 1500 *verstas* from the mouth of the river.

In 1864, he was promoted to rear-admiral, taking the position of the chief of the Baltic Sea fleet headquarters. In 1865, he became the junior chief of the Baltic Sea armored fleet and the commander of the soundly armored sheep squadron. After 1866, he was a member of the artillery division of the Maritime Technical Committee, and from 1867 was the acting squadron-major of His Majesty. In 1868, he went to Germany for medical treatment, where in the resort town Shwalbach he spent his last days.

Fig. 24 A.I. Butakov (1816–1869)

In 1867, the London Geographical Society awarded B. the Gold Medal of the society founder. In the same year, B. became a member of the RGS Board. By proposal of L.S. Berg, the southern cape of the *Barsakelmes Island* (see) was given the name of *Butakov* (see) and his name was also given to the *bay* (see).

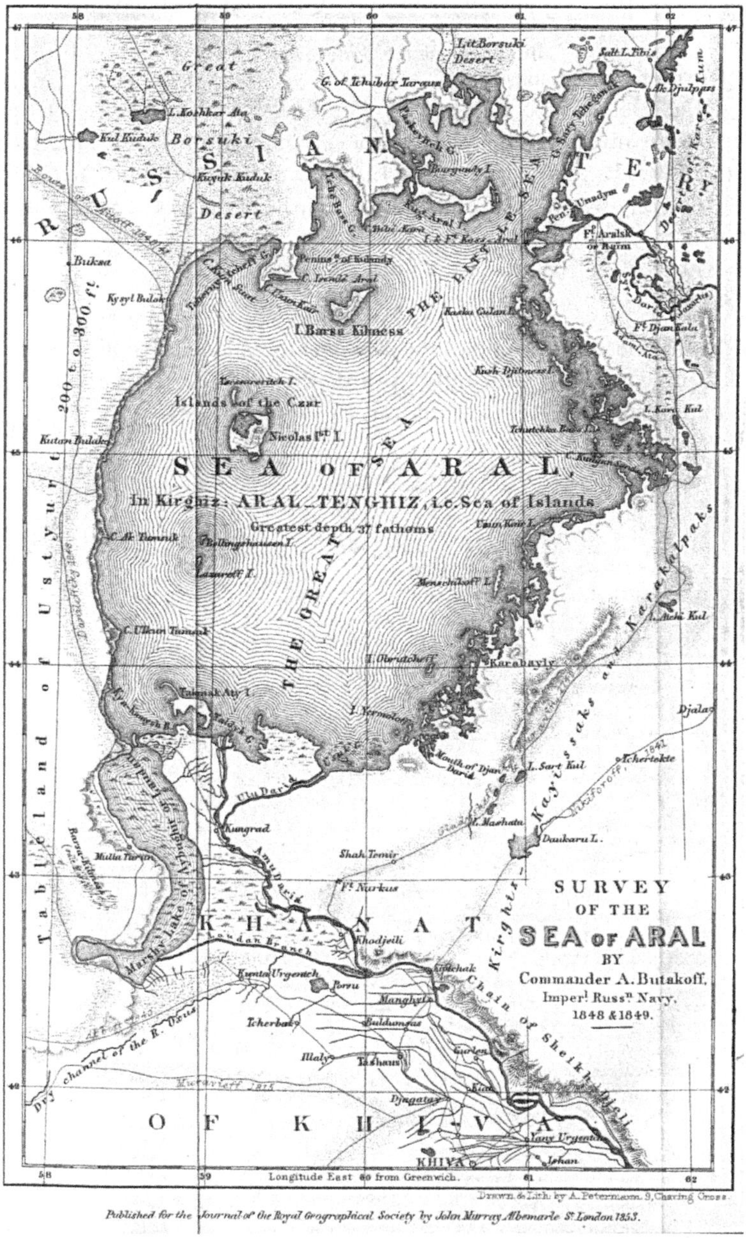

Fig. 25 The map of the Aral Sea by A.I. Butakov (1816–1869)

Among his published works are “Information about the 1848 expedition for the Aral Sea description” (1853), “Notes on meandering of the Syrdarya mouth beds” (1851), “Sailing of the ‘Petrovsky’ vessel over the Syrdarya” (1857), “Recent expeditions for the Aral Sea investigations” (1861), and “Diaries of A.I. Butakov made on the ‘Konstantin’ schooner during the Aral investigations in 1848–1849” (Tashkent, 1953).

Butakov Bay (former Paskevich Bay)* – located in the north of the A.S., demarcated on the east by the Kokturnak Peninsula, on the north by the mainland shore, and on the south-west by the Shubartarauz Peninsula. The entrance into the bay is located between the *Vasily Cape* (see) and the *Tastubek Peninsula* (see). The bay was deep at more than 20 m. The shores are montane: in its northern part, the shores are high and steep, while in the eastern and western parts wide sandy beaches are found. It is named in honor of rear-admiral *A.I. Butakov* (see).

Butakov Cape* – the southwestern tip of the *Barsakelmes Island* (see). Represents a high steep cliff projecting insignificantly towards the sea. The sea is deep near the shore. It is named in honor of rear-admiral *A.I. Butakov* (see).

C

Caique – a big boat without a deck with a carrying capacity of 5–50 tons. When sailing against a current, C. uses a rope that is pulled from the shore. In the early 20th century, C. was widely used on the Amudarya.

Cane (*Phragmites australis*) – before the *Aral crisis* (see), cane thickets covered the greater part of the Amudarya delta, with a biomass reaching 40–60 quintals/ha of dry weight. By the early 2000s, the area of cane thickets shrank significantly from 600 thou ha to 30 thou ha, and their size also somewhat degraded, the cane biomass not exceeding 20–25 quintals/ha and averaging 10–15 quintals/ha. In areas with high water salinity, biomass was only 1.5 to 5 quintals/ha. In the Circum-Aral area, two types of cane were found, aquatic and meadow. Meadow thickets were largely of economic significance because they were widely used for fodder stocking and summer grazing of cattle. One of the main conditions for development of the cane aquatic formations is the level of water salinity. Water salinity from 4 to 10‰ does not interfere with the development of cane, but with salinity above 15–20‰, thicket degradation is observed followed by slow demise. Through-flow of a water body is one of the essential conditions for cane growing. In low through-flow water bodies with the water salinity no more than 8–10 g/l, cane thickets are usually replaced by cattail that does not have wide application. Cane thickets may grow at depths of 1.2–1.3 m, with water bodies of depths no more than 1.0–1.2 m being optimal for cane growing. The main condition for development of the meadow cane thickets is high soil wetting and level of soil salinity. In the Circum-Aral area, these are chloride-sulfate soils with a chlorine ion content of no more than 0.1–0.25% and the sum of salts no more than 0.9%. On more saline soils, cane thickets grow poorly and on highly saline soils (sum of salts over 6% and chlorine ion content over 1.85%), they stop growing. Cane thickets may be restored by seed reproduction. In conditions of abundant light and moisture, cane seeds show good germination and by the end of the first vegetation period they form well-visible roots about 0.5 cm in diameter. In the second year, the cane becomes more than 1 m high and from the third-fourth years, forms normal thickets.

I.S. Zonn et al., *The Aral Sea Encyclopedia*, DOI 10.1007/978-3-540-85088-5_4,

Cane coast – cane-cattail thickets form the coasts of A.S., the Caspian Sea, Balkhash Lake, and others. As the shore zone of these lakes and seas is mostly shallow, the wave effect on the coast is minimized, creating favorable conditions for overgrowing of the coast and offshore zone with *cane* (see) and other hydrophilic plants. Cane thickets here form a band up to 200 or more meters wide with plant heights of 2.5–3 m. Sand and silt are deposited in the thickets and their buildup is conducive to further shallowing and shore development, which was the case on the Aral in the 1960s. This process proceeded at a rate of 15 m a year. First, a meandering C.C. is formed due to protrusion of some parts of the thickets, but bays are filled rather quickly, resulting in realignment of the shoreline.

Carpini Giovanni de Plano (1182–1252) – an Italian traveler, who, in 1245, following the order of Innocent IV, was directed as a head of the diplomatic mission to Mongolia, to the place of the Great Khan. His travel started in Lion, from where, via Kiev and the Lower Volga, he arrived on the northern coast of A.S. and then went along the Syrdarya valley, coming out to the Semirechye and over the southern slopes of the Mongolian Altai, where he reached his destination. In 1246, he arrived to the headquarters of the Great Khan, and in 1246–1247, he returned to Europe by the same route. C. left very valuable descriptions that, apart from purely geographical facts, contained extensive information about the economy, state and military organization, and everyday life and customs of the Tatars and other people.

Caspian Sea–Aral Sea Canal Project (CASPARAL) – a project to regulate the Caspian Sea level, while simultaneously replenishing A.S. to stabilize and restore its level. This project was proposed in 1987. The water should be taken from the Northern Caspian (Komsomolets Bay) and via a canal 540 km long, going along the northern chink (cliff) of the Ustyurt Plateau with the initial discharge of 1700 cu. m/s, reach A.S. Every year, up to 40 cu. km of water should be delivered, which will make it possible to check the water level rise in the Caspian Sea by 10–12 cm a year. Construction of a nuclear power plant to ensure operation of three pumping stations necessary for raising the water in the Caspian to the level of the Aral was planned.

Catfish ordinary (*Silurus glanis L.*) – large freshwater fish of the catfish family (*Siluridae*). Its length was up to 2 m, and its weight was up to 100 kg. The males were larger than the females. It lived in riverbeds and in lakes. In southern seas, it may migrate to the slightly saline waters. Usually, it was represented by a fast-growing and slow-growing (reed) form. The fast-growing form reached maturity on the 4th–5th year of life. It spawned in thickets on the coastal zone at water temperatures no lower than 20°C. Fertility was 11–480 thou eggs. They were large (2–3 mm in diameter) and sticky. Spawning was accompanied by mating games. The male guarded the nest. Already in its first year, S. led a carnivorous way of life. In A.S., it fed on the barbell fries. Being the most common, it had major commercial fishing significance, and was widespread in the

A.S. basin. Aral C. was also met in plain reservoirs. It had nonmigratory and migratory forms living in coastal water bodies. This species may be considered a biomeliorant devouring trash fish. At present, C. does not exist in A.S.

Central Asia – in physiographical terms, this is the part of Central Asia that entered into the Russian Empire and the USSR. It includes the Ustyurt Plateau; the Turanian Depression; the Turgai Plateau; the Kazakh hummocky area; and partially the Kopetdag, Pamir-Altai, Tien Shan, Djungarsky Alatau, Saur and Tarbagatai mountains. In Soviet times, it included Uzbekistan, Turkmenia, Tajikistan, and Kirghizia. Kazakhstan made an isolated subregion that was economically linked with the Siberian subjects of the Russian Federation, though some researchers consider Kazakhstan a traditional part of Central Asia, pointing to the historical, cultural, and language unity of the peoples living in this region.

Fig. 26 The map of the Central Asian States (http://www.lib.utexas.edu/maps/commonwealth/cis_central_asia_pol_95.jpg)

Central Asia – the geographical notion "Central Asia" was introduced for the first time by A. Humboldt in 1843. He defined C.A. as a separate world region. Historically, it was always associated with the nomad peoples populating it and the great Silk Road. On the Eurasian continent, C.A. was understood as a territory of the former Central Asian republics of the USSR – Turkmenia, Uzbekistan, Tajikistan, Kirghizia, and Kazakhstan. Its area is about 4 mln sq. km, with a population of 58.7 mln people (2001). The geopolitical notion of C.A. appeared on January 3, 1993 in Tashkent at the summit of five former republics of Soviet Central Asia – Kazakhstan, Kyrgyzstan, Tajikistan, Turkmenistan, and Uzbekistan, which formed the community called C.A. In hydrographic terms, C.A. is applicable to the basin of the Caspian and Aral seas. The members of this community are solidly linked by dozens of regional bilateral and multilateral international treaties in different fields, including very important political treaties on friendship and cooperation.

Central Asian Research Institute for Irrigation (SANIIRI) – the large scientific center on addressing the irrigation and land reclamation issues of today. In 1924, under the leadership of Professor V.D. Zhurin, Dean of the Engineering-Land Reclamation Department of the Central Asian State University, the project was developed upon establishment of the research hydraulic construction institute in Tashkent for carrying out investigations and training of scientific personnel in its laboratories. The project was approved by the USSR State Committee for Planning and from May 30, 1925, the first Central Asia Research Institute on Irrigation started functioning officially in Tashkent. At that time, it comprised hydraulic and water engineering laboratories and the bureau of field works. In 1926, in view of further development of investigations on water management in Central Asia, a decision was taken to pool all experimental and investigation efforts in irrigation in one organization. Pursuant to Order of July 1, 1926 to the Water Management Department (WMD) of Central Asia, the Experimental and Research Institute of Water Management (ERIWM) was a established, merging the Hydraulic Engineering Institute and Water Gauging and Water Module sections of WMD. This institute comprised the division of the control water gauging net, the experimental irrigation division (with all experimental irrigation stations), the operating and economic divisions, the building material laboratory, the pump water lifting laboratory, the bureau of field works, the bureau of data processing, the chemical laboratory, the calibration station, and others.

As a result of further reorganizations in 1932, the institute was transformed into SANIIRI, the structure of which envisaged integrated investigations in irrigation covering the issues of design, construction and operation of irrigation systems, hydraulic structures, land reclamation, mechanization of construction, and operational works.

From its onset, SANIIRI included into its activities water duty investigations that demanded the opening of new experimental irrigation stations – Pakhta-Aral, Ferghana, Khorezm, Iolotan, Samarkand and Bukhara. The institute was engaged in restoration and further development of the hydrological control net and conducted wide-scale hydrogeological investigation of large land areas. At the institute, there was created the workshop of high-precision tools and models that serviced its laboratories and produced for all Central Asian republics the devices, instruments, and equipment (current meters, water gauging rods, spillways, etc.) that were necessary for investigations and surveys.

SANIIRI conducted research in practically all directions of hydraulic engineering and land reclamation – hydraulic construction and hydraulics, engineering land reclamation, irrigation and operation of irrigation and drainage systems, construction of hydraulic and irrigation structures, and water resources management. The institute developed effective methods of regulation of the water-salt regime in irrigated lands; designs of irrigation, water supply, and drainage structures automatically controlled; and also highly efficient water application practices and new irrigation methods (drip, mist and others). In the recent decades, great attention was focused on integrated management of water resources in the A.S. basin with a view to restore the water surfaces in the Circum-Aral Area.

In 1993 the *Scientific-Information Center of ICWC (SIC ICWC)* (see) was established at SANIIRI.

Central Asian State Shipping Company – established in 1923 following the order of the RSFSR People's Committee of Transport Ways. It was engaged in passages (mostly cargo) along the Amudarya. The management office was in Tashkent, Uzbekistan. Until 1917, steam shipping along the Amudarya was rather insignificant; the passenger traffic from Chardjou to Termez and Chardjou-Petro-Aleksandrovsky (presently Turtkul) was serviced by the *Amudarya military fleet* (see). Cargo was transported along the river on wooden vessels (caique) by haulers. In the Soviet period, the shipping company received motor and non-motor vessels; new ports and quays were constructed. The yards for ship construction and repair appeared in Chardjou. The shipping company imported to the Central Asian republics bread, manufactured goods, chemical fertilizers, forest and petroleum products, etc. and exported to European USSR cotton and cotton oil. Within the basin, raw cotton and cotton seeds were transported. In Takhiatash and Chardjou, cargo was reloaded from railroads to ships and back. In Chardjou, the river shipping college was opened to prepare personnel to work for the shipping company. At its establishment, the company possessed 28 vessels with a total deadweight capacity of 2.23 thou tons, and by 1928 it had 12 vessels and steamers and 23 non-motor vessels with a total deadweight capacity of 3.15 thou tons. In 1934, the management office of the shipping company was transferred to Chardjou. In view of the economic development of the state in 1935, a decision was made to augment the Amudarya fleet. The shipping company received powerful tow and passenger ships, bulk-cargo, and oil barges.

Chagaly Bank* – the former Chagaly Island that was a part of the Tzar Islands. In the early 1990s as a result of scouring it turned into a bank. It was located in the middle of an entrance into the *Vozrozhdenia bay* (see) and stretched from the north to the south. The bank was made of stones and sand.

Chernyshev Bay – projects into the northern shore of the Small A.S. between the Uzynkair Cape and the Bezymyanny Cape located 22 km to the north-northwest of it and representing the southeastern tip of the *Korzhindy Peninsula* (see). It was named in honor of the Military Minister of Russia, Alexander Ivanovich Chernyshev. The eastern shore of the bay formed by the *Kulandy Peninsula* (see) was low and sandy with small dunes. The northwestern shore was high and steep; the height of cliffs increased from the top of the bay to the Korzhindy Peninsula. The shores were deep. Approximately 4 km eastward of the Bezymyanny Cape the bay had a depression 49.5 m deep, the largest in the North Aral. It extended for 8.5 km from the north-east to the south-west. Its width was 3.5 km.

Chimbai – the third largest city in Karakalpakstan. Located on the right bank of the Amudarya in the eastern part of its delta. The Keigli canal runs through the city. There is a cotton gin plant, a brick plant, and other enterprises. A pedagogical college and an agriculture mechanization college are found in this city.

Chink – is a deep, often vertically dipping cliff of a plateau in Western Turkmenistan and Kazakhstan reaching 300–350 m in height. Chinks mostly occur near Southern Mangyshlak, *Ustyurt* (see) (in particular, along the western Aral Sea coast), and eastern shores of the Krasnovodsk Plateau. Chinks often ends in depressions with solonchaks of various origin which are predetermined by tectonic structures.

Fig. 27 The chink along the southwestern coast of the Aral Sea (November 2007). Photo by Pavel Kosenko, http://pavel-kosenko.livejournal.com

Chokolak (Turkmen – a bushy sandy hummock) – high (up to 300 m) sandy hummocks with steep, at times steep, slopes, densely permeated on the top with the roots of plants and thickets (usually of tamarisk). They were composed of wet sand. These forms were linked with permanent groundwater recharge; however, high occurrence of ground water did not guarantee the Ch. formation, which often this led to salinization. They were distinguished from sand massifs and were often characterized by linear extensions. They were most numerous in the Kazakhdarya valley and in the south-west of the Amudarya live delta. Most probable appears the theory that the genesis of such eolina relief forms may be linked with their water-bearing disjunctives. Even small agglomerations of Ch. often coincided with traced tectonic dislocations (to the south of the Sudochie Lake along the chink southward of the Dzhyltyrbas Bay coast and in some other places).

Chumyshkol, Chumysh-Kul Lake – located southward of the Karakol Lake between the Sary-Chaganak Bay and the railroad. It has a rounded form. Its length is 2.8 km, and its width is 2.2 km with a maximum depth of 10 m. By the end of summer, the lake dries out. Its shores are low and even.

Chumyshkol Bay – located in the north of A.S., incised into the top of *Shevchenko Bay* (see), 12 km to the northwest of the Terestubek Bay. The bay was shallow; the depths in its central part was only 1.5 m.

Chushka-Bas Island* – located 10 km southward of *Kuraily Island* (see), in the eastern part of A.S. This was a low, sandy island inundated with water in some places. Sand bars extended from its northern and southern tips.

Circum-Aral Area – the territory around A.S., the vast land areas found in the delta plains of the Syrdarya and Amudarya in particular. In the upper reaches of the Amudarya, this area covered the Prisarykamysh, Akchadarya, and Circum-Aral (often called the Amudarya delta) deltas, originating in the Tuyamuyun narrow from the sea shores and at the foot of the chinks (cliffs) of the Ustyurt Plateau; in the lower reaches of the Syrdarya, it covered the modern and ancient Kazalinsk delta, the lower reaches of the Zhanadarya, Inkardary, and others. Administratively, this area included Karakalpakstan, the Khorezm Region of Uzbekistan, the Kyzyl-Orda Region of Kazakhstan, and the Dashoguz Region of Turkmenistan.

Circum-Aral expedition of the Department of Geography of the Uzbek Academy of Sciences – organized in 1976. In 1977–1979, it carried out field and laboratory investigations in the southern part of the Circum-Aral area. Special attention was focused on studies of the northern part of the Southern Circum-Aral area, which were in the A.S. margins that were most severely affected by the A.S. water level drop. The integrated studies of the natural environment of the Southern Circum-Aral area showed the intensive changes in the geosystems and ecosystems due to regulation of the water regime in the Amudarya delta and the sea level drop in the considered region. The general tendency of these changes have been desertification and aridization of natural complexes. The results of this expedition were included into the monograph, "Water level drop in the Aral Sea and changes in the natural conditions in the Amudarya lower reaches" prepared by A.A. Rafikov and G.F. Tetyukhin (1981).

Coast of the Aral Sea* – In 1960 the length of the coastline of A.S. was 3238 km. L.S. Berg identified three types of coastal topography: smooth even (western), lobate (northern), and embayed (eastern). The coasts differ greatly by their relief. Here one can meet highland plateaus with steep, nearly steep cliffs as well as low-lying sand coasts broken by many bays, and *kultuks* (see) overgrown with cane and reed. The sands of the *Greater Barsuki* (see), Smaller Barsuki, and *Circum-Aral Karakums* (see) come close to the northern coast, while the sands of the Kyzylkums (see) to the eastern coast, the desert *Ustyurt Plateau* (see), comes close to the western coast. The greater part of the southern coast is covered by the Amudarya delta.

The coastline in the north of A.S. is broken by many bays, gulfs, and kultuks. Many of these are shallow and partially overgrown with reeds. Five great bays are found in this part of the sea: the *Greater Sarychaganak* (see), *Butakova* (see), *Shevchenko* (see), Tsche-Bas, and *Chernysheva* (see). Far into the sea there four peninsulas: *Konturnak* (see), *Shubartaruz, Kapatyun* (see), and *Kulandy* (see).

The northern coast of A.S. is mostly high and steep, except for small low-lying coastal parts in the Greater Sarychaganak and Butakova Bays, in the eastern part of the Shevchenko Bay, to the south of the Torangly Cape, and in the eastern part of the Tsche-Bas Bay.

The vegetation on the northern coast is rather sparse, and in some places some shrubs and thin grass cover is found.

steep shores usually go deep into the water, with depths near them often reaching 10 m; the low-lying sandy shores are often shallow.

The western coast of A.S. is high and steep. The spurs of the desert Ustyurt Plateau come close to the coast along its whole run. From the Karatamak natural complex tract located in the northern part of the western coast the height of the cliffs tends to increase, reaching 190 m over the sea level. To the south of this point, the cliffs become lower, their height near the *Adjibai Bay* (see) being no more than 90–100 m. The western coast is broken, but insignificantly. There are no large gulfs and bays here.

The vegetation on this coast is very sparse. On the slopes of cliffs and in splits some shrubs are found.

The whole western coast goes deeply into the sea. At some distance from the coast, a deep trough, the deepest area (69 m in 1960) of A.S, runs parallel to the coast.

The southern coast is composed of sediments brought here by the Amudarya, which has changed its channel many times. Practically the whole coast is overgrown with reeds. For the greater part it is low-lying, strongly dissected, and several bays, such as *Adjibai* (see), *Muinak* (see), *Rybatsky* (see), and *Djiltyrbas* (see), extend into it.

The southwestern part of the sea is covered by reed thickets separating the Adjibai Bay from the low-lying Aibugir depression where the large shallow *Sudochy Bay* (see) is found.

From the Djiltyrbas Bay to the eastern shore of A.S., the coast is snaky; many kultuks and bays indent into it here. The Kazakdarya River, being one of the arms of the Amudarya, flows into the southwestern part of the Djiltyrbas Bay.

The eastern shore of the sea may be referred to as the so-called "Aral-type" *bay coasts* (see). These are low, sandy shores, heavily indented by numerous shallow bays and gulfs (*kultuks*, see) that at times extend far into land, like the Asche-Bas, Ak-Saga, Big-Tau bays, among others. Many small islands separated by a narrow straits, the *uzyaki (see)*, may be found near the eastern shore.

Numerous islands and islets are located in the southeastern part of the sea. They form the *Akpetkinsky Archipelago* (see) that covers an area of more than 2 thou sq. km. It was formed in the early 20th century when A.S., due to water level rise, projected deeply (for 40–50 km) into the Kyzylkum sands. As a result, many shallow lakes appeared on this coast that were linked with the sea via the narrow uzyaki; some of them were separated from the sea, but did not dry out because they were permanently replenished with water seeping through sandy bay-bars.

Fig. 28 The southwestern coast of the Aral Sea. Photo by Dmitry Soloviev, June 2008

The whole eastern coast of the Aral Sea from the Akpetkinsky Archipelago to the Syrdarya mouth, inclusive, was covered with thickets of reeds growing not only near the water edge, but offshore where depths may reach 2.5–3 m. Due to this fact, the configuration of the eastern coast was uncertain, which prevented an accurate delineation of the coastline.

Fig. 29 The sea foam along the southwestern coast of the Aral Sea (November 2007). Photo by Pavel Kosenko, http://pavel-kosenko.livejournal.com

Committee for Sudochie Lake Management (KUOS) – a nongovernmental organization, the associated members of which are JSC "Karakalpakbalyk," the State Committee for Nature Conservation of the Karakalpakstan Republic, the Ministry of Agriculture and Water Management of the Karakalpakstan Republic, and the Department for Nature Preserves and Hunting Grounds. The main function of KUOS is management of the Sudochie Lake in cooperation with the associated members and public supervising commissions established in settlements located nearby the lake. The KUOS Council, consisting of representatives of the associated members and public supervising commissions, makes decisions to ensure strict adherence to the rules and instructions on management of the Sudochie Lake. Decisions of the KUOS Council are binding for all associated members and representatives of the public supervising commission. Taking into consideration the need to maintain the minimum standards of water quality in the Greater Sudochie and Begdulla Aidyn Lakes, every year an agreement is made concerning additional maximum admissible water supply via the system of irrigation canals Raushan and Ustyurt. Every year, the required quantity of additional fresh water is approximately 600 mln cu. m and is negotiated between *BVO "Amudarya"* (see) and KUOS, which is authorized by the Council of Ministers of the Karakalpakstan Republic (Resolution No. 263/12 of November 1997) to undertake general management of the wetland areas of the Sudochie Lake.

Convention on the protection of transboundary water courses and international lakes – signed on March 17, 1992 in Helsinki, Finland. The Convention consists of the Preamble, 28 clauses (arranged in 3 parts: the first part – provisions common for all parties; the second part – provisions concerning the littoral parties; the third part – organizational and concluding provisions) and 4 annexes. In the Preamble, the parties, understanding that protection and management of the lakes are key and urgent issues the effective addressing of which may be ensured only through close cooperation, undertake all appropriate efforts to prevent, restrict, and reduce any transboundary impacts. Here, the parties are guided by such principles as taking cautious actions, "the polluter pays," and management of water resources without incurring damage to future generations. Littoral parties cooperate on the principles of equality and reciprocity, in particular, by entering into bilateral and multilateral agreements with a view to elaborate coordinated polices, programs, and strategies regarding respective watersheds or their parts to prevent, restrict, and reduce transboundary impacts and with a view of ensuring protection of the environment of transboundary waters or the environments influenced by such waters, including marine environments. This Convention formed a basis for preparation of various agreements among the Central Asian states on joint management and protection of the flow of the transboundary rivers Amudarya and Syrdarya.

Coordination Metrological Center (CMC) – The Regulations on CMC were approved by a Resolution of MKVK on February 11, 2000. Its main functions: provision of a single methodology for water measurement and improvement and updating of water metering devices installed on water management systems; metrological support of programs and resolutions of MKVK; development, improvement, and metrological support of devices for measurement of parameters of water resources; and participation in activities of interstate bodies on the issues of metrology and standardization in water management. In addition, CMC is responsible for supporting interstate programs of water resources management and protection, application of automation and metrology in the A.S. basin in particular. The center is located in Bishkek, the Kyrghyz Republic.

Cotton (*Gossypium*) – the "visiting card," brand of Central Asia. This is a plant of the mallow family having 35 varieties, including 5 cultural. Perennial woody and shrub wild plants are found in tropical and subtropical countries. They are cultivated in tropical and subtropical zones and in the south of the moderate zone. When cultivated these plants may be as high as 1–1.5 m. C. is the principal world fiber plant. Its fruits are balls containing seeds covered with long (25–50 mm) and short (3–15 mm) fibers – cotton. At ripening, the balls open, exposing fibers. The long fiber is used for manufacturing of cotton yarn, while the short one is used for manufacturing silk, cotton wool, lacquers, paper, explosives, and other substances. Cotton oil is produced from cotton seeds and is used for food and technical purposes. The stems are used for production of bast fiber.

C. is a warmth-loving plant. It develops well in dry weather and wet soil, thus it is largely cultivated under irrigation. The optimal temperature for growth is 25–30°C. It does not like frosts and survives them badly. Among the best cultural varieties of C. are Egyptian and Soviet fine-fiber varieties, Mexican ordinary, and Indochina.

It is believed that C. came to Central Asia via Persia from India, although the opposite is quite possibly true. In India, C. was already cultivated about 5000 years ago, a fact confirmed by digs in Mokhendjo-Doro.

C. was mentioned by chroniclers who accompanied Alexander the Great in his campaigns to the East, including India. From that time, European merchants brought, among other things, cotton fabric. As evidenced by Titus Livius in 63 B.C., Lentull covered the whole Rome forum with a tent of cotton fabric. Thirty years later, Caesar made a tent over the whole street leading from its palace to the Capitol. At the beginning of our era, in the Kushanian time, the peoples of Central Asia were already well acquainted with C. Herodotus, who lived in the 5th century B.C., wrote that the Indians wore clothes made of fibers growing on trees. The Greeks called such fabric the "wooden wool." In Europe, they knew and highly praised cotton fabric for a long time; however, they had quite a vague notion about the raw material from which it was derived. Various rumors, some rather fantastic, about cotton being half-plant and half-animal circulated. C. was called a Scythian lamb.

After the discovery of America, the European market was flooded with cotton fabric (perhaps the Mexicans and Peruvians were acquainted with cotton making still before the opening of America). In 1793, the American Whitney invented a machine to separate fiber from seeds, which caused a revolution in cotton manufacturing. By the late 20th century, the manufacturing of cotton fabric increased by more than 10-fold. The major manufacturers were China, USA, India, and Japan.

In the Soviet period, the objective was to ensure cotton independence of the country which spurred intensive development of the lands in Central Asia. To meet the objectives for the increase of cotton production in the late 1940s and early 1950s, gardens were cut off and vineyards were destroyed to clear the best lands for cotton growing. In the late 1970s and early 1980s, the rural settlements were enlarged for the same purposes. Small camps (kishlak) were leveled by bulldozers, and cotton fields appeared in their place.

In 1964, Uzbekistan produced 4 mln tons of raw cotton. The next target was 5 mln tons and then 6 mln tons. The latter was unattainable because all lands that could be withdrawn from crop rotation had been already in use.

By the 1980s, Uzbekistan became one of the world leaders in raw cotton production. Every year this republic produced 5–5.5 mln tons of cotton.

In the years of independence, due to a reduction of export possibilities because of the low quality of C.A. cotton and the progressing shortages of water resources, the lands under C. were gradually withdrawn from production. Cotton ruined the Aral Sea.

Fig. 30 The cotton (http://news.ferghana.ru/photos/2008_01/paxta_kuprin5.jpg)

"Creeping environmental problems and sustainable development in the Aral Sea basin" – the only collection of works by Russian and Uzbek scientists published abroad in 1999 by the Publishing House "Cambridge University Press" (ed. by M.H. Glantz (USA)). For preparation of this book, M.H. Glantz invited papers from Russian scientists who started working in the Aral from 1960 and were witnesses of the transformations of the sea proper and its environment. The book addresses the following issues: landscape composition changes (Ye.A. Vostokova); alteration of water level and salinity of the Aral Sea (V.N. Bortnik); desertification (A.A. Rafikov); climatic fluctuations and changes in the Aral Sea basin (A.N. Zolotokrylin); Pryaralie ecosystems and creeping environmental changes (N.M. Novikova); public health (L.I. Elpiner); impact of political ideology on creeping environmental changes in the Aral Sea (I.S. Zonn); changes of river flow (K.V. Tsytsenko, V.V. Sumarokova); fish population (I. Zholdasova); creeping environmental changes in the Karakum Canal's zone (N.S. Orlovsky); environmental changes in the Uzbek part of the Aral Sea basin (A.N. Krutov); and creeping changes in biological communities (N.V. Aladin).

Curative Aral muds – extensively occurring in the Amudarya lower reaches. Nearly all saline lakes abound in mud permeated with saline solutions. Their largest accumulations are found near a salt deposit in the *Barsakelmes Lake* (see). They are represented by black liquid and viscous muds with inclusions of small crystals of halite, gypsum, and fine-grained carbonate. A mud layer thickness varies from 0.1 to 0.45 m, and a mud zone has a width from 100 to 1100 m. In Karakalpakstan, mud of the highest quality used for mudtherapy is found in Karaumbet Lake. Here, seasonal therapeutic mudbaths are found where in summer the local population often goes for mud treatments under medical supervision. Therapeutic mudbaths are located near railway crossing point Kunhodja. Between *Muinak City* (see) and *Ushsai village* (see), the dried out *Sorkol Lake (Surgul*, see), the bottom of which is composed of black mud that by its composition and curative properties is similar to that in the Karaumbet Lake but of smaller scale, can be found. Curative mud is also found in other saline lakes.

D

Dambal Island* – located near the eastern coast of the Aral, 15 km to the south of the Kokarev Cape, concealing *Karatma Bay* (see) to the west. A wide strait formed between *Biiktau* (see) and D. Islands leading to Karatma Bay. The island is low, in some places inundated with water.

Darya, Derya – a big full-flowing river, sea (Persian). Great rivers and sometimes bays – Syrdarya, Karadarya, Amudarya and others – are referred to in this way in Central Asia.

Daryalyk, Kunyadarya – one of the western dried riverbeds that is clearly seen and traced along the whole Amudarya delta. The D. riverbed is more recent, and has a definite form of a meandering hollow that cuts into the Sarykamysh depression like a canyon up to 50 m deep through which the Amudarya waters flowed. In 1878, the waters broke through for the last time, resulting in the formation of a lake 8 m deep.

Dashoguzsky Velajat (former Tashauz Region) – one of 5 vilajats in Turkmenistan that was renamed in 1999. Belongs to the zone of the Aral ecological crisis or Turkmen *Circum-Aral area* (see). Located in the north of the country. On the north, north-west, and north-east, it borders on the Republic of Uzbekistan; on the west, it borders Balkansky vilajat, on the south it borders Akhalsky vilajat, and on the east it borders Lebansky vilajat. Its area is 73.43 thou sq. km, and has a population of 1196.7 thou people (2002). The center of D.V. is Dashoguz (former Tashauz). Practically the whole territory of the vilajat is covered by the Karakum Desert. In the north-west, the greatest part of Sarykamysh Lake is located. The Amudarya River flows along its eastern border. Being not far from A.S., this vilajat has experienced all of the consequences of the sea's drying. The greater part of its population is Uzbek.

The climate here is sharply continental and arid. Annual precipitations is less than 100 mm. The average annual temperature varies from 11°C to 13°C. The coldest month is January with an average temperature of –6°C (the absolute minimum is –36°C), and the hottest is July with an average temperature ranging from 27°C to 32°C.

I.S. Zonn et al., *The Aral Sea Encyclopedia*, DOI 10.1007/978-3-540-85088-5_5,

The share of D.V. in the country's industrial product is not so large as yet and by many per capita indices is lower than the country's average. Its specific weight in the total industrial production of the country is 9.2% (2000).

The area of irrigated and drained lands is 2711.5 thou ha, including priority development lands on 1300 thou ha. The key branch is distant grazing animal husbandry (73 thou sq. km). Farming is practiced here on 100% irrigated lands. The main crops are cotton and grains. The source of irrigation water is the Amudarya. In low-flow periods, the shortage of irrigation water is felt. Some *etraps* (districts) of D.V. suffer from a deficit of drinking water of good quality. This is due to high pollution of surface waters, transboundary transfer of *collector-drainage waters* (see) from Uzbekistan, and the narrowness of fresh groundwater lenses under the channel.

Degradation of lands – reduction or loss of biological and economic productivity and a complex structure of rainfed arable lands; irrigated arable lands or pastures; forests and forested areas in arid, semiarid, and dry humid regions as a result of land use or the action of one or several processes, including those related to human activities, such as: (1) wind and/or water soil erosion; (2) deterioration of physical, chemical and biological or economic soil properties; (3) loss of the natural vegetation cover for a long period. The three principal factors leading to degradation of soils in arid areas are overgrazing (34.5%), deforestation (29.5%), and existing agricultural practices (28.1%).

Deigish, Degish (Turkmenian) – descructive scouring by a river of low loose banks causing avalanches, partial collapse of bank deposits, and displacement. This is especially true of the *Amudarya* (see) in its lower and middle reaches. In this way in 1949 the city of Turtkul (the former capital of Karakalpakia) was destroyed and the capital was moved to Nukus. The main causes of D. are occurrences of easily scoured alluvial soils in the banks, a high flow rate in a river, and great fluctuations of water discharges. During many years, the riverbed has altered significantly; since 1943, bank displacement was as large as 2.5–4.0 km.

The D. phenomena affects mostly the stretch from Tashsak to Djumurtau and banks here must be kept under constant surveillance to prevent likely scouring.

In the coastal strip and the likely zone of river meandering, the settlements, quays, and water intakes of canals are found. The coastline may be stabilized with the help of structures made of strong and reliable modern materials. Bank fixation works should be carried out along the whole stretch of a scoured bank.

Delta of the Amudarya* – located to the north of the line connecting the Takhiatash Cape and the eastern tip of an offspur of the Ustyurt Plateau (northward of the Buten-Tau upland). The *Near-Sarykamysh delta* (see) is located to the south of this line. The western border runs along the eastern chink of the Ustyurt as far as the Aral, while the eastern border runs along the right bank of the Kuanishdjarma. Within the aforementioned limits, D.A. area is equal to

19 thou sq. km. It represents a monotonously flat plain composed of intercalating sands, sandy loams, loams, and clays. It extends for 360 km to A.S. One side of this plain smoothly lowers down to the north to A.S., while the other extends to the *Sarykamysh depression* (see). Against this generally flat relief are several prominent residual uplands, including Kubetau, Djumurtau, Kulaknatau, *Beltau* (see) and others. Their relative elevation over the delta plain is 60 to 80 m. Water abundance, rich vegetation, and A.S. proximity influence the climate of D.A. The sums of negative temperatures in the wintertime vary from –300°C to –500°C in the north. Winter lasts here for 3 months. The absolute maximum of temperatures over the greater part of the territory is 41 to 43°C. The sums of positive temperatures over 10°C make 3800° in the north of the delta and 4250° in the south. Annual precipitations here range from 80 to 110 mm; summer precipitation composes 10–15% of the annual amount. In D.A., draughts and dry winds are observed quite seldom, which may be attributed to good air moistening due to evaporation from the surface of the river and its tributaries as well as rich vegetation.

Modern D.A. covers nearly the whole southern coast of A.S. in its former borders (up to 53 m abs.). Until recently, it was second only to the Volga delta by its size, productivity, and biodiversity. In the course of its development, D.A. constantly migrated following the buildup of alluvial deposits and the changing direction of the main river bed.

The most recent Amudarya deltas were formed in the Paleozoic depression more than 140 m deep, the northern part of which is occupied by A.S. During the Quaternary period, three deltas were formed here – *Sarykamysh* (see), *Akchadarya* (see), and the modern Circum-Aral. In the period of A.S. stability, D.A. expanded mainly due to sediments forming a protrusion delta. The average annual delta increment was 13 sq. km. In the period of sea level drop, the delta grew, largely due to extension of the dried sea bottom. In the last 35 years, the delta area has increased from 14 to 21 thou sq. km. The relief and modern structure of the D.A. landscape was formed under the influence of geology, hydrological regimes, climatic conditions, and anthropogenic factors.

In the period of the ordinary delta regime, the average annual input of suspended matter into the delta was 3–4 tons a second. And during a year the sediment layer in the floodplain increased by 7 mm. The deposition intensity in the regions of highest active accumulation was as high as 1 mm/day during a flood period.

As a result of the annual cycle of self-leveling, a hydrographic network of the delta composed of fine-grained sands was formed. Natural levees along the riverbed were characterized by interbedding deposits represented in the foot by thick horizons of fine-grained sands and loamy sands overlain with loamy sandy and loamy horizons. The near-bed floodplains were composed mainly of alternating loamy sands and loams. Internal floodplains were formed in the lowest parts of the delta. They were mostly occupied by marshes and lakes. At present, many lakes and marshes have dried out and salinization and takyr-formation processes are observable.

At present, many different landscapes may be found in the delta; they were shaped due to different combinations of moistening, salinity, and soil composition.

The near-mouth terrain and water-filled arms of the Amudarya are covered by special types of vegetation communities: *tugai forests* (see); tamarisk communities (found on dried riverbeds); and delta depressions moistened by collector-drainage waters which are overgrown with reeds.

In the upper part of the delta, oases of cotton, corn, white durra, alfalfa, melon crops are cultivated under irrigation. Most marshlands are used for rice growing.

In the pre-mouth area and in the dried bottom of A.S., solonchaks and saline sands prevail. This territory is characterized by low productivity with thinned communities of annual thistle. With further drying out of A.S., this type of landscape moves more and more to the north. In water abundant years, after the periodic flooding of these territories, the vegetation productivity here increases significantly. In the south of the dried sea bottom, the areas of artificially watered irrigation and disposal lakes gradual expand.

The delta starts below Takhiatash and represents a slightly inclined plain. The first right delta channel going out from the Amudarya is Yerkendarya. The Raushan channel originates nine kilometers downstream; in 1943 it was dammed. The water is supplied here by floating pumping stations (discharge is up to 100 cu.m/s). Downstream, at the head of the Raushan channel, the Amudaria is divided into two branches that form an island. Near the northern edge of the Kyzyldjar upland the Priemuzyak channel branches off westwards from the Amudarya. Downstream of this place, the river bed breaks into two channels: *Kipchakdarya* (see) (left) and *Akdarya* (see) (right). At a distance of 20 km from A.S., these channels join together forming a single bed, Inzheneruzyak, that at a distance of 3 km from the sea is broken into mouth arms. Large channels, such as Taldyk, Kipchakdarya, Akdarya, and others flowed over the delta. Until recently, it had many small lakes, marshes, hydrophilic vegetation, and mostly reeds. Tugai thickets extended along the channels. Now everything is dried out, and the delta surface is barren. Only those lakes have survived that received collector and drainage waters from irrigated fields. Here, residual hills composed of parent rocks, the largest of them being Kyzyldjar, with its relative height of 50 m, are found.

Delta of the Syrdarya* – a plain slightly inclined towards A.S., low-elevated, alluvial, having a triangle shape with the apex to the east of *Kazalinsk City* (see). At the head of the delta, its uplifted part is 65–67 m high, while the peripheral parts are lower than 53 m. In the middle of the delta, the Syrdarya River flows, the width of the delta plain on the right bank of the river varying from 10 to 70 km and on the left bank from 5 to 50 km. The total delta area was approximately 7,000 sq. km. In spite of the ideally flat relief of the delta, it has many channels and arms, near-bed levees, and inter-bed lowerings. The river floodplain is not well visible: in some places it can be traced as a narrow strip up to several dozen meters wide, while during spring floods it is covered with water and disappears completely. The live (Aksai) and dead (Karauzayk, Karaaryk, Keigushken, Birkazan, Sarykol, and others) arms branch off from the right and left banks of the Syrdarya. Some of the arms are used as irrigation canals, while others are used as disposal headers. Along the riverbed, narrow, elevated levees with gently-rolling microrelief are clearly visible in the delta head where they rise by 2–3 m over the peripheral plain. Their width here reaches 2.5–3.0 km. The closer the sea is the lower the levees, and consequently, the narrower is their width. In the delta head in many places with lower elevations of near-channel levees, earth dams are constructed to protect the nearby territories from flooding. The near-bed levees are usually composed of loamy sands and sands whose layers may be as thick as 4 m. The inter-bed depressions occurring between the channels are places where flood waters accumulate and this facilitates formation of lake-marshy complexes here.

The Syrdarya lower reaches confined to the desert zone are characterized by a sharply continental climate. In winter, the temperatures vary from $-10.7^{\circ}C$ (in January at the Karak station) to $-13.4^{\circ}C$ (at the Aral Sea station); in summer, temperatures range from $+27.6^{\circ}C$ to $+26.1^{\circ}C$, respectively, in July. The absolute maximum of temperature is observed in July (43–46°C); the absolute minimum is registered in January (–38°C to –42°C); the average annual temperature varies from 7.0°C to 8.8°C. The cold period with average daily air temperatures below 0°C lasts for 120–130 days, while the non-frost period lasts for 170–180 days.

The average annual precipitations range from 104 to 132 mm. The prevailing wind direction during the year is northeastern. The wind velocity tends to increase in the northwestern direction (i.e. towards A.S). The average annual wind velocity over much of the territory is 3 to 5 m/s, and the number of days with strong winds (15 m/s) varies from 8 to 22.

The average annual water flow in the river on the line of Kazalinks City (1912–1966) was 476 cu.m/s (with the extreme values being 378–670 cu.m/s). The average many-year maximum flows were as high as 865 cu.m/s.

The delta zone abounds in lakes. In terms of origin of their depressions, shape, and water composition, they form 5 groups. The first group includes oxbow lakes having a sickle form. They were usually found within the near-bed levees and represented sites separated from the main riverbed or arm. The depth of these lakes usually varied from 2 to 3 m, while their width were up to 200–300 m, and their length was up to several kilometers. The lakes contained fresh waters that were replenished during floods on the Syrdarya. The second group includes lakes of inter-bed depressions, the so-called floodplain lakes. They were distinguished by their greater size, often reaching several kilometers in their cross-section. They were characterized by vague, uncertain outlines, marshy banks, small depths – only up to 2 m. Such lakes were inundated with freshwaters during floods on the Syrdarya, but by autumn, many of them would dry out completely. These lakes appeared when deflated depressions formed within spot sand terrains and filled with water. They were filled with ground waters backed up due to hydraulic head from the flooded areas. The third group includes lakes with brakish waters that were cut off from sea lagoons. The largest of these groups were Kamyshli-Bash, Chumyshli-Kul, Rimsky, and others. They are dammed lakes.

D.S. is composed of the following soil types: meadow (deltaic), marshy (deltaic), solonchaks, takyr-like, and others. These soils contained small humus reserves in the upper horizons (1–2%) and within a 50 cm-layer (57–106 t/ha), and no progressive buildup of salts was observed in them.

The vegetation cover in the delta was formed due to the influence of such factors as relief, soils, their salinity and moistening, ground waters, depth of their occurrence and salinity, human direct and indirect impacts, and others.

The near-bed levees of the Syrdarya and its live tributaries were overgrown with tugai forests extending in strips up to 1–3 m wide along the riverbed. Among the named trees were oleaster, poplars, willow as well as the thickets of salt trees and tamarisk. The inter-channel depressions that are filled with water during floods are covered with reed thickets, and on the lake shores cattails are growing. The near-channel levees are usually covered with weed vegetation. Scarce halophyte thickets are found on saline soils and solonchaks.

Regulation of the Syrdarya flow, drastic reduction of the Aral water area; resultant climate changes in the region, making it more arid and continental; growing salinity of irrigation water; deposition of salts drifted from the exposed A.S. bed; and deficit of irrigation water led to the drying out of the Syrdarya delta.

Department of the Amudarya Irrigation Canals (DAIC) – one of the oldest regional water management organizations. In 1927, on the basis of the parity commission located in Tashkent, there was established the Department of the Amudarya Delta Irrigation Systems (DADIS) in Novo-Urgench. This organization was assigned to construct, rehabilitate, and refurbish irrigation and drainage systems, as well as to establish an equitable inter-republic division of water. With the extension of cultivated areas, irrigation and drainage systems were developed, increasing the activity of DADIS. The irrigation canals were linked. After the separation of the water management organizations of Khorezm and Tashauz from DADIS, the latter received a new name, the Department of the Amudarya Irrigation Canals (DAIC). For satisfaction of the needs of water users, DAIC operated 350 km of interstate irrigation canals, 60 major hydropower plants, and 110 gauging stations.

When the Amudarya water resources could no longer fully cover the needs of water users, a very difficult water situation was witnessed even in years of average water availability. The problem with water supply was aggravated by the absence of a single water management body.

Pursuant to the decisions of the October (1985) Plenum of the CPSU Central Committee, in 1987 the basin water management associations (BVO) *"Amudarya"* (see) and *"Syrdarya"* (see) were established under the system of the USSR Ministry of Water Management. Their tasks included the apportioning of water resources among the republics and operation of water intakes and hydraulic structures. They were located in Urgench and Tashkent. Later on within their framework, the

Kurgan-Tyube, Chardjou, Urgench (DAIC), Nukus (BVO "Amudarya") and Gulistan, Uchkurgan, Chardara, and Chirchik (BVO "Syrdarya") Territorial Production Departments were formed. They engaged in the regulation of water resources and the operation of water intake structures.

Today, DAIC has undertaken operation of 11 river water intakes, 52 hydraulic structures on main canals; maintains and operates 386 km of main canals; and controls water intakes on the river stretch from the Tuyamuyun hydraulic unit to the Kipchak gauging station, a river stretch of 167 km. Three large irrigation systems – Tashsakinsky, Klychniyazbaisky and Kipcha-Bozsuisky – are subordinated to DAIC.

Desertification assessment and mapping in the Aral region – the section in the World Atlas of Desertification – II (second edition, 1997). Prepared by Doctor of Biology G.S. Koust (MSU). It contains descriptions of desertification in the Southern and Eastern Aral Region, including cartographic materials illustrating such desertification parameters as causes, vectors, risks, degrees, rates, and depths. Among the maps are basic ones illustrating the prevailing causes and vectors of desertification and supplements showing the areas affected by various directions and impacts of desertification. The studied region covered 127 thou sq. km, and the mapping base contained 1500 contours identified by the results of analysis and interpretation of satellite images for the period from 1975 to 1989.

Desertification control and Sanation of Solonchak deserts in the Aral Sea region – BMBF-GTZ/CCD Project (Germany). Experiments on the development of new methods for saline soils of heavy lithology were conducted from 1998 to 2000 within the framework of BMBF Project 10 km to the southwest of the Syrdarya mouth by the Aral Institute of Agroecology and Agriculture (Kyzylorda). In 2002–2004, the experiments were continued. The representatives of the Bielefeld University (Germany) as well as Kazakh organizations: the Institute of Botanics and Phytointroduction (Almaty), the Aral Institute of Agroecology and Agriculture, the Scientific-Production Forestry Center (Kokshetau), and enterprises of "Syr-Tabigaty" (Kyzylorda) took part in this project. During two seasons (November 2002–March 2004), the following salt-resistant varieties of local flora were sown and planted: black saxaul (*Haloxylon aphyllum*), tamarisk (*Tamarix elogata, T.ramosissima, T.laxa, T.hispida*) and *Halocnemum strobilaceum*. Trial plots were established on soils of different mechanical composition and salinity level on the eastern shore of A.S. in the Kozzhetpes natural area within the sea belt dried in the 1970s. The plantings were made both with a tree-planting machine (252 ha) and manually (9.5 ha). The experimental results were used for preparation and publication of the "Recommendations on assortment and technology of cultivation of halophytes: shrubs and trees on the dried out Aral Sea bottom" (2003).

Desertification map of the southern and eastern circum-Aral area – prepared by G.S. Koust in 1993 at a scale of 1:500,000 within the framework of the UNEP/USSR Project, "Assessment of Desertification in the Southern and Eastern Aral

Area." The map shows the causes of desertification, main trends in desertification development, and their manifestations. The map was used as a basis for preparation of the comprehensive *"Desertification Atlas of the Southern and Eastern Circum-Aral Area"* (see).

Desertification map of the southern circum-Aral area – prepared in 1988 at scale 1:200,000 by the Geography Department of the Uzbek Academy of Sciences under editorship of A.A. Rafikov. Much attention was paid to consideration of the desertification factors of both a natural and anthropogenic nature. An attempt was made to assess the relationship between natural and anthropogenic factors, providing particular examples when the development of anthropogenic desertification is aggravated by natural factors. The map shows a classification of desertification causes and desertification types. In particular, *desertification types* identified are vegetation degradation, deflation, water erosion, soil salinization, technogenic causes, desertification related to the water level drop in A.S., desertification related to groundwater level lowering, and related soil salinization; also identified are *causes of desertification*, such as (a) undergrazing; (b) overgrazing; (c) cutting of trees and shrubs; (d) low efficiency of collector-drainage systems (KDS) and poor leaching of saline soils; (e) high occurrence of ground waters and low efficiency of the existing KDS; (f) hydromorphic regimes of irrigation on the basis of well-developed drainage systems; (g) irregular motion of motor transport, crawler and wheel tractors and others, intensification of research, drilling, construction and other kinds of works; (h) drying and salinization of the beds of water bodies due to a water level drop in them; (i) degradation of the vegetation cover due to termination of regular water supply of the delta plains; (j) cultivation of steep slopes and inadequate erosion-control efforts; and (k) insufficient seepage-control measures.

Desertification of the circum-Aral area – a comprehensive analysis of the environmental situation in the region has shown that the key factors of desertification are human activities changing ecosystems on the A.S. dried territory, salinization and waterlogging of irrigated lands, cutting out of forests in the zones of the river flow formation and saxaul vegetation in deserts and semi-deserts, loss of vegetation cover of pasturelands due to overgrazing and transport denudation, growing areas of solonchak deserts in drainless zones, and irrigation erosion of irrigated lands.

Rough estimates have indicated that in the period 1960–1990, the area of natural deserts in Central Asia increased by more than 100 thou sq. km, or nearly 8%. Over 50% of irrigated lands were affected by salinization. In the lower reaches of the Amudarya and Syrdarya, where irrigation was mainly practiced, the land salinity varied from medium to strong, though in the recent 25 years it had doubled.

Until the mid-20th century, the wet and dry periods had alternated, and as a result, due to riverbed erosion, one and another area of the delta was always either wet or dry. With the establishment of grand projects on irrigation development and desert conversion, Central Asia moved to a new stage – sustained desertification in river deltas and further on in the Circum-Aral area in general. Large headworks regulated the flow of the Amudarya and Syrdarya Rivers, disturbing its annual and seasonal dynamics. In conditions of extensive water consumption, the low efficiency of irrigated networks and water application practices led to the increase of irretrievable water losses. Beginning in the 1960s, cyclic climate aridization also affected river flows in Central Asia. As a result, in 1961 a permanent reduction of flow to A.S. was witnessed.

In the 1950s, the first rough features of desertification onset and the further intensification of aridization in the eastern part of the Amudarya delta were observed. This development was connected with a sharp increase of water intake by the Karakum Canal, the appearance of large areas of newly irrigated lands, etc. At the same time, large floods and inundations ceased and many arms in the delta dried out. Also witnessed were the gradual decrease of evaporation from the delta surface, some drawdown of the groundwater table, substitution in uplifted areas of the flushing water regime of soils with the exudative regime, soil salinization, soil drying, and rough successions of hydrohalophytic and haloxerophytic vegetation.

In the 1960s, the processes of overall aridization and continental desertification were developing in the delta. This was associated with a reduction of water inflow into the delta and a dropping of the Aral Sea level. In this period, the drying of many secondary river arms, drawdown of the groundwater table, and a growth in groundwater salinity were observed. The zone of favorable climatic impact of the sea shifted following its regression into the newly-dried zone. Deeper spreading of aridization and drying of hydromorphic landscapes and mass successions of hydrohalophytes were witnessed in the central and western parts of the delta. The area of cane wetlands shrank drastically, replaced with tamarisk and thistle growths.

The lowering of the base level of erosion spurred erosion processes in riverbeds, the prevailing exogenous processes changing from waterlogging to deflation and salt accumulation. As a result, the zonal landscapes of sandy deserts with eolian reliefs began forming in the eastern part of the Amudarya delta.

From the early 1970s, the aridization and transition of some landscapes to zonal (continental) desertification was observed. The ongoing reduction of water inflowing into the delta and general intensification of economic activities in the "live" delta of the Amudarya against the continuing climate aridization became typical events. Most intensively, the desertification processes increased in 1974–1977 when excessive water intakes combined with a dry period and the insufficiency of water in rivers. In this period, "chemical" desertification started in the delta, connected with the growing salinity of the river water and the growing level of pesticides that drained into waterways with the runoff. The water level drop in the sea increased enormously, the area of the dried seabed widened, and the puffy solonchaks that lead to salt-dust storms formed. The hydrological role of rivers changed from feeding to draining, and nearly all channels and arms of the Amudarya dried out. Deltas and especially their depressions accumulated salts from the whole Aral basin. At the same time, a considerable amount of salts fell with atmospheric precipitations or drifted with wind (up to 1 t/ha a year).

In the Syrdarya delta, the beginning and rate of desertification were somewhat different because the moist Syrdarya delta consisted of two isolated massifs; the desertification process did not affect them simultaneously. Already in the 1910s the lake-wetland landscapes of the Kyzyl-Orda part of the delta were first affected by drying associated largely with intensive agricultural development, land reclamation development, and to a lesser extent with the reduction of the Syrdarya flow. Still earlier (in the late 19th century–early 20th century), the hydromorphic landscapes in the middle reaches of the Kuvandarya River were affected by desertification. In the lacustrine-wetland, depressions composed of meadow-bog soils formed sagebrush-thistle communities on the meadow-desert soils as well as paleohydromorphic relics like puffy solonchaks and black saxaul growths. In the Kyzyl-Orda delta and on the Kuvandarya plain, the desertification process went slowly. By the 1920s on the Kuvandarya plain, the zonal desert landscapes were formed. The Kyzyl-Orda delta (as well as the Khorezm ancient delta of the Amudarya) was used nearly all for irrigated farming. Complete drying of the lacustrine-wetland had occurred by 1950–1960.

Simultaneously with the Amudarya "live" delta that had passed similar stages of desertification, the landscapes of the "live" Kazalinsky delta of the Syrdarya River were also transformed. In general, the desertification processes led to still more intensive changes of hydromorphic complexes in the Kazalinsky delta compared to the Amudarya delta, which may be attributed, perhaps, to its lower watering.

Desertification also affected the landscapes of sandy eolian plains in the Circum-Aral area, in particular in the east and southeast. Desertification processes were spurred here by a drop of groundwater level (in some places to 8–10 m), changes in the hydrostatic backup of the sea, climate alterations toward aridity, intensification of eolian processes, and anthropogenic activities (cattle grazing, etc.).

Desertification of the southern circum-Aral area – a map at scale 1:200,000 prepared by the Department of Geography of the Uzbek Academy of Sciences. Its editor was A.A. Rafikov.

Dictionary on desertification – prepared by Doctor of Geography I.S. Zonn. It was published in 1995 (1st edition) and in 1996 (2nd edition). It contains about 400 terms on arid and semiarid territories, their study, mapping, and ways of desertification control. It describes widely the role of UN organizations (UNEP, FAO, UNESCO, WMO and others) and international governmental and nongovernmental organizations in desertification control. The dictionary has been translated into English, Japanese, and Chinese.

Djalpakkair Strait* – located to the south of the Akpetki and Karabaili Islands. On the northwest it is covered by the Orussengir and Karashokhe Islands.

Having a general eastern orientation, it extended for over 11 km. The low-lying sandy islands *Karashokhe, Shagyl, Annabai, Estai, Kuvat, Erdjan* (see) are found to the west of the strait along the southeastern coast of the sea.

Djaltyrbas, Djiltyrbas, Djaltyr-Bas Bay – see *Zhiltyrbas bay*.

Djaksyklych Lake (Kazakh – *Jaksykylysh*) – located in the north of the Circum-Aral area, 15 km north-east of the city of Aralsk in Kazakhstan. The absolute elevation of the lake bottom is +43 m, which is nearly 10 m lower than the maximum (+53 m) Aral level. Terraces visible on its shores reflect the stages of periodical drying of the lake basin. D.L. may be divided into two large basins: Northern (17.6 sq. km) and Southern (57.6 sq. km) that are surrounded by numerous small saline lakes. The origin of salts is connected, on the one hand, with the evaporation of A.S. waters that flow via a narrow strait during high water levels, and on the other, with ground waters that come to the surface not only in the periphery, but in the bottom of the lake basin. The large Djaksyklych salt deposit is found here.

Djeikhun – an Arabic name of the Amudarya.

Djut (Mongolian) – lack of fodder for cattle due to coating of pastures with ice. This phenomenon occurs quite often in the A.S. region.

Does the Aral Sea merit heritage status? – this article, written by Michael H. Glantz and Robert M. Figueroa, Philosophy Department of the University of Colorado, Boulder, USA, was published in December 1997 in the journal "Global Environmental Change." The article was written for the occasion of the passing in 1972 by the UN of the Convention Concerning the Protection of World Cultural and Natural Heritage Sites in view of the need to acknowledge the global significance of protection of cultural and natural heritage. According to the authors, A.S. satisfies many criteria of this Convention as an object deserving such status. This article pursued three purposes: first, to discuss the notion "heritage" and its applicability to the world heritage status; second, applicability of this notion to the A.S. basin in an attempt to find an answer to the question posed in the article title; and third, a discussion of "environmental fairness" issues that may be raised as a result of such status.

Dolgy Island* – covers *Kishi-Karatyup Bay* (see) on the north–west. This island is low-lying and sandy, with small barkhans 2–3 m high. Several small bays extend into its coast.

Domalak Cape* – the northern tip of *Kokaral Island* (see) and the end of a low sandy bar overgrown with shrubs that stretches for 4 km northward of the northeastern shore of the Kokaral Island.

Dying and dead seas: climate versus anthropic causes (Edited by J.C.J. Nihoul, P.O. Zavialov, Ph.P. Micklin. NATO ARW/ASI Series, Kluwer Acad. Publ., Dordrecht, 2004) – the proceedings of the colloquium with the same name that was held at Liege University, Belgium, in 2003 within the framework of the 35th

International Liege Colloquium on Ocean Dynamics under the auspices of the International Year of Fresh Water declared by UNESCO. This collection contains 15 articles that consider the problems of A.S., Caspian Sea, Dead Sea, Chad Lake, Issyk-Kul Lake (Kyrghyz Republic), and Corangamite Lake (Australia).

Dynamics of the Aral Sea from 1957 to 1989 with forecast till 2000 – a map at scale 1:1,000,000 prepared in 1990 by Z.D. Tkacheva and V.M. Sigalov in Moscow.

E

Eastern (Aral) Sea – the remaining part of A.S. located to the east of the merged chain of the *Vozrozhdenia* (see), *Lazareva* (see) and *Komsomolsky* (see) Islands is referred to in this way in scientific literature.

Ecological-geographical database of the Aral Sea basin – developed by the Laboratory of Land Ecosystem Dynamics of the RAS Institute of Water Problems for monitoring of the natural ecosystems in the A.S. basin. The database was prepared using the results of expedition research and also data of stationary and semistationary observations received for the period 1978–1997. The data from the published works of foundations and archives for 1947–1994 were also included into this database as literary sources. The relevant hydrometeorological information received by hydraulic and meteorological stations located in the Amudarya and Syrdarya basin was also used. All data were subject to preliminary analysis and processing and were converted into a standard form for input, storage, and further utilization for addressing a wide spectrum of environmental issues.

The database includes several thematic sections reflecting the condition of the ecosystem's basic components – soils, vegetation, climate, surface, and groundwaters. All sections are interrelated and provided with a reference-information system. The geographical database is based on the scheme of landscape-typological zoning with a multilevel hierarchical system of units. This scheme was used for identification of the Amudarya and Syrdarya basin and the drainless areas in the Karakums and Kyzylkums in the territory of the A.S. watershed. Each river basin is divided into functional parts: watershed, floodplain, mountain, plain, and delta.

Ecological map of the circum-Aral area – published in 1991 in Almaty.

Ecological refugees – people whose lives were at risk or whose quality of life was seriously affected and were forced to leave their residences due to environmental disturbances. Three categories of E.R. are distinguished: first, people who leave their residences temporarily due to environmental stress (natural events); second, people who leave their residences for good and resettle to new places (the result of construction of large reservoirs, etc.);

I.S. Zonn et al., *The Aral Sea Encyclopedia*, DOI 10.1007/978-3-540-85088-5_6,

and third, people or groups of people that migrate from their residences, either permanently or from time to time, to a new area within a country or abroad in search of better living conditions (in ecological terms). Due to the Aral's drying out, a part of the Karakalpak population became Ecological Refugees.

Ecological research and monitoring of the Aral Sea deltas – interdisciplinary project of UNESCO initiated in 1994 with the financial support of the Federal Ministry for Research and Technology, Germany. The project united 130 scientists and specialists on the Aral region.

Ekzek Bay* – juts for 6.5 km into the western shore of the *Small Aral Sea* (see) 20 km to the south-west of the Tarangly Cape. The bay is shallow. The fishery area Shemyshkol (now closed) was found on the southern shore of the bay.

Encyclopedia of the Aral Sea – book by I.S. Zonn and M. Glantz (edited by A.N. Kosarev and A.G. Kostianoy) and published in the Russian edition in 2008.

Encyclopedia of Islam – the work of *V. Bartold* (see) published in German in the early 20th century. It includes 111 historical-geographical entries describing a vast territory from the Crimea and Povolzhie to Mongolia. Among them are articles on the "Aral Sea" and "Amu Darya." This work was translated into Russian and presented in the book, "Works on Historical Geography," by V.V. Bartold, and published in "Oriental Literature" by the Russian Academy of Sciences in Moscow in 2002.

Environmental crisis – the global change of the environment leading to disturbances of the natural conditions for the existence of modern forms of life on the Earth and the established dynamic equilibrium in the biosphere in general. The report of the UN Environmental Program of 1992, referring to the Aral Sea crisis, states, "we can hardly find any other region on our planet, except may be Chernobyl, where a more severe environmental crisis affected such a large territory and interfered with the life of such a large number of people."

Erzhan Island* – located to the west of the *Djalpakkair strait* (see) along the southeastern coast of A.S. in the center of the *Zhiltyrbas Bay* (see) and to the southeast of the Uzynkair Cape. The island is low and sandy. Due to sea water level fluctuations the coastline is subject to sharp changes.

Executive Committee of the International Fund for saving the Aral Sea (IFAS) – established in 1997 as the permanent executive and managing body of IFAS in Tashkent having branches in all states of Central Asia. The Executive Committee is a legal entity with the status of an international organization. The IFAS Executive Committee has its own running and settlement accounts in national and foreign currencies for supporting its activities and collects contributions from founding states, international organizations, donor countries, grants, charities, as well as other donations from legal entities and private persons for the implementation of programs and projects and the distribution of urgent aid to the population in the A.S. basin.

F

Farming (Historical) in the circum-Aral area – the initial demarcation of states in the region coincided, in many cases, with irrigation zones and irrigation systems, which appeared in the Aral basin in the first quarter of 1000 B.C. This period was preceded by the large-scale unification of local tribes in 7th–6th centuries B.C.

The most developed state in the Circum-Aral area in the 6th–4th centuries B.C. was ancient Khorezm. There, on the territory of the Sarykamysh delta of the Amudarya, large irrigation systems were constructed, the length of main canals in them reaching 100–150 km, with a width of 10 to 30 m and a depth of 2 to 3 m. Plough farming was also intensively developed, applying iron agricultural implements (hoes, shovels) which enormously improved labor efficiency and enabled the population to transform desert lands into cultural oases.

The irrigation systems in ancient Khorezm were rather ramified and, in general, their structure was as follows: river – head water intake – main canal (*arna*) – primary and secondary distributory canals – irrigation ditches – field.

In the Khorezm antiquity, the following periods are distinguished: the archaic (6th–5th centuries B.C.); the Kangyuisky; and the Kushansky (4th century B.C.–4th century A.D.), the latter being the period when ancient irrigation systems flourished, shaping the "Great Silk Road." During this time, the irrigated areas in the Khorezm oasis expanded enormously, the irrigation water intake and head structures were improved significantly, water application practices were perfected, and general irrigation schemes became more intricate; however, investigations have shown that only gravity irrigation was applied in those times, so farmers were fully dependent on the hydrographic changes of delta arms. Despite the enormous total area of irrigated lands, covering 4.3 mln ha of the Aral basin flatlands (out of which up to 2.5 mln ha were serviced by large irrigation systems), the land use factor did not exceed 10–12%. The lands that became saline shortly during several years of their utilization were abandoned, which is confirmed by a small thickness (15 to 20 cm) of the irrigated layer in agricultural lands. At that time, a fallow farming system was practiced. In the mid-1st century A.D., the process of irrigated farming escalation in ancient Khorezm ceased due to an acute economic and social crises.

I.S. Zonn et al., *The Aral Sea Encyclopedia*, DOI 10.1007/978-3-540-85088-5_7,

At present, soils in the Circum-Sarykamysh area, upon which irrigation ceased in the 5th century and which were no long irrigated as cultivated land, are more subject to desertification and at the stage of takyr formation. The nearby ancient delta territory that was irrigated only from time to time (or was not irrigated at all) is now in transit from takyr formation to deflation.

The specific feature of these landscapes is the presence of leveled, rather lengthy sections of *takyrs* (see) and takyr-like soils in an area that corresponds to the area of the ancient fields and is crossed by sandy ridges running largely in perpendicular directions. The areas of cattle routes are characterized by a higher sand content, greater breakdown of road edges, and nascent vegetation successions. The territories adjoining the large waterways are affected by submersion and solonchak-formation processes.

Irrigated farming in the Lower Syrdarya originated in mid-1000 B.C. (approximately 300 years later than in the Amudarya). Traces of primitive irrigation and semi-nomadic horticulture-farming-fishery occupation dating back to the bronze age have survived. At one time, basin-type irrigation was practiced; it used flood waters regulated with the help of leveeing. Such irrigation systems existed along the Kuvandarya and Zhanadarya river channels. The following scheme of irrigation was applied: river-oxbow (reservoir) – irrigation ditch – field. During an irrigation season, the irrigation layer of agricultural lands, notwithstanding the periodical renewal of irrigation, didn't have sufficient time for formation, which is demonstrated by the presence of intercalating single thin takyr-like interbeds in the profiles of post-irrigation soils.

Desertified soils in the Kuvandarya delta abandoned in mid-1000 B.C. are at the takyr-forming stage at present, which coincides with the period of general aridization of this territory. Landscapes with dominating takyrs and takyr-like soils and fragmentary sandy soils interchange with areas of drifting sands and sandy broken takyrs in the cattle grazing zones.

Up until to the 8th century, the Khorezm oasis was in decay – irrigation systems were not operational and became dilapidated. During the Afrigid dynasty, farming was of a spot nature. In the 8th century, however, elements of the ancient irrigation systems were used for occasional irrigation.

Extensive irrigation construction began again in the 11th century. The organization of irrigation was radically transformed and permanent irrigation emerged. On the main Amudarya channel head structures (*saka*), main canals (*arna*), large distributory canals (*yab*), secondary aryks, chigir pits, and Persian wheels (chigirs) were constructed. Chigir water-lift irrigation or "Persian wheel" made possible the delivery of irrigation water to the uplands to increase the irrigated areas and enhance the land use factor by up to 30–40%.

Irrigation systems developed in the Middle Ages were, of course, more efficient than ancient ones; a complete lack of a collection-drainage network, however, made it actually impossible to control secondary salinization.

At the same time, farmers applied various local land reclamation practices (sand application, loosening, fertilizer application) that, together with silt deposition, facilitated formation of thicker agricultural irrigation layers (up to 50 cm).

The conquests of Genghiz Khan and his sons and the ruinous wars of Timur (13th–15th centuries), confirmed by the results of archeological and topographical investigations of the Khorezm expedition in the 1930s, led to irrigation decay in Khorezm in the 13th century. The post-irrigation landscapes abandoned in those times have evolved at present to a stage of takyr formation.

The main occupation of the population in the Syrdarya lower reaches, which in the 12th–13th centuries belonged to Khorezm, was cattle breeding.

Farming was of a spot (non-regular) nature. Irrigation systems were primitive and, as in ancient times, they used flood waters. Irrigation was practiced mostly on submerged near-channel territories. Permanently irrigated areas were found along Zhanadarya, Inkadarya, and, partially, Kuvandarya, and they covered in total 700–800 thou ha. The area of intermittent irrigation was 400 to 500 thou ha.

Massifs of takyr and takyr-like soils has formed in the Zhanadarya delta in place of irrigation systems abandoned in the 13th century. They alternate in the landscape with the territories used for intensive grazing, are distinguished by a high sand content and the availability of multiple deflation sources.

The total area occupied by medieval irrigation systems in the delta plains of the Amudarya and Syrdarya lower reaches is approximately 1 mln ha.

In the post-Mongolian time and up to the early 20th century, irrigated farming developed in separate localities, covered small areas, and was practiced largely in the deltas of main channels of the Amudarya and Syrdarya Rivers. The lands in the Circum-Aral delta that were abandoned in the 15th–18th centuries and turned into deserts are at present in transition from aridization to takyr-formation.

In the 19th–20th centuries and today, the desertified territories of the ancient delta plains are used as pastures.

Films about Aral – The early period the A.S. drying and the emergence of related problems, such as water pollution; dust and salt drift; loss of biodiversity and resultant unemployment; growing death rate among the adult population and children; etc. stirred great anxiety and, at the same time, the interest of the wider public. Film-makers from many countries could not avoid this environmental disaster. Thus, in 1995, the American producer, Christopher Hooke, made a 14 minute documentary film, "Women of the Aral." The film was shot in the former fishery port of Aralsk, Kazakhstan, where fishermen were left without work, and the women were forced to take up the burden of survival, organizing a cooperative for tailoring garments and making carpets from camel, goat, and sheep fur.

In 2000, Turkish cinematographers ("Academi Production"), with the support of the former chairman of the Soviet-Turkish Council on Business Cooperation (Nikhat Gekyingit, 1988–1998), made a documentary called "Aral" (producer Kemal Uzun) about the environmental disaster in the region.

Leading Russian scientists *N.F. Glazovsky* (see), I.S. Zonn, N.V. Aladin, and D.Ya. Ratkovich took part in this film with their commentary.

In 2004, Dutch producer Jakob Gotteschau shot a 29 minute film called "Predictable Catastrophe: the Aral Sea" (Late Lessons from Early Warnings).

In the same year, one more film, "Aral, Fishing in an Invisible Sea" (52 minutes), was made by Uzbek and Italian producers Saodat Ismailova and Carlos Casas. This film tells about the fate of three generations of one family of fishermen living near what remains of the Aral.

In 2005, the film "Killing the Aral Sea: Catastrophe by Design" was made.

One more film about the Aral called "The Poisoned Land, the Dying Sea" was awarded the prize for Full-Length Films and TV Films at the Monte-Carlo International Festival.

Fisherman of the Aral – newspaper published in the 1960s in Karakalpakia.

Fisheries on the Aral Sea – the history of fisheries in the lower reaches of the Amudarya goes as far back as the Neolithic tribes of primeval hunters and fishermen. Archeological excavations of the encampments of the peoples of the ancient Khorezm prove the existence of an ancient fishery. Later, it was identified that the bones found in the Djambaskal encampment 60 km to the north–east of Turtkul belonged to several fish species, mostly pike and common carp. In excavations of other settlements in the lower reaches of the Amudarya from the period of the flourishing of ancient Khorezm (3rd century A.D.–11th century) fish remnants belonging to 5 fish families were found. The species of fish that were caught by the people in ancient Khorezm were similar to those of the present – bastard sturgeon, roach, rudd, Turkestan barbell, asp, and others. The people of ancient Khorezm caught fish in rivers and lakes. Such mixed fisheries (lake-river) existed until the present day.

After the downfall of Khorezm and up until the establishment of regular relations between the Khiva Khanate and Russia, the fishery in A.S. developed very slowly because of the difficulty of commercial distribution. The significance of commercial fisheries on the A.S. was not high (about 5% of USSR fishery); however, in catches of such valuable fish as bream, common carp, pike perch, barbell, and others, the basin took the first place in the USSR. The share of catches of these fish species in A.S. was as high as 93% of the total catch, while the shares of the Caspian and Azov Seas were 18% and 9%, respectively. While in the period of the optimal sea regime (1958–1960), the fish catches in the basin were maintained at 46 thou tons, already by the late 1970s, they were only 6.7 thou tons and in early 1980 the total catch dropped to 1 thou tons. By this time, the share of valuable fish species dropped sharply. At present (2008) there are no fisheries on the Large Aral Sea.

G

Geller Samuil Yulievich (Yevelievich) (1906–1972) – Soviet geographer who was one of the pioneer researchers studying the Karakum Desert. He graduated from the Geographical Department of Leningrad University, Doctor of Geography (1947). In the 1920s, he took part in the expeditions led by A.Ye. Fersman and D.I. Scherbakov. Their investigations focused mainly on desert geography and saline water desalinization. In 1930, the USSR Academy of Sciences (USSR AS) sent G. to investigate the western part of the Northern Karakums, after which he worked in the Kara-Bogaz-Gol area, then studied the Southern Mangyshlak Plateau, the Karyn-Djaryk depression, and Ustyurt. As a result, G. developed his theories about the utilization of deep drainless depressions for hydropower generation and raw chemical materials production. In 1934, together with V.N. Kunin, he published a review of the projects on watering of the Turkmen Karakums. Here, he drew attention to the utilization of Kelif Uzboy for transit of the Amudarya waters to the west.

His hydrological studies of deserts resulted in the formulation of an engineering-geographical practical proposal on deriving fresh waters from saline water by their natural freezing. G. built a trial site on which saline water was frozen in a layer-by-layer mode over a winter. At freezing, salts in the form of brine precipitate from ice that after subsequent thawing becomes fresh water.

In his work, "Some Aspects of the Aral Sea Problems" (1969), G. disagreed with those who advocated for the calculation of an optimal water level in the Aral. Relying upon economic estimates, G. argued that the benefits from expansion of irrigated farming would outstrip by 100-fold losses due to degradation and the interruption of the fishery activities that were the key economic sector for this water body.

Considering the effects of wind drift of salts from the drying Aral Sea, the stock of which is evaluated at 10 bill tons, G. comes to the conclusion that "wind drift of salts precipitated from waters of the Aral Sea is not hazardous for the irrigated lands of Central Asia and Kazakhstan because only calcium salts and quite insignificant quantities of sodium sulfates may be transferred by wind." G. proposed using a part of the dried out sea for irrigated farming.

During his last years, G. devoted much time to the problems of the Caspian Sea in view of a steady drop of the sea level.

I.S. Zonn et al., *The Aral Sea Encyclopedia*, DOI 10.1007/978-3-540-85088-5_8,

At the Institute of Geography of the USSR Academy of Sciences, G. headed research on likely changes in the physiographical conditions in the A.S. basin due to construction of the Karakum canal. Unfortunately, a comprehensive summary devoted to this problem was published only after his death.

Since 1956, G. was a member of the editorial board of the journal "Proceedings of USSR AS. Geographical Series." For several years, he headed the editorial board on geography of the Foreign Literature Publishers in Moscow.

Geology of the Aral Sea – monograph by I.V. Rubanov, D.P. Ishniyazov, M.A. Baskakov, and P.A. Chistyakov published in Tashkent in 1987. It describes the geological structure and history of the Aral depression; provides data on stratigraphy and lithology of the Paleogene, Neogene, and Quaternary deposits in the Circum-Aral area and in the Aral depression; details bottom and beach sediments and also the delta deposits of the Amudarya and Syrdarya; and studies the issues of Holocene and modern lithogenesis. The book studies the lithology and salinity level of deposits formed as a result of the disastrous drop of the sea level as well as scales of ground deflation due to drying out and desertification of the Circum-Aral territory. Brief information is given about mineral deposits of the region.

Glazovsky Nikita Fedorovich (1946–2005) – well-known scientist, geographer, geologist and ecologist. From 2001, he was a Corresponding Member of the Russian Academy of Sciences (RAS). In 1964, he graduated from the Geological Department of the M.V. Lomonosov Moscow State University (MSU) which was also his first place of work; here, he defended his candidate (in 1972) and doctoral (in 1985) academic theses. Worked at MSU and RAS Institute of Soil Science and Photosynthesis. From 1988 and until his death, he was deputy director of the RAS Institute of Geography. In his scientific and organizational activities, apart from fundamental investigations in physical geography and the geochemistry of landscape, G. paid much attention to the problems of arid territories in Central Asia. Many times, G. went on the centerboard boat "Mega" over the Aral. In 1980, he published his monograph *"Aral Crisis"* (see), which suggested a solution of the A.S. problem. Quicker accession of the Russian Federation to the UN Convention on Desertification Control and Alleviation of Draughts may be credited to Glazovsky. In 1995, he became a director of the Regional International Program, "Leaders in Environment and Development," and he energetically engaged young Central Asian specialists and scientists in activities of this creditable organization. G. was awarded the Silver (IInd degree) Medal "For Russian Nature Conservation" and the Honorary Diploma of Russian Environmental Movement "Green" "For Environmental Education Development in the Russian Federation." For a long time, G. was an active member of the editorial board of the international scientific-practical journal, "Desert Development Problems."

G. is the author of a concept on a conjugate analysis of matter migration in the Earth's landscape shell. He determined the correlation of regional matter flows in the biosphere, defined specific features of the biogeochemical cycle in various natural zones, and identified salt balances in some arid regions and landlocked seas. G. introduced new notions of technogenesis geochemistry, and developed an original methodology and estimated the technogenous geochemical load on the biosphere and its particular regions. He also made a comparative assessment of the efficiency of natural resource management in the world and elaborated proposals on the improvement of various nature management mechanisms. G. supervised the process and directly took part in preparation of ecological maps for Russia and other countries; he also proposed a classification of environmental situations based on a conjugated analysis of the states and dynamics of nature, population health, and economics and social relations.

In the 1980s, he was one of the initiators of the State Environmental Expertise, and after its establishment, he became a member of the Board of the RSFSR State Environmental Expertise. After 1990, he was chairman of the nature management section of the Higher Environmental Council at the State Duma of Russia.

In 1991–1992, G. was Deputy Minister of Ecology of the Russian Federation, and in 1991–2000, he was a member of the board and scientific-technical council of the Ministry of Ecology and State Committee for Ecology. In 1992–1994, he was a member of the Council on Critical Situations and Analysis of Governmental Decisions for the President of the Russian Federation.

In 1997–2000, he was elected the Chairman of the Board of the Federal Ecological Fund of the Russian Federation, and in 2004, he became Chairman of the Trust Board of the Russian Foundation for Wild Nature Protection. G. was actively involved in international activities addressing the problems of nature conservation and sustainable development. He was a research manager of the USSR-UNEP Projects "Reclamation of Saline Irrigated Lands" and the Russia-UNEP project on desertification control, the UNEP expert on the project "Agriculture Impacts on the Natural Environment," and a UNEP consultant on desertification issues. In the period 1989–1997, he was a member of the Advisory Board of the World Resources Institute (Washington). In 1990–1994, he represented Russia at the Interstate Council on the Aral Sea

Fig. 31 N.F. Glazovsky (1946–2005)

Basin Problems. In 2000, he was elected, and in 2004, re-elected the Vice President of the International Geographical Union. He was a member of editorial boards of various Russian and foreign journals addressing environmental protection issues. G. was a Fellow of the Environment Academy (1994) and a Fellow of the Natural Science Academy (1996). In 1996, he was awarded the title "Honored Ecologist of the Russian Federation."

Global Environment Facility (GEF) – established in 1991 as a grants program aimed at financially supporting the implementation by developing countries of commitments made under international agreements on environment protection. The GEF also sponsors activities on the implementation of the Convention on Climate Change and the Convention on Biological Diversity adopted in 1992 in Rio de Janeiro at the UN Conference on Environment and Development. GEF activities are supervised by the World Bank, UNDP, and UNEP. GEF actively participates in financing the Aral international projects.

Grass carp *(C. idella (Valenciennes)* – the introduced fish belonging to *Ctenopharyngodon* (*Steindachner*) genus, carp family (*Cyprinidae*). It was brought here in the early 1960s from China and let go into the sea and rivers. Its body length is up to 120 cm, and its weight is up to 30 kg. It prefers living in arms and lakes. It reaches maturity at 7–8 years, reaching a length of 65–70 cm. Spawning is fractional in June-July and at water temperature of 26–30°C. Eggs are pelagic, spawning occurring during water level rise. The fertility rate is 29–816 thou eggs. Larvae are carried into the coastal zone where they feed on zooplankton and algae. In spring, full-grown fish migrate to floodplains and lakes where they feed extensively (they are voracious) on water plants and on-land immersed soft vegetation. It was recommended to use this attribute of G.C. for biological cleaning of canals and headers. It is acclimatized to Central Asia. It is widely used in pond farming and acclimatization. In 1960, G.C. was let go into the Karakum Canal where it naturalized well and populated the canal and the Amudarya River.

Fig. 32 Grass carp (or White Amur) *(C. idella (Valenciennes)*

Greater Amudarya Shovel-Nosed Sturgeon *(Pseudoscaphirhynchus kaufmanni)* – fish of the Sturgeon family (*Acipenseridae*). At the end of the snout it has from 1 to 9 sharp thorns; 2 sharp thorns are also found both between the eyes and at the occiput. The upper blade of the caudal fin ends with a long filament. It has 10–15 dorsal bony plates, 28–40 lateral plates, and 5–11 pelvic bony plates. The body between the bony plates is covered with bony grains. The mouth is large. The eyes are very small. The length (without caudal filament) is up to 60 cm, and the weight is up to 1 kg. It lives in the fresh waters of rivers with sandy bottoms. It reaches maturity at the age of 6–7 years (males) and 7–8 years (females), having a length about 25 cm. The spawning period is April. Fertility is 1000–1900 eggs, 1.9–2.4 mm in diameter. The largest specimens lay up to 37 thousand eggs. After hatching, the fries move downstream, and feed on aqueous larvae of insects, eggs, and fish fries. Specimens older than 10 years are met quite seldom. Lives mostly in the Amudarya. A valuable fish species; due to its low population, however, it has no commercial significance. It is also found in the Mississippi River (USA).

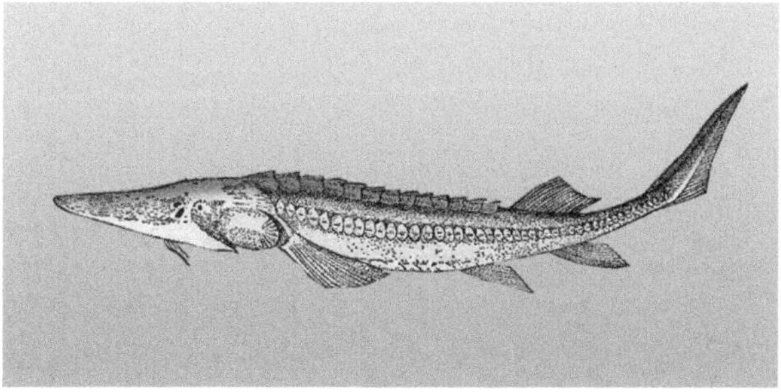

Fig. 33 Greater Amudarya shovel-nosed sturgeon (*Pseudoscaphirhynchus kaufmanni*)

Greater and Smaller Barsuki sands – sandy deserts extending meridionally as two long and narrow strips from A.S. and divided by the parent relief band. G.B. is approximately 200 km long, while S.B. is about 100 km long. Their elevations reach 100 m. In some localities, the sands are deflated, forming hummocks, ridges, and barkhans. The vegetation is not uniform here: on flattened terrains, xerophytic shrubs, wormwood, thistle, and ephemers prevail, while the slopes of hummocks and ridges are covered with the thickets of calligonum, sandhill wattle, milk vetch, kundym, and salt trees.

Greater Chushka Island* – located near the eastern coast of A.S. It is one of the islands that cover the *Aksaga Bay* (see) on the west. The island is low and sandy, and its elevations vary from 1 to 3 m.

Greater Sarychaganak, Sarychabanak Bay* – located in the northeast of A.S. The bay is large. The depths in the center reach 12–13 m. Two kinds of currents are observed: the southwestern current generated by north and northeastern winds and the northeastern current. Severe wind-generated water-level fluctuations are recorded in the bay, with range reaching 3 m. They are generated largely by the southern, southwestern, and western winds.

Greater Sudochie Lake – a part of the water-marshy lands of *Sudochie Lake* (see), composing approximately 22.4% of their total area. The lake area is 98 sq. km, its maximum length reaches 10 km, and it has a width of 4.5 km. Its depth does not exceed 1.0–1.2 m, and its coastline is 32 km long. During the disastrous low-water period in 2000–2001, the lake dried out completely.

H

History of Aral Sea investigations – as *V.V. Bartold* (see) noted, the areas nearby the Aral "had been involved in historical life rather late." First allusions to the A.S. existence were found with Greek authors. Strabo (64–63 B.C.–23–24 A.D.) pointed out that the territory to the east of the Caspian Sea was populated by nomadic tribes *daev* who came here from the area located "behind Tanais and Meotide," meaning near the Syrdarya and the Aral Sea.

The first data about A.S. were obtained in 138 B.C. from Chang Chien who was delegated by the Chinese government as an ambassador to Central Asia and who wrote about finding there "a big lake without high shores." Chang Chien meant "Northern" Sea, but to other Chinese sources this was the "Western" Sea. As is known, in 97 B.C., Chinese military commander Ban Chao with his troops approached A.S., while the Byzantine ambassador who was directed in 568 to Turkey brought among other things information about A.S. Nevertheless, Ptolemy showed on his map in the 2nd century and published in Europe in 1490 that the Amudarya and Syrdarya flowed into the Caspian; A.S. was not shown at all, making it clear why the first attempts of the Russians in the early 18th century to make an accurate map of the territory to the east of the Caspian were reduced, first of all, to identifying the places of inflow of the major Central Asian rivers.

In the 4th century, Greek historian Ammian Marcellin was "the first and only of the ancient authors whose words contained clear indication to the existence of the Aral Sea."

Mostly correct data about A.S. could be found in the treatises of the Arab scientists of the 9th–10th centuries. They provided data about the size of the sea and described its shores. In the mid-10th century, Arab writer Ibn Ruste mentioned A.S. in his works. At the same time the first cartographic presentation of the "Khorezmian Sea," Istakhri, had appeared. The Istakhri map, descriptions of Ibn Ruste, and other Medieval investigators proved that at those times A.S. had the same size and countours as in the early 10th century (Berg, 1908).

Up to the 16th century there was no new information about the sea.

In 1552, Ivan the Terrible ordered subjects "to measure the terrain and make a drawing of the state," in this way spurring the development of cartography not

I.S. Zonn et al., *The Aral Sea Encyclopedia*, DOI 10.1007/978-3-540-85088-5_9,

only in Rus' but in cross-border territories as well. Work on the "drawing" contemplated by the tsar gradually progressed, and during the reign of Boris Godunov, new things were added to it. At last, in 1627 its description was complete. It was named the "Book to the Great Drawing," and contained, among other things, information about the Aral, which was called "Blue Sea" there.

Meanwhile, West European scientists had no notion about the existence of the Aral up to the late 17th century (for the first time it was shown on the map of Vitzen (Ides) in 1704 under the name "Since"). On the world map prepared by Fra Mauro (1457) from Venetia, a lake was shown to the east of the Caspian Sea, but it had no name.

In the late 18th century, the outstanding Russian geographer and historian S.U. Remizov wrote a comprehensive work, the "Drawing Book of Siberia," one of the drawings in which showed rather correctly A.S. with its tributaries, the Amudarya and the Syrdarya.

The first scientific investigation of the nature of the Circum-Aral area was the expedition led by Duke A. Bekovich-Cherkassky and directed in 1715 to the eastern shore of the Caspian by Peter I. The expedition found out that the Amudarya flowed into A.S. Present-day cartographers praise the sophisticated materials of this expedition, which were helpful for the preparation of a map depicting the Aral Sea quite accurately. The correct outlines of the sea could be provided only by people who have navigated over it or at least who have seen it. In 1716, Peter I ordered Bekovich-Cherkassky to go on a new expedition, this time to study the Amudarya, but the members of this expedition met a tragic death. In the decree that Peter I handed to A. Bekovich-Cherkassky before the expedition, the sea into which the Amudarya flowed was called the "Aral" for the first time, though "Aral" in the Turk languages meant "island." The most probable explanation fir this is that the sea was like a blue island among a boundless sandy desert.

In the 1720s, thanks to the Russians, Western European cartography was enriched with new data about the vast Aral-Caspian area. Meanwhile, A.S. cartography was verified further. The first accurate topographical and geodetic data about its northern shores were obtained when, in 1731, a Russian legation, among the members of which were two officers–land surveyors, was sent to the Kazakh khan who ruled over the Aral shores. In that period, nearly a third of the Aral coastal area remained unsurveyed; it had only been mapped by the results of interviews with local people.

In the mid-18th century, after the establishment of a stable relationship between Russia and Kazakhstan, targeted surveys of the Circum-Aral area began. In 1740–1741, on the initiative of F.I. Soimonov and P.I. Rachkov, the Russian government directed to the Aral a hydrographic expedition headed by I. Muravin whose task was to carry out reconnaissance studies for construction of a city at the mouth of the Syrdarya. The members of this expedition studied the eastern shore of A.S. with the Syrdarya delta, and on the basis of the results of instrumental surveys, they made a landscape map. Many of the features that were marked on it disappeared later on: some cities were destroyed, and the channels of some rivers were displaced or dried out.

Then there was rather a long interval in the A.S. investigations that lasted up to 1825 when the expedition headed by Colonel *F.F. Berg* (see) went to the western coast to describe it and to carry out leveling surveys of the plateau Ustyurt.

In the 1830s, the A.S. coast was investigated by the well-known zoologist and professor of Kazan University, E.A. Eversman. His detailed description is now of great value because it helps to reconstruct the nature of those areas 150 years ago. E.A. Eversman gave the geological and physiographical characteristics of the Aral coast and made suggestions about the drying of the sea.

In 1831, A. Levshin made a map of A.S. on the basis of the data from the archive of the Orenburg Frontier Guard Commission and "reconnaissance data and results of interviews with Russian engineers or Quartermaster officers who were in the Cossack steppes in 1820–1821, 1824–1826."

In the history of A.S. investigations, a considerable contribution was made in the 1840s by the Khiva expedition led by Colonel G.I. Danilevsky and natural scientist F.I. Baziner. They supervised the topographic works and astronomical determinations of the points in some coastal areas. The large-scale map of A.S. and the Khiva Khanate prepared by F.I. Baziner was also included by A. Humboldt in his book, "Central Asia." The Expedition Report, "Description of the Khiva Khanate," published in 1851, contained a detailed story about the climate and relief of the Circum-Aral and A.S. regions.

In 1848–1849, the first marine "dangerous expedition" was organized on the schooner "Konstantin," commanded by naval officer *A.I. Butakov* (see). In fact, this expedition was the first to prepare an accurate map of the whole A.S. Apart from reconnaissance surveys, measurements were made over a vast area, and as a result, the maximum (69 m) sea depth was determined and the direction of the "permanent" current, which unlike other seas in the Northern Hemisphere is not counterclockwise, but clockwise, was identified. During two years of investigations, the expedition carefully studied the relief of the sea bottom, the configuration of the coastline, the currents, ground samples, salinity levels, and the color and turbidity of the water in A.S. It turned out not an easy task to publish the results of this expedition, however. The reason was that Butakov fell into disfavor with the tsar due to indulgences of the "strict behavior code" assigned to *T.G. Shevchenko* (see), the Ukrainian poet and artist who had been exiled to Orenburg as a soldier but was later included on this expedition. This circumstance turned out a serious obstacle to publishing this fundamental scientific work. Only a communication about the expedition was published, the full text of the report not being released for another 100 years.

A detailed description of the investigations carried out in the A.S. area was prepared by Ya.V. Khanykov (1851), who detailed in particular the "cartographic materials of the area."

By the mid-19th century, thanks to the efforts of A.I. Butakov and his predecessors, the A.S. hydrography was already rather well studied. At that time, however, hydrological knowledge about the A.S. remained poor. Only K. Sharngorst in 1871, J. Grimm in 1873, and E. Pratz in 1874 took a few water samples of the surface water in the summertime and measured its temperature.

Investigations of the Circum-Aral area and A.S. were reanimated to a great extent in 1873 when Khiva was joined to Russia. Already in the following year, two expeditions worked in this region: the Aral-Caspian Expedition organized by the Petersburg Society of Natural Scientists investigated the western and northern Aral coast, while the Amudarya Expedition sent by the Russian Geographical Society studied the southern and southeastern coasts. Worth mentioning is that a participant of the latter expedition, the well-known land surveyor A.A. Tillo, carried out an accurate leveling of the sea level and placed a benchmark that was used as the datum in further determinations of its level. Outstanding Russian zoogeographer N.A. Severtsov (1874) focused on the Aral drying process, which was quite visible even at that time.

From 1874, occasional foot-gauge observations were carried out on the sea shores. Soon it was found that its level was subject to perceptible fluctuations: after a very low level in the 1880s, it rose rather sharply and quickly (over 10–15 years by nearly 3 m) until it stabilized by the 1850s.

Under the guidance of A.V. von Kaulbars (1881), all arms in the Amudarya lower reaches were investigated, the delta maps were prepared, and a rather detailed description of the delta was given.

In 1889, Strelbitsky determined for the first time the morphometrical characteristics, particularly estimations of the area of its surface and islands, of A.S. In 1884 the hydrometeorological station "Aral Sea" was opened in the northern part of the sea on the shore of the Sarychaganak Bay. Not long before this event, the first biological investigations of the sea's flora and fauna began. The expedition of Butenkov in 1841, which included natural scientist A. Leman; the expedition of Butakov in 1848–1849; and, lastly, the Aral-Caspian expedition in 1874, the members of which included zoologist V. Alenitsyn, whose materials were interpreted by I. Borschov (1877), N. Andrusov (1897), K. Kessler (1877) and others, deserve special mention.

In 1897 in Tashkent, the Turkestan Branch of the Russian Geographical Society was created, an event that played an important role in further investigations of A.S. Although having at its disposal only meager finances, it immediately initiated studies of the steadily rising water level of the Aral, a phenomenon that began in the last quarter of the 19th century. Successful implementation of this plan required a man of great erudition and persistence in accomplishing the established objective, however, and such man was found in L.S. Berg, a young geographer who had just graduated from university. He organized expeditions to A.S. during which hydrological observations were conducted and rich collections of botanic, ichthyologic, geological specimens were made. When the expedition of 1900-1903 worked near the hydrometeorological station "Aral Sea," L.S. Berg organized the first foot-gage observations of the Aral water level. The results of these physiographical investigations became the basis for Berg's essential monograph, "Aral Sea" (1908), which immediately received wide publicity and was translated into many foreign languages. Investigations carried out by Berg made a whole epoch in cognizing A.S., and even now offer a creditable example of a comprehensive analysis of natural events.

In 1899–1902, Girschfeld conducted statistical hydrological investigations in the Amudarya delta. He found out that from the late 1840s to the late 1880s, the Amudarya did not, in fact, deposit sediments to the sea because they were all deposited on the floodplain. Beginning from the late 1880s, the Amudarya scoured one channel, after which great quantities of silt found their way into the sea. He also noted a considerable rise of the Aral Sea water level beginning from the late 1880s, attributing this to the enhanced air humidity in Central Asia and a changed structure of the Amudarya delta. He was the first to formulate a hypothesis about the effect of floods in the Amudarya delta on Aral water level fluctuations.

In 1905, during the construction of the railroad Orenburg–Tashkent, the coastal settlement Aralsk appeared on the route along A.S. Beginning from 1911, the first regular observations over the A.S. water level were began here.

From 1906–1911, the water level in A.S. rose. The sea actively encroached upon the land, filling depressions and forming bays deeply incised into the shores. For example, during this time a great salt water bay formed in place of the Sudochie Lake. From then on, investigations in the Circum-Aral area increased. In 1910, the Hydrometric Unit of the Land Management Division of the Turkestan Territory was also founded. It undertook the identification of water reserves and all water sources in this territory.

From 1912, the Resettlement Department of the Ministry of Agriculture worked diligently. On the basis of their survey materials, a "Land Use Map of the Aboriginal Populations of the Amudarya Unit," at a scale of 8 *verst** to one inch, was prepared in 1914. After 1912, the soil and geological surveys also began under supervision of N.A. Dimo, V.V. Nikitin, and L.L. Nozhin. From 1913–1915, geological investigations were conducted by *A.D. Arkhangelsky* (see) and B.N. Semikhatov. In 1931, Arkhangelsky published his most interesting first geological map of this area (10 *verst* to one inch).

In Soviet times, investigations of the natural conditions of the Aral increased significantly to a major scale. In the early 1920s, with the introduction of tougher requirements to navigate the Aral, it became evident that "the Aral Sea had no appropriate coastal pilot." In order to cover this gap, already during the 1921 navigation season hydrological works were conducted in the A.S., the Syrdarya, and the Amudarya deltas. Data obtained enabled A. Malinin to prepare and publish within a rather a short time the "Brief Pilot of the Aral Sea and the Amudarya Delta."

But there were other problems related to the Aral. In the early 1920s, a period of severe hunger struck, and the whole country awaited fish from this sea. At that time, the sea had sufficient fish resources. For example, in the early 1930s,

---

* One *verst* – 3.500 feet.

when the fishery was first established in A.S., the annual fish catch was approximately 500 thou quintals. Also at this time, the Main Fishery Department organized the Aral Research Expedition, led by F. Spichakov, to study the fishery potential of the Aral Sea and to find ways of its sustainability and possible intensification. During the year and a half from the summer 1920 to the end of 1921, they succeeded in collecting a rather extensive amount of material, reflecting rather completely the fishery conditions in the Aral and its natural-historic (including hydrological) situation.

In 1929, the VNIRO branch of the Aral Fishery Station was opened in Aralsk. L.S. Berg, who headed a 1925 fishery expedition to A.S., wrote that the VNIRO Aral Branch "was also addressing, apart from hydrological studies of the Aral Sea, the fishery issues." In 1932, he organized two expeditions on the motor vessel "Dekabrist" over the Aral to prepare a fishery map of the sea. Extensive materials collected during these expeditions provided the basis for the major monograph, "Hydrological and Hydrobiological Materials for Preparation of the Fishery Map of the Aral Sea," written by A.L. Benning (1934). In 1933, works were carried out on the expedition vessels "Aralets" and "L.S. Berg," the results of which were included into the second volume of the monograph (Benning, 1935).

In the 1930s, attempts were made to study in more detail the currents in A.S. In 1931, L.S. Berg measured them with a Woltman current meter, which was lowered from the deck of an anchored vessel. In 1932–1933, A.L. Benning also conducted observations over currents, and in the spring of 1936, M.Ye. Zhdanko continued studying the coastal currents near the HMS "Aral Sea." In 1936–1937, he managed to carry out up to 130 observations. From 1933, over 80 determinations of currents were made at HMS "Aral Sea," which initiated systematic observations. After generalizing of all these materials, M.Ye. Zhdanko prepared and published a complete work on the A.S. currents in 1940.

Summing up available materials on the Aral, N.M. Knipovich (1932) wrote that "its nature is described in general, but investigations conducted thus far are not sufficient, and, in particular, they fail to cover the whole annual cycle of seasonal changes." In response, when the supervision of all marine works in the VNIRO Aral Branch was assigned to G.V. Nikolsky in 1935, regular seasonal investigations of the A.S.'s hydrological conditions began at permanent stations. He summed up this work in his monograph, "Fish of the Aral Sea," which became, by expert appraisals, "the second most comprehensive book after the Berg monograph (1908) on the hydrology of the Aral Sea."

During the World War II, observations of the A.S. continued, but episodically. Thus, in 1943, a combined expedition of the "Aralrybvoda" and the Aral Fishery Station studied spawning waters of the Syrdarya and Amudarya deltas. Much time was given to interpretation of the pre-war materials, and after the end of the war, a research work was published that focused on the creation of a new equilibrium level in A.S. in connection to its diminished inflow (Zaikov, 1946).

In 1950, after the governmental resolution on hydraulic construction and land irrigation in the A.S. basin, the need emerged for wider-scale and more diverse investigations of the sea, targeted not only to the study of the existing hydrological and hydrochemical regimes but also to the identification of changes in the physiographical images of the sea that would appear after completion of hydraulic construction and the beginning of regular withdrawal for irrigation of a part of the waters feeding the sea.

In the 1950s, hyndrological investigations of A.S. continued, producing, among other works, "Marine Hydrometeorological Yearbook," the publication of which was initiated by UGMS of the Uzbek SSR on the basis of the data of six hydrometeorological stations: "Aral Sea," "Bayan," "Uyaly," "Tigrovyi," "Lazarev," and "Barsakelmes." Beginning in 1955, hydrometeorological observations became systematic, but only during the period of navigation and not covering the central part of the sea. They were, however, carried out every year in the same months and even on the same days: May 7–9, August 13–15, and October 20–22.

All scientific materials were systematized, mainly by GOIN, which was engaged in thematic studies of A.S. At the same time, sedimentation conditions in A.S. (N.G. Brodskaya), stratigraphy, tectonics, and mineral deposits (A.L. Yanshin) were also investigated.

Unlike L.S. Berg (1908), who considered the Aral to represent the typical lake basin because of its chemical composition, L.K. Blinov tried to demonstrate that A.S. was a typical sea basin, although special (as in, differing from the basins of the World Ocean) correlations of the chloride content and salinity of water should would have to be obtained. He focused attention on the fact that the Aral sea-lake was completely isolated from the ocean and under a particular effect of land runoff, so the permanent ion composition typical of oceanic waters was disturbed. Therefore, "Oceanographic Tables" elaborated for oceanic waters would give errors in hydrological estimations for A.S. GOIN began development of A.S. specific tables, and all necessary estimates were carried out in 1952–1953 by L.K. Blinov and A.P. Tsurikova. For the first time, the newly established relationships between the chloride level, salinity, and density of A.S. waters were presented in a monograph devoted to the A.S. hydrochemistry (Blinov, 1956).

In the late 1950s, V.S. Samoilenko specified the issues on thermal exchange in the sea's active layer in relation to the Aral conditions. Applying yearly hydrometeorological observations by means of square-based calculations, he made maps of the "external" and "internal" annual heat cycle in the Aral and also a map of the differences of these heat cycles. He showed that the differences represent the individual features of separate sea regions (coastal, shallow-water, and central deep-water): the effect of currents, turbulent exchange, ice formation and thawing, etc.

Although ice conditions in A.S. were first investigated in the 1930s, only in the 1950s was this problem addressed on a higher level – forecasting the ice regime.

Seiches (standing waves with large periods) in A.S. were also studied. L.S. Berg (1908), and later V.N. Obolensky (1920), having pointed to the existence in the Aral of longitudinal single-node seiches with periods 22.7 and 28.0 hours, assumed that sieches were possible in other periods. In the 1950s, seiches were studied on a specially built laboratory model of A.S., and as a result, "there were found single-, double- and triple-node longitudinal and single-node transverse seiches... [and] the objective criteria were calculated for the location of mareographic stations in the Aral Sea water area to ensure a more rational study of its seiches."

Significant work was conducted in the Amudarya lower reaches in 1951–1954 during surveys for the project of the *Main Turkmen Canal* (see) and the partial transfer of the Amudarya flow to the Caspian Sea. Historical geomorphological investigations were conducted under the guidance of S.P. Tolstoy and with the participation of A.S. Kes'. The principal stages of development of the A.S. water area and paleogeography of the Circum-Aral Area were also defined (Kes', 1969). G.V. Lopatin (1957), who studied the structure of the Amudarya delta in these years, had found that the Aral delta itself was formed over 7 thousand years, while the Khorezsm-Sarykamysh was formed over 10 thousand years. In the same year, M.M. Rogov published the comprehensive book, "Hydrology of the Amudarya Delta".

The Aral investigations attracted still more attention in the 1960s in connection with the economic development and irrigation of large land areas in the Amudarya and Syrdarya basins and the construction of unique main canals and reservoirs with large storage capacity. As a result, attention was also drawn to the Aral proper, its declining levels due to intensive withdrawal from both the Amudarya and Syrdarya for irrigation, and the first appearance of concerns about the likelihood of its complete disappearance. Several works were published, including one that described the economic significance of the Aral (Geller, 1969).

Important investigations were carried out in the late 1960s by V.I. Lymarev (1967), who studied the A.S. coasts. He was the first to identify the transgressive and regressive stages of the Aral.

The climate and related water availability in the Aral-Caspian basin were studied by V.P. Lvov (1965) and A.V. Shnitnikov (1968). They attributed the fluctuation of the water levels in the Caspian and Aral to the rhythmic activity of the sun, which enabled a conclusion to drawn about the presence of century-wise, many-century, and larger rhythms of wetting of this region.

In 1969–1971, VNIIMORGEO of the USSR Ministry of Geology, under the leadership of I.G. Vainbergs, organized a sea expedition that brought new data on the conditions of modern sedimentation and sediment composition in the Aral.

Marine terraces (Gorodetskaya, 1978), the condition of the dried zone, and its geomorphology (Bogdanova, Kostyuchenko, 1977, 1981) as well as the Aral history (Kes', 1985) were investigated.

In general, in the period from 1976 to 1978, attention was focused on variations in the hydrological, hydrochemical, and hydrobiological regimes of the sea; identification of the specific features of natural environment degradation in the Circum-Aral area; and assessments of the socio-economic consequences of the changes. In the next two years (1978–1980), scientific justification of actions on maximum prevention of negative ecological and socio-economic consequences of the A.S. water level drop were pursued.

By the 1980s, it became clear that there was not a single component of natural conditions and not a single branch of agriculture or industry in the Circum-Aral area that, in the future, could develop independently of the A.S. water level drop and the man-made desertification of the nearby territory. In response, the key research and design-survey work on the Aral issue included study of the changes that had occurred in the Aral and the Circum-Aral area and their socio-economic consequences, as well as a forecast of further changes and the development of the scientific basis for actions aimed at alleviating and arresting the negative consequences of man-made desertification in the area.

Investigations into various aspects of nature in the lower reaches and delta of the Amudarya and of the drying A.S. were conducted quite purposefully. These were geographical and ecological (Kuznetsov, 1977, 1980, 1991), soil (Zhollybekov, 1983), natural-reclamation (Rafikov, 1981, 1982, 1984), suspended sediments, changes in the cycle of biogenous elements, carbonates and humus in the "Basin-Aral sea" system (Klyukanova, 1985, 1986), ecological-geobotanical (Bakhiev, 1985; Bakhiev et al., 1978; Kurochkina, 1984, 1991, 1996; Novikova, 1996–2002), geological (Rubanov et al., 1987), and climatic (Molosnova et al., 1987). Later on, remote methods of nature studies and monitoring found wide application in special landscape investigations in the Circum-Aral area (Popov, 1990; Ptichnikov, 1994, 1996; Glushko, 1995, 1996). From 1977 to 1993 the integrated hydrological and hydrochemical investigations of A.S., Syrdarya, and Amudarya were conducted under supervision and with direct participation of A.G. Tsytsarin. From 1977 to 1994, V.N. Bortnik investigated water circulation and changes in the hydrological and hydrochemical regimes (seasonal and many-year) in A.S. The 1990s were years of wide-scale investigations of consequences of the A.S. drying, such as desertification (Rafikov, 1988, 1994, 1995; Kust, 1992, 1994, 1999), salt and dust transfer (Rozanov, 1992–2000), and salt balance (Glazovsky, 1995). With enviable consistency, N.V. Aladin, I.S. Plotnikov, and A.A. Filippov carried out studies of zoobenthos and zooplankton as well as general variations of the sea ecosystems affected by man's activities in the Aral. In 1994, V.I. Kuksa published a work that, on the one hand, summed up the variability in space and time of the key hydrometeorological, hydrochemical, and hydrobiological processes observed in A.S. in the preceding four decades, and, on the other, elucidated the relationships between the natural and anthropogenic factors determining the variability of the sea regime parameters.

A considerable contribution into the development of project proposals on, first, environmental protection, conservation, and restoration of the Aral Sea and then of the Circum-Aral area (Amudarya delta) was made by workers of SANIIRI and then by NIC MKVK, headed by V.A. Dukhovny (1990–2000).

From 2000 to 2006, under the supervision of A.G. Kostianoy (P.P. Shirshov Institute of Oceanology, Moscow, Russia) and S.V. Stanichny (Marine Hydrophysical Institute, Sevastopol, Ukraine), integrated satellite monitoring of both the Large and Small Aral and the nearby region was carried out. During monitoring, the following parameters were controlled: surface and volume of the sea, sea surface temperature, sea level, ice cover, vegetation index, desertification, etc.

Since 2002 P.O. Zavialov (P.P. Shirshov Institute of Oceanology, Moscow, Russia) organized a series of complex expeditions on motor boats in A.S. that included hydrological, hydrobiological, and hydrochemical surveys.

These works continue in studies of the causes and dynamics of the seal level drop and changes in the hydrology and coastline of A.S. (V.N. Mikhailov, V.I. Kravtsova, 2000, 2001, 2006).

Several contributions into A.S. studies, specifically in generalizing the available investigative materials and suggesting their own visions of the problem, which improved awareness of the foreign scientific community about the Aral environmental crisis, were made by the following scientists: from USA – Ph. Micklin (1991–1996), M. Glantz (1993–2006); from France – M. Mainguet, R. Létolle (1992–1994); from Japan – Ishida (1995–1996), Ogino (1995–1996), Tsutsui (1992–1996); from Australia – W.D. Williams (1993–1996).

The number of investigations devoted to the Aral problem is enormous. Only in the recent two decades did it exceed 1000 publications, and more than two-thirds of those were published in the late 1990s.

Hydrochemistry of the Aral Sea – a classic monograph of L.K. Blinov, published in 1956. This book amassed investigations of the sea chemistry carried out over five years by the author and his team at the laboratory of the State Oceanographic Institute. It considers both generalizations and criticisms of the findings of previous investigations along with the enormous amount of material he collected during his expeditions and experimental works. This publication describes not only the present sea regime, specific features of the composition and physico-chemical properties of sea waters, and salinity and biogenous matter defining sea productivity, but also makes a well-grounded estimate of future changes in the chemical nature of the Aral and the conditions of its biological productivity on the basis of a comprehensive consideration of its geographical and hydrological specifics. This is the first attempt of integrated investigations targeted not only to the study of the existing physiographical conditions of a particular feature, but also at elaboration of a scientifically-validated forecast of their changes in the future as a result of economic activities. This work also considers some of the theoretical issues involved in the formation of the chemical composition of the sea waters over time and the alteration of the hydrochemical regime due to the flow regulation of rivers running into the sea.

Hydroecological monitoring – the regular accrual of targeted data of hydrochemical and hydrobiological studies describing the state of various aqueous ecosystems and the analysis of these data for decision-making on improvement of this state. H.M. is also needed for addressing such problems as assessment of water quality in sources used for drinking, technology, and domestic supply; assessment of the possibility of using river flow or ground waters of available quality for irrigation of different agricultural crops and industry development; collection of the hydrochemical data necessary for construction and operation of various hydraulic structures (reservoirs, canals, headers); development of recreational and fishery sectors; elaboration of actions aimed at conservation of various aqueous ecosystems in terms of their chemical composition; utilization of the flow of both transborder rivers and those from which transfer of some flow to other basins is planned; and fulfillment of hydrochemical zoning and development of geoinformation system (GIS) maps of different scales and applications. H.M. may have different orientations and content depending on the purpose of water utilization.

Hydrogen sulfide zone of the Aral – in recent years, a hydrogen sulfide zone began appearing in A.S., which could be explained by the fact that the more salinated and denser waters of the eastern basin of the Large Aral have moved along the northern slope of the western depression, reaching its isopycnic level at a near-bottom layer and forming a sharply haline and dense stratification accompanied by anoxia in that layer, which has resulted in its contamination with hydrogen sulfide. This phenomenon was revealed for the first time by the expedition of the Institute of Oceanology of the Russian Academy of Sciences in November 2002 when hydrogen sulfide was found at depths more than 22 m. In autumn 2003, the concentrations of oxygen and hydrogen sulfide were measured, and their vertical distribution was demonstrated. The upper quasi-homogeneous layer (about 15 m thick) was saturated with oxygen in concentrations up to 7 mg/l, but already at the 20 m-horizon, the oxygen disappeared and hydrogen sulfide appeared in its place. Its content increased with greater depth, and near the bottom (40 m) reached 80 mg/l. This figure is an order of magnitude greater than H_2S concentrations in the Black Sea. In this period, hydrogen sulfide has also been recorded in the Chernyshev Bay in the north of the western basin of the Large Aral. Unlike the hydrogen sulfide zone of the Black Sea, the presence of hydrogen sulfide in the Aral Sea is not permanent. Its appearance and disappearance depend on its density stratification. Thus, in spring 2004, the whole water thickness was ventilated due to winter convection processes, and there was no H_2S. Hydrogen sulfide was also not found in the summer of 2004 when the density stratification was insignificant. In autumn 2005, however, the presence of hydrogen sulfide was again registered, but in small concentrations and at horizons over 35 m. If hydrological conditions conducive to anoxia in the Aral have been established, then the biochemical mechanisms predicting such quick buildup of H_2S may become the subject of future investigations.

Hydrology of the Amudarya mouth Area – a generalizing work written by State Oceanographic Institute specialists M.M. Rogov, S.S. Khodkin, and S.K. Revina and published in 1968. It provides a hydrographic description of the Amudarya delta area, hydrological characteristics of the river proper and its delta, dynamics of the perennial flow, hydrochemical indices, and the effects of water intake on irrigation. Forecasts of A.S. level fluctuations in the future are especially focused on.

Hydrology of the Caspian and Aral Seas – a fundamental work of Doctor of Geography and Professor A.N. Kosarev (Oceanology Chair, Geographical Department of the M.V. Lomonosov Moscow State University), published in 1975. It covers the principal aspects of the hydrological regime of the Caspian and Aral seas, and generalizes practically all observation materials amassed by the early 1970s, providing estimates based on different schemes. The book describes the principal hydrological processes contributing to the formation of a structure and the state and dynamics of the natural appearance of the Caspian and Aral. This publication is one of the most comprehensive generalizations of the natural stable regime of the hydrometeorological conditions of A.S. existed before 1970s.

In the introduction, Kosarev stresses that the modern conditions of A.S. are a reference level by which future changes will be assessed. The analysis of the year-by-year alterations of the hydrological conditions used data obtained mainly from field studies in the 1950s and 1960s when such studies were most widespread. The book is divided into three sections, preceded by a brief historical sketch about the development of the studied landlocked seas. It underlines that in geological terms these water bodies differ radically by both their genesis and age.

The first section analyzes basic factors that contributed to the formation of the hydrological regime of the Caspian and Aral Seas and reveals their similarities and differences. In this section are found: the morphological characteristics of seas (by the mid-1970s); data about flow and water balance; thermal balance; synoptic conditions; severity of winters; and the hydrometeorological factors defining many-year fluctuations of the water levels in the Caspian and Aral. A joint analysis of year-based water level fluctuations in the Caspian and Aral confirm, first, their dependence on the hydrometeorological processes occurring in the basins of these seas, and, second, the lucidly demonstrated heterochronic (non-coincidental) nature of these fluctuations.

The second section of the monograph is devoted to Caspian Sea hydrology.

The third section describing the Aral Sea includes an analysis of currents, temperatures (over two thousand observations), salinity, water density, vertical stability, oxygen level, ice conditions, and seiche oscillations. This section studies in great detail the convective mixing processes in A.S. as well as the formation of bottom waters. Already in 1975 the author of this monograph stressed the irreversible changes in the hydrological conditions of A.S. caused by the reduction of the river flows. Emphasized is that due to the reduction of the sea depth and surface area, the climate in the coastal regions will be more continental.

I

Ichthyofauna of the Aral Sea – ichthyofauna is made up of the Asian mountain, Ponto-Caspian, Turkenstan, and other fauna complexes. Up until the 1960s, I.A.S. comprised 20 fish species belonging to 7 families. The most abundant was the carp family, which included 12 species (bream, common carp, sea roach, *Chalcalburnus*, Aral and Turkestan barbell, asp, white-eye, sichel, rudd, ide, and crucian carp). This family made up 60% of the whole ichthyofauna. The next most abundant was of the perch family living in lakes, including the pike perch, perch, ruff; the sturgeons (bastard sturgeon); salmons (Aral salmon); catfish; pikes (pike); and sticklebacks (stickleback) were each represented by 1 species. Due to insufficient population, newly introduced species had no commercial significance; however, they influenced the biological regime of the Aral. Among the invaders, only the plant-eating fish had some commercial significance. In the 1980s, the main commercial fish were bream, common carp, sea roach, pike perch, barbell, asp, *Chalcalburnus*, catfish and others. By the end of 2002, only 2 fish species survived – flatfish and Aterina – and only in the western part of the Large Aral Sea. At present (2008), I.A.S. has disappeared (except for in the Small Aral Sea), the main cause of this outcome having been the increasing water salinity.

Ide (*Leuciscus idus*.) – commercial fish of the carp family (*Cyprinidae*). Its length reached 70 cm, and its weight was 6 to 8 kg. In A.S., a subspecies, the Turkestan ide, was found. This fish usually lived in small plain rivers, in lakes, and in reservoirs. It reached fertility at the age of 4–6 years when its length was 25 cm and more. Spawning was in April–May in floodplains, and at times in bars at water depth 0.5–0.7 m and temperatures of 3–4°C and higher. Fertility was 39–114 thou eggs. It fed on insect larvae, small mollusks, worms, algae, and higher vegetation. It was not inclined to long-distance migration: Spending winter in rivers, in spring it ran to small tributaries and floodplain lakes for spawning. After flood recession, it returned to the rivers.

Information about the Aral Sea and lower Amudarya from the ancient times unitl the 17th century – one of the principal geographical works of the outstanding Russian Oriental specialist *V.V. Bartold* (see) published in 1902 by the Turkestan Branch of the Russian Geographical Society in the series, *"Scientific results of*

I.S. Zonn et al., *The Aral Sea Encyclopedia*, DOI 10.1007/978-3-540-85088-5_10,

the Aral expedition" (issue 2) (see). Being involved in integrated studies of A.S., V.V. Bartold critically examined numerous written sources, many of which had been made available to scientists for the first time. He succeeded in proving that historical sources decisively observed that from the 13th to the 16th century, the Amudarya flowed into the Caspian Sea. Such conclusions were made in spite of the prevailing opinion of geologists and geographers. More recent investigations, in particular of S.P. Tolstov, confirmed, in general, the data contained in the historical sources given by Bartold, with some reservations concerning the period and volume of flow along the Uzboi channel, however. This work stirred great interest – in 1910, it was published in German and later in English. This work was highly praised by *L.S. Berg* (see).

Institute of socio-economic problems of the circum-Aral Area, Karakalpak Branch of the Uzbek Academy of Sciences – established in 2000 on the basis of the Computer Center. The Institute contends with the regional socio-economic problems of the Circum-Aral area and econometric modeling. Today, the Institute comprises the following divisions: the division on modeling socio-economic processes, the division on the study of production forces, the division on population and social problems, and the division on environmental problems.

The Institute is a member of the International Society for Ecological Economics (ISEE), is included in the Bureau on the Coordination of Economic Studies of Eastern Europe (Germany), and maintains contacts with the Levy Economics Institute of Bard College (USA) and others.

Since its founding, the Institute has implemented research projects relevant for the region. In recent years, completed projects have included: the advancement of the concept of sustainable development in the Circum-Aral area; the elaboration of a program on socio-economic development for the Circum-Aral area from 2000 to 2010 (using materials from the Republic of Karakalpakstan); the implementation of economic instruments for regional sustainable development; the application of econometric methods in the study of demographic processes; and the development of mathematical modeling of climate responses to changing ecosystems in the Southern Circum-Aral area.

Within the framework of the Institute, the scientific workshop, "Problems in the Sustainable Development of the Circum-Aral Area," functions.

Integrated hydrometeorological atlases of the Caspian and Aral Seas – prepared and published by the Research Institute of Aeroclimatology in Leningrad in 1963, and edited by V.S. Samoilenko. The Atlases comprise maps of atmospheric pressure, resultant winds and their stability, solar heat flux, effective radiation, radiation balance, and wave height; maps of contact heat exchange, evaporation, and heat fluxes; maps of water and air temperature, recurrence of air temperature, atmospheric events and visibility; maps of cloudiness, quantity and intensity of precipitations, and recurrence of clear and gray skies, mists, and precipitations; maps of visibility and recurrence of weather conditions; maps of an average rate of wind and mixed waves and recurrence of swell waves; maps of dominating winds; and maps of absolute humidity for estimating components of the radiation balance.

Integrated Institute of Natural Sciences of the Karaklpakstan Branch of the Uzbek Academy of Sciences – located in Nukus.

Interstate Coordination Water Commission (ICWC) – established in 1992 under the Agreement among the Republics of Kazakhstan, Kyrghyzstan, Uzbekistan, Tajikistan, and Turkmenistan on cooperation in joint management and protection of interstate water resources. ICWC is a collective parity body of the Central Asian states, acting on the basis of equality, justice, and consensus of opinions. Pursuant to the Resolutions of the heads of Central Asian states on March 26, 1993 and April 9, 1999, ICWC and its divisions were included into the International Fund for Saving the Aral Sea and acquired the status of international organizations. The ICWC executive bodies are Basin Water Management Associations (BVO) "Amudarya" and "Syrdarya" and the Scientific-Information Center (SIC).

ICWC and its executive bodies ensure strict observance of water release regimes and water consumption limits, implementation of efforts on rational and wise management of water resources, sanitary water flushes along river channels and irrigation systems, and supply of guaranteed quantities of water to the Circum-Aral area and A.S. for improvement of the environmental situation and maintenance of the water quality level in accordance with the agreements. The decisions taken by ICWC concerning observance of the assigned water intake limits and rational management and protection of water resources are binding for all water users. At its meetings, ICWC approves the annual limits of water intake from interstate water sources (classified by vegetative and inter-vegetative periods) for member countries with regard to the predicted water management situation and the assigned water releases to A.S.; considers and makes decisions on correction of water intake limits from the actual situation (BVO "Amudarya" and "Syrdarya" are permitted operative corrections of water intake volumes within 10%); annually mandates the program of BVO activities and finances the operational and other costs. The decisions made by ICWC concerning regulation, utilization, and protection of water resources are binding for all water users regardless of their state or departmental affiliation and forms of property.

One of the clauses of this Agreement imposes on ICWC the power to define the water management policy in the region, elaborate its directions with regard to the needs of all economic sectors, integrate and rationally manage water resources, and develop a prospective program of water supply for the region and act on its implementation. Among its other functions, the ICWC also develops and approves yearly water consumption limits for each state and the region, matching, in general, the regimes of reservoir operation and their correction on the basis of verified forecasts depending on the actual water availability and the established water management situation.

The ICWC structure includes the Secretariat, Scientific-Information Center, Coordination Metrological Center, and BVOs "Amudarya" and "Syrdarya".

International Commission on Irrigation and Drainage (ICID) – one of the largest nongovernmental organizations, it was established on the initiative of the Indian Government in 1950 in Simla, India as the International Commission on Irrigation and Canals. At the 1st Congress on Irrigation that was held in New-Delhi (India) in 1951, the Commission acquired its present name – ICID. At this meeting, the ICID Statutes were adopted to define its purposes and tasks. The Statutes defined ICID's purpose as to comprehensively assist the development and application of achievements in science and technology to irrigation, drainage, flood control, and riverbed regulation in technical, economic, and social disciplines. The scope of the issues addressed by the Commission included all problems related to the planning and financing of efforts on land reclamation, flood control, riverbed regulation and design, and construction and operation of respective engineering structures. Later, the scope of issues was extended to include the study and analysis of all factors contributing to successful irrigated farming development.

Within the first 50 years of its existence, the ICID membership increased from 11 to 87 countries (2000). The Soviet Union joined ICID in 1955. After the disintegration of the USSR, ICID was joined by Uzbekistan (1996) and Tajikistan (1997). ICID's highest management body was the International Executive Council, comprised of the president, 9 vice presidents, and the general secretary (selected for a term of 3 years), and representatives of its national committees. In 1972, at the 8th ICID Congress Ye.Ye. Alekseevsky, the USSR Minister of Land Reclamation and Water Management, was elected the ICID President. The ICID Executive Council performed its activities via several committees, and the ICID Secretariat was located in New-Delhi (India).

The meetings of the Council were convened every year. The 26th Meeting of the ICID Executive Council was held in Moscow (USSR) in 1975 within the framework of the 9th Congress on Irrigation and Drainage. In 2004, Moscow hosted the 55th Meeting of the Executive Council. The European and Afro-Asian Conferences were also organized. On the USSR's initiative, in 1976 Tashkent became the venue of the Afro-Asian Conference on Irrigation and Drainage.

The tasks formulated in the ICID Statutes included improvement of the exchange of scientific-technical information among the national committees; the convening of international congresses, symposia, and ad-hoc sessions; organization of joint researches and experiments; publication of congress transactions, papers, world reviews, and other materials; and the promotion of cooperation with other international organizations. For the 55 years of its activities, the ICID organized 19 congresses to discuss the most burning issues on irrigation and drainage. The issues for discussion at the coming congresses were selected taking into consideration the interests of a host country and also the significance of the problem for the majority of the ICID member countries.

The first publication of ICID was the review, "Irrigation and Drainage in the World – A Global Review," which was re-published three times (1969, 1981 and 1983). All in all, ICID issued more than 90 publications, including special issues, world reviews on relevant issues, guidelines, memorial publications, and technical memoirs. A large contribution of ICID in the development and unification of terminology in irrigation and drainage was publication in 1967 of the "Multilanguage Technical Dictionary on Irrigation and Drainage," which contained over 10 thousand terms and definitions. In 1996, an enlarged and revised edition of this Dictionary was circulated. The Dictionary was translated and published in 14 languages, including Russian. In addition, ICID published a Bibliography (from 1954 – annually), Newsletters (from 1986), monthly News Updates (from 1993), and the ICID Bulletin (from 1952). In 1997, ICID opened its Internet-site.

International cooperation on the Aral Sea problems – large-scale cooperation began in 1993–1994 after the breakdown of the Soviet Union when the difficult period of reforms and alienation from financial sources, material resources, etc. made the young independent states of Central Asia dependent on international aid for addressing such serious and complicated problems as the Aral Sea problem, water resources management in the A.S. basin, and others. One of the first proposals to the world community with a view to raise financial resources for addressing the Circum-Aral problems was submission to the World Bank in 1992–1993 of the "Aral Sea Basin Program." In it, *SANIIRI* (see) and other co-authors of this program included ideas that had been developed by scientists for nearly the entire preceeding decade. In early 1994, this program was presented to the Summit Meeting of the Central Asian countries, which approved it in the form of 8 items (see PBAM). In the same year, these items were submitted to the donors' meeting, at which they were approved as a first-stage endeavor and US 40 mln was conferred for their implementation. From this time, scientists and designers have actively cooperated with foreign consultants and financial organizations, reflecting the growth of ideas and methodological approaches proposed by Central Asian research organizations.

The main projects adopted for implementation related primarily to water resources management, agriculture improvement, environment protection, and, to a less extent, health improvement. Many UN organizations (UN University, UNDP, UNESCO, UNEP, UNIDO, FAO, WMO, UN High Commissioner for Refugees, and the International Labor Organization); financial organizations (World Bank, Asian Development Bank, European Bank for Reconstruction and Development, International Monetary Fund, Global Environment Facility); European Union Programs (TACIS, INTAS, INCO-Copernicus, OSCE, TEMPUS); international nongovernmental organizations ("Doctors Without Borders"); regional organizations (International Fund for Saving of the Aral Sea, Interstate Coordination Water Commission, Commission on Sustainable Development, Central Asian Economic Community); and bilateral organizations (US Agency for International Development,

Soros Foundation (USA), Konrad Adenauer Foundation, Friedrich Ebert Foundation, Germany Agency for Technical Cooperation (Germany) NOVIB (the Netherlands) NATO Program "Science for Peace," JAIKA, Global Infrastructure Fund Research Foundation (Japan) and others) were involved in the implementation of many hundreds of projects.

Apart from these organizations, experts, consultants, scientists, academicians, and others from more than 30 countries took part in the study and preparation of project proposals on the Aral problems. Needless to say, from 2000 more than 30 international projects devoted to various aspects of problems in the Aral Region were elaborated within the framework of the International Programs INTAS and INCO-Copernicus. Dozens of Eastern-European, Russian, and Central Asian institutions and laboratories were also involved in comprehensive investigations. And, of course, ministries, local authorities, institutes of the Academy of Sciences, and national nongovernmental organizations of all Central Asian countries participated in this international cooperation.

International Fund for saving the Aral Sea (IFAS) – interstate organization established in 1993 by the heads of Central Asian states – Uzbekistan, Turkmenistan, Tajikistan, Kazakhstan, and Kyrghyzstan. In 1997, after merging with ICAS, the final organizational structure of IFAS was shaped. The main tasks of IFAS are raising funds in the 5 Central Asian states and through international donors to financially support the *Aral Basin Program* (see); implementating joint environmental and research-practical projects on saving the sea and on environmental improvements in the regions affected by the Aral disaster; financing joint fundamental and applied investigations and research-technical developments on restoration of the environment balance; and rational management of natural resources and environmental protection. The IFAS Executive Committee was established to ensure the general guidance of the Aral Program. Branches of the IFAS Executive Committee were organized in Almaty, Bishkek, Dushanbe, Dashkhovuz, and Nukus. The Agreement signed by the heads of the states on April 9, 1999 confirmed the following division of duties among regional organizations:

- *IFAS Board*, comprising the deputy prime ministers of 5 states, is the highest political level of decision-making and finalizing approval;
- *IFAS Executive Committee* is a permanent body including the chairman and two representatives from each state and in charge of realizing the decisions adopted by the IFAS Board via the IFAS national committees. At the same time, the Executive Committee may organize, on behalf of the Board, and implement various other projects (international or donor).

The presidents of the Central Asian states are appointed Foundation Chairs for 2 year terms on a rotating basis. IFAS is the main instrument of collective influence on the environmental, social, and economic situation in the A.S. basin.

Interstate Commission for Sustainable Development (ICSD) – established pursuant to the Agreement of July 19, 1994 within the IFAS (initially – Interstate

Commission for Socio-Economic Development and Scientific, Technical, and Economic Cooperation). It is charged with coordinating and managing the regional cooperation for environmental protection and sustainable development of the Central Asian countries, including organization and coordination of development of regional strategies, programs, and plans of sustainable development; management of regional programs, plans of action, and projects in environmental protection and sustainable development; organization of expertise and preparation of regional projects; coordination of efforts on execution of the commitments of the Central Asian countries concerning implementation of the transboundary nature conservation conventions; promotion of the unification of the legal and methodological base in environmental protection; and facilitation of interstate information exchange and creation of a regional information databank on environmental protection and sustainable development, including preparation of the Regional Agenda-21 and the Convention on Sustainable Development. ICSD comprises 15 members – 3 representatives from each state appointed by the governments of member countries. The ICSD executive bodies consist of the Secretariat and the Research-Information Center (SIC ICSD) which has branches in all IFAS member states. ICSD activities are managed by the Minister of Nature Protection of each Central Asian country on a 2-year rotation.

Interstate Council for the Aral Sea Basin (ICAS) – established in 1993 in Kyzyl-Orda at the Summit of the five Central Asian states. Among other divisions, working of the ICWC were assumed by the Council. In 1997, ICAS was abolished and its functions were assigned to the reorganized *IFAS* (see).

Invaders – alien organic species introduced, either intentionally or accidentally, into new habitats from adjacent or remote water areas through human shortsightedness or negligence. Plans for new fish introductions into the Aral Sea were elaborated since 1920, while the beginning of transition coincided with the renewed biological investigations.

The first attempts at the introduction of valuable feed species were in 1929–1930. These attempts unsuccessfully tried to introduce the Caspian "shad," which died en route. Other attempted introductions were the larvae of two other herring species, *A. kessleri* and *A. volgensis,* both from the Volga delta and foredelta. The Aral aqueous fauna was replenished with at least 7 species of free living animals and at least 5 species of saline-water aqueous parasites: the Caspian stickleback and its specific saline-water parasite (*Trichodina*), one of the first intermediate hosts of the sturgeon specific parasite (*Cystoopsis acipenseris*); the Caspian zebra mussel (*Dreissena caspia*); 2 species of the Caspian *Cerastoderma ornate* and *Cardium edule rusticum*; the Caspian *Theodoxus pupus*; and the hydrobiide mollusk *Caspiohydrobia*, the first intermediate hosts of the Caspian saline-water trematodas.

The principal places of Caspian herring hibernation are in the Southern Caspian where the water temperature does not usually drop below $+10^{\circ}$C. By the average many-year data, however, February temperatures in the deep troughs of the Aral (up to 60 m) dropped to +1.0–1.4°C, while in the north-eastern and southern shallow areas, the drop was even to negative temperatures (–0.2 to –0.5°C). This temperature difference is why the introduction of the Caspian herring did not result in their acclimation. Out of 8 million transported larvae, only a few specimens of the two-year shad (1931) were ever caught.

In 1933–1934, specific a monogenetic fluke of sturgeons, *Nitzschia aff. sturionis,* was introduced together with the stellate sturgeon (*Acipenser ratzeburgii*) from the Volga delta. The resulting 1936 epizootic outbreak became one of several well-known consequences of uncontrolled inter-basin fish transfers. In the summer of 1936, on each of the dying bastard sturgeons were found hundreds of *Nitzschia* (approximately 600 species max). Feeding on the blood of their hosts, they sucked nearly all blood from the sturgeons, thus causing their mass death in the population.

In 1948–1963, after another unsuccessful transfer of sturgeons from the Ural delta, less than 6 species of hydrobionts were introduced, including 5–6 species of freshwater bullheads: Berg's bullhead (*Hyrcanogobius bergi)*; *Knipowitschia lencoranica; Apollonia melania; Neogobius pallasi; N. gorlap,* and, possibly, *Proterorhinus semipellucidu*; and freshwater silverside (possibly, *Ichthyotaenia gobiora*) together with intermediate hosts of plankton crustaceans (*Cyclopidae* and *Diaptomidae*).

The widespread opinion that in the 1940s–1950s, introduction of Ural sturgeons (stellate sturgeon, sturgeon, bastard sturgeon) was carried out only by fertilized eggs was not true. In 1948–1956, the larvae (of stellate sturgeons – 2–10 days of age) and fries (of sturgeons 2–4 weeks of age) caught in the Ural lower reaches were also transferred here. Later, the fertilized eggs of the sturgeons were brought to the fishery base of "Aralrybvoda Tastak"; however, out of the myriad-strong stellate sturgeon population that allegedly appeared in the Aral from the (detached eggs), by 1967 dozens if not single specimens of stellate surgeons and sturgeons were fished here. After the abnormally cold winter of 1969–1970, the fishing of Caspian sturgeons ceased and they again had to be re-introduced in the late 1970s.

In 1954–1956, together with the Baltic herring (*Clupea membras*), at least three species, such as *Gammarus locusta, Diacyclops bisetosus*, and *Mesorchis denticulatus*, were introduced into the Aral.

In the same period of fruitless attempts to acclimate Mediterranean gray mullets (golden mullet (*Liza aurata*) and little mullet (*L. saliens*)) from the Bekovich Bay (Krasnovodsky Bay, South Caspian), the following species were introduced into the Aral: one species of macrophytes (*Cladophora aff. fracta* – specific spawning substrate of the Caspian silverside and at least 17 species of animals (including 8 parasitic): 2 species of the Caspian bullheads – long-tail *Knipowitschia longecaudata* and *Neogobius niger*); Caspian silverside (*Atherina caspia*); 2 species of Black Sea cockles (*Cer. maeotica* and *Cer. picta*), which were introduced into the Caspian along with Mediterranean hydrobionts transferred in 1930–1940; Caspian *Evande angusta*; Caspian *Calanipeda aquaedulcis*, *Palaemon elegans*; holoeurihalinic Caspian hydrobiides (including 3

Aral "endemics" *Caspiohydrobia behningi, C. kazakhstanica and C. sidorovi* that survived in the brines of the Gorkaya River (the Baskunchak Lake basin) since the Khvalyn Time); 2 or 3 species of parasitic infusoria: *Trichodina meridionalis* and *Tr. puytoraci*, specific of bullheads.

The indispensable condition for young mullet survival during their first winter is that the water temperature does not rise higher than 7–8°C. As water cools to + 5°C, they die in 1–2 days, which is why in the Caspian in winter they flock in the southernmost bays near the Iranian coast, and the attempts of their introduction in the Aral Sea were absolutely hopeless. Only a small part of holoeurihalinic hydrobionts could assimilate there. And Some of them (e.g. Caspian freshwater shrimp *T. priscus*, Black Sea *Ath. pontica,* and Ural sturgeons) were frozen out during the cold winter of 1969/1970. Attempts to introduce holoeurihalinic boreal fish (Baltic herring and flatfish) were more fruitful.

In 1958–1967, after transfers of opossum shrimps (the reasoning was to prepare them for introduction to the Volga and Caspian shrimps from the Don delta) and *Monodacna colorata* from the Taganrog Bay, 2–3 species of opossum shrimps *Paramysis baeri, Par. intermedia* and *Par. ullkyi,* being endemic of the Caspian basin, were introduced.

Despite assurances of wide environmental plasticity and the eurihalinic nature of the so-called Ponto-Caspain mysids, they proved incapable of surviving even oligohalinic spaces, while at the Amudarya mouth, they managed only with human help. That is why the opinion about introduction of Don opossum shrimps in the Aral was exaggerated. Negative results of *M. colorata* transportation to the Aral and salt lakes of Central Asia and Kazakhstan vividly prove the inability of this euxinian relic to multiply even in the oligohalinic water.

In 1959–1963, a transition of the mollusk *Abra ovata* and polyhaete worm *Neanthes succinea* from the Azov Sea was proposed.

In the 1960s, in the course of the planned introduction of two fish species (grass carp – *Ctenopharyngodon idella* and white silver carp – *Hypophthalmichthys molitrix*) from the rivers of North China and the Amur basin, at least 14 taxons of Far-Eastern fish were introduced into the Aral rivers, including the Amur bullhead (*Rhinogobius similes*, snakehead (Chinese – *Channa argus* or Amur – *Ch. warpachowskii*), Amur pseudogudgeon *Pseudogobio rivularis*, three-lips (Chinese *Opsariichthys bidens* or Amur *Op. amurensis*), Amur *Pseudorashora parva*, white Amur bream *Parabramis pekinensis*, bitterling *Rhodeus ocellatus* (may be represented by several species), Japanese *Oryzias latipes* from the Amur or Chinese *O. sinensis* from the Yangtze, *Micropercops cinctus*, ordinary sawbelly *Hemiculter leucisculus*, one-color *Nemacheilus labiatus* and spotted *N. strauchi, Aristichthys nobilis*, and the black Amur *Mylopharyngodon piceus*. The parasitic fauna of commercial fish was added with at least 12 new species: *Balantidium stenopharyngodonis (Peritrichida), Dactylogyrus aristichthys, D. chenshuchenae, D. ctenopharyngodonis, D. hypophthalmichthys, D. lamellatus, D. magnihamatus, D. nobilis, D. scrjabini, D. suchengtaii, Diplozoon bychowskyi (Monogenoidea) and Bothriocephalus opsariichthydis (Trematoda)*. The ratio of planned invasives to accompanying ones (no less than 1:13), usually undesirable, proves once more the hazard of such acclimatization efforts. Perhaps the free-living Far-Eastern invertebrate and algae were introduced into the Aral basin.

The Far-Eastern fish, similar to Don mysids, did not usually go beyond the confines of the freshwater zones in the Aral rivers; however, Amur bullhead and *Oryzias latipes* were found in Sudochie Lake (1999–2000) at water salinity up to 30–35 mg/l.

In 1965–1971, after transfer of *Calanipeda* from Azov brackish lagoons, *Calanipeda aquaedulcis and Rhithropanopeus tridentatus*, which before the beginning of trans-Atlantic shipping lived near the North-American coast, were introduced, as were, perhaps, the holoeurihalinic marine *Popella guernei* with parasitic nematode *Con. septentrionale* or *Con. rudophii*.

The habitat of freshwater *cal. Aquaedulcis* in the Aral was limited by river mouths. The holoeurihalinic twin-species belonging to the balakhansky complex (*relics* (see) that were accidentally introduced during transfer of gray mullets from kultuks of the Krasnodovsky Bay) propagated here. It was found, together with *Acanthocyclops viridis auct.* and other unnamed forms of the Aral zooplankton, for the first time at water salinity over 50 g/l in kultuks in the north-east of the sea in the summer 1955.

In 1971, in a vain attempt to introduce *Heterocope caspia* from the Volga foredelta, *Limnomysis brandti* and one more species of Caspian *Evande trigona* were introduced.

Heterocope caspia is a stenohalinic freshwater species. In the Volta, it is found up to Saratov, while beyond the zone of the Volga water transit, it is eliminated, which is why its transfer to the saline Aral waters was useless and quite logically ended in a failure.

From 1978 to 1980, the last transfers of sturgeon and stellate sturgeon fries (largely hybrid forms) from the Volga sturgeon farm were made. Analyzing the results of the introduction, it was found that sturgeon fries (7.7 cm, 1.4 g) may live in the Aral waters with salinity level up to 16.2 g/l, while at salinity 19.5 g/l about 87.5% of the fries die. In 1978–1980, water salinity in the Aral Sea was up to 17 g/l and greater salinity was quickly approaching. Thus, introduction could not provide and did not provide any positive economic results.

In 1979–1987, with the introduction of *Platichthys flesus* from the Azov Sea, one more of Black Sea *Cer. Glauca* and *Halicyclops rotundipes* were transferred here.

In 1984–1986, an attempt to introduce mussels, sand gapers (mia), and *acarcia* from the Azov Sea was made.

Before wide-scale introductions, the Aral took a leading position in economically valuable fish catches among the internal waters of the former USSR. Fish catches (benthos feeders – 30 thou tons and predatory fish – 3 or 4 thou tons) were formed mostly by the bottom trophic chain. The fish that fed mostly in the pelagic zones provided an additional 3 thou tons to the catch every year. In the water abundant period (1942–1960), numerous populations of freshwater aboriginal fish were formed, and the catches were record high, as with the 42 thou tons caught in 1957. In the early 20th century, this regularity was unknown. It seems that dwindling native fish fauna, unaccustomed to the full production capacities of the water body, were to be augmented with valuable fish species; however, not only were the conditions of the recipient water body not studied well enough but neither was the ecology and parasitic fauna of the invasives for studied for acclimatization purposes. Not enough attention was paid to taxonomic studies, while efficacious conclusions often contradicted even well-known facts. Hence, quite logically, introduction efforts had mostly negative results. One cannot name a single water body in the world where introduction have led to improved fish production. That is why there is a special law in North America and Western Europe, though it is not upheld in international conventions, that bans introduction.

The autochthonous biota of the Aral basin was destroyed as a result of its total "reconstruction"; its economic value dropped long before its wide-scale hydraulic construction. As a result of epizooty, the catches of the Aral bastard sturgeon, being the most valuable representative of the commercial fish fauna, dropped from 300 – 400 t/year from 1928–1935 to 13.8–53.8 t/year from 1936 to 1940. A complete ban on bastard sturgeon fishing and transfers of the Ural sturgeons during many years did not improve the situation. The maximum catch of bastard sturgeon in the 1950s was 6.0 t/year (1952), while the average did not exceed 2.5 t/year (i.e. the catches of sturgeons that appeared in the *Nitzschia* period had increased only by 1% of the natural level). In the 1960s, when the Aral water level dropped and the water salinity increased, the last sturgeons (less than 60 tons) were caught. After introduction of short-cycled plankton-eating fish like the Baltic herring, silversides, and bullheads, the invasives devoured the remaining, poor as it was, Aral plankton on which only the larvae and fries of freshwater fish had once fed. In the early 1960s, only 1 fry out of each 10 was referred to as a commercial species, which is why in spite of the growing salinity and fish concentration in rivers, a further drop in the catch was witnessed.

Inzhener-Uzyak* – the westernmost of the three main arms of the Amudarya that flows into *Rybatsky Bay* (see) of A.S. Over this arm, vessels with a draft of 1.2–1.4 m sailed to Ushsai port. The entrance into the arm is surrounded by a bar with the depths of about 1.6 m. From time to time, bottom dredging works were carried out here.

Irrigation management for desertification control in the Aral Sea basin – a collection of materials published in Tashkent in 2005 on the basis of the results of the research project, "Management of Agricultural Crop Irrigation for Control of Anthropogenic Desertification in the Aral Sea Basin," edited by Professor L.S. Pereyra, Professor V.A. Dukhovny, and engineer M.G. Horst. The project was implemented jointly by scientists and specialists from Portugal, France, Uzbekistan, and Kyrghyzstan. 20 chapters of this work combine the results of scrupulous investigations of desertification control through the management and regulation of a system of irrigated farming in the A.S. basin in the Central Asian countries that face water deficit, limited material resources, and population growth. It considers different viewpoints and approaches in the theory and practice of irrigated farming development in Central Asia (both "Pros" and "Cons"), and provides an assessment of the problem and offers proposals that, in the authors' opinion, may curb and even reverse the ongoing negative environmental processes and facilitate the application of new practices in irrigated farming that would lead to sustainable yields.

Islands of the Aral Sea – More than 1100 islands were found in A.S., most of which belonged to the *Akpetkinsky Archipelago* (see) and were located near the eastern shore. The quantity of islands in the Akpetkinsky Archipelago and near the eastern shore as well as the configuration of their coastline changed permanently because they depended on the sea level. Most of these islands were very low and were separated by multiple shoals and bars. During water level rise, some islands were inundated, while the configurations of the coastlines of the most elevated ones changed consistently. Their total area was 2235 sq. km, or 3.5% of the total sea area. The largest islands were Kug-Aral (273 sq. km), Vozrozhdenia (216 sq. km), Barsakelmes (183 sq. km), and Muinak (127 sq. km). The islands near the eastern shore were smaller in size: Kaska-Kulan, Kuzzhetpes, Uyaly, and Djudeli. Many of these islands had an area from dozens of square meters to several square kilometers.

Izendyral Cape* – the southeastern end of *Kulandy Peninsula* (see). Its bank is steep and high (up to 14 m).

J

Jenkinson, Anthony (?–1611) – a British tradesman and diplomat, the envoy of Queen Elizabeth, and the agent of a company founded in England for trading with Moskovia. Traveled over Europe, Asia, and Africa. In the period from 1557 to 1572, he visited the Russian state four times. From 1558 to 1559 and again from 1562 to 1564, having permission from Tzar Ivan IV, the Terrible, he traversed Russia (Kazan and Astrakhan) to Persia and Central Asia (Khiva and Bukhara) to study anew the trade path to China. J. succeeded in obtaining approval from Ivan IV for the British trade monopoly (1567–1569), provided only that a union treaty was signed between England and Russia. After the refusal of England to sign this agreement, Ivan IV abolished the monopoly (1570).

J. prepared a description of his trips that became an important source of knowledge on the history of the relationship between Russia and England, Bukhara, Khiva, and Persia. Of special importance was the map prepared by J. and published in 1562 (reprinted several times later). This map was made on the basis of his own observations and the results of interviews and showed Central Asian cities that were absent on previous maps. It also gave geographical latitudes of the Mangyshlak Peninsula, Kunya-Urgench, and Bukhara.

Concerning the hydrography of the Aral basin, L.S. Berg noted that "Jenkinson's report and map were so confused and unreliable that practically no useful data can be taken from them." Jenkinson also wrote that the "Water that irrigates all of Khorezm country is supplied via canals made from the Ox (Amudarya) River, causing much damage to the mentioned river. For this reason it does not flow as before into the Caspian Sea."

Joint statement of the heads of the water management bodies of the Central Asian Republics and Kazakhstan – this document was adopted by the ministers of land reclamation and water management of the Central Asian republics at a meeting held in October 1991. In this Statement, which was passed unanimously, the participants recognized the indissoluble interrelation of the interests of all republics in the region in joint management of water resources of the A.S. basin and the need to equitably realize for all republics principles for the just regulation of water use with regard to the needs of all peoples living in the region. It was acknowledged as advisable to establish new organizational

I.S. Zonn et al., *The Aral Sea Encyclopedia*, DOI 10.1007/978-3-540-85088-5_11,

structures for coordination of such joint activities in view of the breakup of the former economic ties. The participants also arranged to undertake development and correction of the inter-republican quotas for water intake and water use by years and by sources with regard to ensuring the guaranteed supply of water to the Circum-Aral Area and the Aral Sea; to ensure full exchange of information on water resource management; to abstain from taking unilateral actions that might have negative consequences for neighboring states; and to settle all disputed issues through the heads of relevant organizations in the republics and with representatives of non-interested party.

The Statement was published in the newspaper “Pravda Vostoka” on October 13, 1991 in Tashkent.

K

Kaitpa – deep pits in the Amudarya riverbed that are erodeded as the water flows in the river meanders.

Kamyslybas, Kamyslybasi Lake – the largest water body in the *Kamyslybas lake system* (see). Located in the Syrdarya valley, in the depression between bald mountains. It extends from the west to the east. Its length is approximately 30 km, and its width reaches 9 km in some places. The maximum depth is 10 m, and the surface area is about 178 sq. km. Its coastline is 116 km long, and its elevation is 58 m above sea level. The water level in the lake is subject to fluctuations dependent on the quantity of water from the Syrdarya.

The eastern coast is flat and sandy. The railroad Kyzyl-Orda – Aralsk runs along the shore. The southern shore is elevated, in some places rising steeply (up to 23 m), and is flat, composed of gray marls and loams, near the bay. The northern shore is high and steep (in some places up to 50 m) nearly along its whole run and is composed of loams, with some outcrops of gypsum. The southwestern part is flat; a lake bay up to 4 km long connects here via a channel with the Syrdarya.

The ground is silty (gray silt) closer to the shore, sandy in some places, and stony (gravel and pebble) in others. In the reed thickets, the silt is black, containing plant remnants and slightly smelling of hydrogen sulfide. Silt deposits in the central part of the lake may be up to 150 cm thick.

The lake was populated with fish (1960): common carp, pike roach, sea roach; pike; catfish, perch, and others. The shores were nesting places of birds: ducks, geese, sandpipers, and others. Muskrats lived in the cane thickets. In the southern part of the lake, nearby the Koszhar settlement, a fish farm supplies fish fries to the Kamyslybas, Akshatau, and Karakol Lakes. Before 1968, up to 70 thou quintals of fish were caught here. After 1977, to maintain a certain level in the lake, water was pumped here from the Syrdarya. Thus, in 1977 the maximum depth of K.L. was 6 m, while in 1978 it had already reached 10 m. These efforts, to some extent, helped to maintain the average fish catches and the average level of fish farming.

Kamyslybas Lake system – found on the right bank of the Syrdarya, 60 km from the mouth. The system includes 5 lakes: Raimkol, Djalanash, Kayazdy, Laikol,

I.S. Zonn et al., *The Aral Sea Encyclopedia*, DOI 10.1007/978-3-540-85088-5_12,

and *Kamyslybas* (see), all of which are linked by small channels. In the 1980s, these lakes were full mainly during the Syrdarya flood periods, when water flowed to the lakes along the Taupzharminsky canal. The other Sovetsky canal also fed water to these lakes, but after Syrdarya regulations were imposed it became shallow and then ceased to exist.

Kantyubek – a small port that once existed on *Vozrozhdenia Island* (see).

Kara-Karnau Cape* – located to the east of the Berg Strait, 2.4 km south of the bar that run to the south-southwest of the *Sarykamysh Cape* (see). It restricts entrance into *Bugun Bay* (see). The cape is low, flat, and sandy.

Karabaily Island* – entering into the *Akpetkinsky Archipelago* (see), it is located 11 km to the south-southwest of he Tailakdjegen Islands. The 1.5 km long *uzyak* (see) separates it from nearby Akpetky Island. K.I. is low and sandy.

Karadjarsky Lakes – comprised of Mashankol, Khodjakol, and Ilmenkol Lakes, all of which are found on the left bank of the Amudarya delta. Because of their water exchange, they are referred to as periodical through-flow lakes. These lakes are supplied with water via the Raushan canal that breaks the nearby Moshanaul settlement into two branches, the Sudochjyab canal and the Raushan arm. The latter goes into the Mashankol Lake, from which, along local arms, water flows into the Khodjakol and Ilmenkol Lakes. A water-distribution structure is built on the Raushan canal near Moshanaul. It is designed for river water transfer via the Main Drain (formerly the Ustyurtsky) to the *Sudochie* (see) wetlands. In late 2002 and early 2003, because of increased water supplies along the Raushan canal, the water regime of all K.L. was restored and excessive waters were transferred towards to the Sudochie wetland.

Karakalpak branch of the Academy of Sciences of the Uzbek Republic (KKB AS UR) – established in 1959 on the basis of the Karakalpak Integrated Research Institute of the Uzbek Academy of Sciences. KKB AS UR unites 5 research institutes, a botanical garden with the status of a research division at the Presidium of KKB AS UR, a fundamental library, the editors office of the scientific journal "Vestnik" (published since 1960), and a branch of the Research-Educational Center "FANUM" of AS UR.

Karakalpak language (Karakalpak Tili) – the language of the Karakalpaks, the official language of the Autonomous Republic of Karakalpak. It is in the Kipchak subgroup of the Turk languages (Tartar, Bashkirian, Karachaevo-Balkarsky, Kumyksky, Karaimsky, Crimean-Tartar, Kazakh, Karachagsky, and Nogaisky). Together with the Nogaisky, Kazakh, and Karachagsky languages, it is referred to as the Kipchak-Nogai branch. Before 1928, written Karakalpak used the Arab alphabet. In the period between 1928 and 1940, however, the Latin alphabet was used, and later on the Cyrillic alphabet. After the declaration of independence of Uzbekistan in 1991, it was decided to return to the Latin alphabet, and at present this decision has gradually been realized in Karakalpakia. Approximately 400 thou people speak K.L., largely in Uzbekistan. Approximately 2000 native speakers of K.L. live in Afghanistan, with smaller diasporas existing in Russia, Kazakhstan, Turkey, and other countries.

Karakalpak Oasis – located on the right bank of the Amudarya. Only its Khodjeily and Kungradsky regions are found on the left bank of the Amudarya within the confines of the modern delta. The total irrigated area is over 350 thou ha. The main agricultural crops grown here are rice and cotton. In the east, the oasis borders the *Kyzylkum Desert* (see), while to the west is the waterless *Plateau Ustyurt* (see), and to the north is the A.S. In the Amudarya delta, vast areas with the traces of ancient irrigation systems that are mostly buried under sands can be found. The Amudarya flows over several million hectares of territory, out of which about 3 mln ha are suitable for irrigated farming. Before 1917, two irrigation systems existed in the oasis, Shurakhansky and Chimbaisky. In 1935, the Kyzketken main canal (25 km long) was constructed with concrete headworks that had a discharge capacity of 210 cu. m/s. The canal commanded an area of 200 thou ha in 5 littoral regions of the Karakalpak Republic. In 1940, the V.I. Lenin canal (110 km long) with headworks that could discharge 240 cu. m/s was constructed to irrigate lands on the left bank of the Amudarya in the Khodjeily and Kungradsky regions. Later, this canal was reconstructed. Its route now flows over the lands of the Khodjeily, Leninabad, and Kungradsky regions, originating near Takhiatash and ending near Kungrad. The purpose of the reconstruction was to increase the carrying capacity of the canal to supply more water for rice growing.

On the right side of the Amudarya, the Pakhtaarna canal, built in 1936 in the Turtkulsky and Birunyisky regions, was also refurbished. After improvement, the canal stretched to 78 km long with a headworks that could discharge 75 cu. m/s. Such large canals as Nazarkhan, Mangit-Kipchak, and Keneges-Kchil were also built. The length of irrigation canals here reaches 30 thou km, including on-farm irrigation canals of 4.1 thou km long.

K.A. is distinguished by deep groundwaters that resulted from the fallow farming that was widely practiced here in the past. Many lands with abandoned irrigation systems are especially found on the right bank of the Amudarya in the Turtkulsky and Birunyisky regions and on the left bank in the Kungradsky region. Well known is the larges Shumanaisky ancient irrigation area, which may be equated now with the whole irrigated area in Karakalpakstan. The Chartambaisky area is overgrown with cane on the right bank of the Amudarya and is suitable for irrigation development.

To ensure the reliable water intake for the irrigation systems on the right and left banks, the *Takhiatash headworks* (see) were constructed in 1975 downstream of the main canals (named after Lenin, Sovetjyab, and Kyzketken).

Such large main drainage canals as Kungradsky, Khodjeily, KS-1, KS-2, KS-3, KS-4, and others were constructed to maintain and improve the condition of irrigated lands. The length of the drainage system is 6 thou km, including on-farm drains of 2.1 thou km. This has enabled improvements on more than 500 thou ha of irrigated lands.

At present, plans are underway on radical improvements of water supply of irrigated lands, flow regulation, looping of irrigation systems, utilization of ground waters for irrigation, improvement of the concrete chute network, and seepage-control lining of canals. In addition, the surface of irrigated areas will also be improved as well as the practices of vegetation and leaching water applications.

Karakalpaks (self-named Karakalpak) – descendants of the eastern Pechenegs and referred to in the Russian chronicles as “black klobuks,” they are Central Asian people inhabiting mostly the Republic of Karakalpak. Their language, Karakalpak, is of the Kypchak-Nogai group of the Western branch of the Turk languages. Their religion is Islam (Sunni), and Sufism is also widespread. Remnants of pagan beliefs also survive. Among the most ancient predecessors of K. are the Sako-Massaget tribes that lived from the 7th to the 2nd centuries B.C. on the southern shores of the Aral. From the 6th to 8th centuries A.D., Turkish tribes partially mixed with the local population, and from the 8th to 10th centuries, the formation of the K. people began in the Pechenegs and Oguzs environment. Some Pechenegs gradually merged with the Kipchaks that came here from the Irtysh basin and took up their language. In the 14th and 15th centuries, the K. ethnogenesis was influenced by their links with the Nogai people. From the late 16th century, K. were mentioned in Central Asian publications, and from the 17th to mid-18th century, K. inhabited largely areas in the middle and lower reaches of the Syrdarya. In the mid-18th century, the greater part of K. moved to the Zhanadarya, the western arm of the Syrdarya delta. By 1811, K. were conquered by the Khiva Khan and resettled to the Amudarya delta. In 1873, the territory on the right bank of the Amudarya inhabited by K. became a part of the Russian Empire. After the 1917 revolution, as a result of Central Asia separation from Russia, the Karakalpak Autonomous Area was formed, receiving in 1932 the status of Autonomous Soviet Socialist Republic. The traditional occupation of K. is semi-sedentary cattle breeding (mostly large-horned cattle) combined with irrigated farming and fishing.

Karakalpakstan, Karakalpak Assr, Karakalpakia – see Republic of Karakalpakstan.

Karashokat Cape* – the tip of the low-lying peninsula that confines *Akkol Bay* (see) on the south. The cape’s coast is flat and sandy.

Karasholan Bay* – located to the east of Berg Bay, 9.3 km to the south-west of the *Kara-Karnau Cape* (see). This is a shallow bay that intrudes to the southwest. The eastern coast of K.B. is low and sandy, and the southern and western shores are formed by deposits transported from the Syrdarya delta. It is connected with the sea via a narrow strait.

Karateren, Kultuk (see)* – traverses the northern part of the Akkol Bay and is connected with it via a narrow strait, the Karateren Path. Its depths approaching kultuk vary from 2 to 3 m, while in the Karateren Path, its depth is 2 to 2.5 m.

Karateren Lake* – found near the western foot of the *Beltau upland* (see) at an absolute altitude of 40 m in the northeastern part of the southern Circum-Aral area. It belongs to the wetlands of Sudochie Lake, covering 2.1% of its water area. Its width is 1.25 km, and its area is 9.3 sq. km, extending for approximately 6 km from the north to the south. It is a relic depression in the Tertiary deposits, surviving from the times of the sea basin and, obviously, was recharged with groundwaters. At present, it is permanently replenished with drainage waters from irrigated lands. Its maximum depth is 2 m. It is one of the unique natural features of Uzbekistan.

Karatma Bay* – found in the eastern part of A.S., directly to the south of the Biktau Island and extending for 24 km along a meridian. The bay has low, sandy shores. Its depth is 3 to 4 m, and its maximum depth is 9.5 m.

Karatobe Cape* – the southern tip of the Karatyup Peninsula. It is the eastern entrance of the Tsche-Bas Bay. The cape is elevated and has steep shores that extend deep into the water.

Kashkynsu Bay* – extends into the eastern coast of A.S., 25 km to the south of *Bozkol Bay* (see). The entrance into the Bay is limited by a low sandy island on the north and the mainland coast on the southeast. The eastern coast of the bay is shallow and broken.

Kaskakulan, Kaska-Kulan Island* – located near the eastern shore of A.S., 1.5–2 km eastward of the southern part of *Uzynkair Island* (see) and separated from it by a shallow strait. K.I. is low and sandy, and overgrown with shrubs in some parts. Its eastern shore is more elevated. Here, the settlement Kaskakulan, which extends along the shore, and a fish farm are found.

Kazakh Sea – see Small Aral Sea.

Kazakhdarya River* – an arm of the Amudarya River, it flows into the southern part of *Zhiltyrbas Bay* (see). It was once one of the largest Amudarya arms. Its mouth is at the top of the Iske-Kun-Uzyak Bay. On the right bank the fishery settlement Kazadarya is located.

Kazakov, Nikolai Ivanovich (1829–1903) – Major General of the Fleet Navigator Corps (FNC). In 1841, he entered as a cadet into the 1st Navigation Half-Crew from which he graduated in 1850 with the rank of FNC guard. In two years, he became a warrant officer on the frigate "Pallada," on which he went to the Pacific where, in 1854, he took part in a hydrographic expedition near the Korean coast. Then he was promoted to second lieutenant. In 1855, he returned to Petersburg from Ayan via Siberia. In 1857–1858, in the rank of a senior navigation officer on corvettes "Voevoda," "Boyarin," and "Novik," he again sailed over the Pacific Ocean. In a year from Ayan he returned to

Kronshtadt and on the corvette "Vol," he went to the Mediterranean. From 1860 to 1875, he served in the Aral fleet where he commanded the vessels "Perovsky," "Aral," and *"Samarkand"* (see). In 1865, he carried out surveys on a stretch of the Syrdarya River from the mouth to 300 km upstream. In 1873, on the vessel *"Samarkand"* (see) in the formation of the Aral fleet, he took part in the battle near the Akzal fortress in Khiva. In 1874, he was promoted to Lieutenant Colonel and transferred to the Baltic fleet where he continued his service. In 1884, he was given the rank of FNC Major General, after which he resigned. In his honor, a cape on the eastern coast of the Korean Peninsula was given his name.

Kazalinsk (in Kazakh – *Kazaly*) – a city in the Kyzyl-Orda District of Kazakhstan; the administrative center of the Kazalinsk Region. Population of about 6 thousand people. Located on the right bank of the Syrdarya River. The railroad station "Kazaly" (formerly "Kazalinsk," built in 1905) is on the Aralsk–Kyzyl-Orda line 12 km north to the territory of the Aiteke-Bi settlement (formerly Novokazalinsk, population: 33,000).

Before the degradation of A.S. and Syrdarya delta in the second half of the 20th century, the main occupation of the local population was fishing and fish processing. Here, the fishery and brick plants were functioning. This city is officially included into the Aral zone of environmental disaster. The Russian military fortification that was in place at the site of the modern city was founded in 1853 and was officially called Syrdarya Fort No. 1. This was the second Russian fortification established in this region, after fort Raim (1847), which was further downstream. In 1867, the settlement was given the status of a city and named K. The name Kazaly (Kazala), which is the same name as the natural area, had been known for a long time before.

By the late 19th century, the population of the city was up to 5 thousand. The national composition was very diverse: Kazakhs, Russians, Cossacks, Tatars, Khivins, and others. Before the revolution, K. was an administrative center of the Kazaly Division of the Syrdarya Region. K. was known as a center of the Syrdarya Ural people, which they called themselves. These were descendants of the Ural and Orenburg Cossacks, Old Believers who organized their settlements near the Syrdarya delta at least half a century before Russian expansion. After joining the empire, they were assigned at different times to the Ural, Orenburg, and Semirechie troops, but due to their remoteness from troop centers, enjoyed relative autonomy. They engaged in fishing and hunting, and lived in isolation among the Kazakh population that settled along the lower reaches of the Syrdarya. In their everyday life, they distanced themselves from the Russians who were adepts of the official Orthodox Church.

Kazalinsk delta – see Syrdarya Delta.

Kazaly Bay* – located in the northwest of A.S., at the northeastern part of Chernyshov Bay, 23 km to the north-northeast of the *Uzynkair Cape* (see). It is confined by the *Kulandy Peninsula* (see) and by the sandy bar running from it for 2.2 km to the north-northeast. A fishery was developed on its eastern shore.

Kegeily – the center of the Kegeilinsky Region of Karakalpakstan, Republic of Uzbekistan. Located in the Amudarya delta, 42 km to the north of Nukus and 60 km from the railroad station "Khodjeily" on the Dashoguz–Kungrad line. Irrigated farming, cotton growing, rice growing, and cattle breeding are practiced here.

Keinchiyak Cape* – protrudes from the western shore of the sea, 38 km to the south-southwest of the *Aktykendy Cape* (see). Near this cape, the shore gradually steps down in the form of terraces, ending with a steep cliff.

Kendyrli Island* – located in the western part of the Small Sea, 6 km to the south-east of *Peschany Island* (see). This is largely a low island with some small uplands in its central part, ending on the northern side with a cliff. The shores of K.I. are infringed by a shoal of up to 2 m deep that runs for 2.5 km in the northwestern direction.

Kendyrly Island (former Tyulkeli, Menshikova)* – a part of the *Akpetkinsky Archipelago* (see). It lies 20 km to the south-west of *Uyaly Island* (see). K. island is low and sandy, and in some places subject to flooding. K.I. extends in a narrow arch from the north to the south. Small underwater bars stretch to the north and south of the island.

Ken-Kamysh Bay – see Adjubay Bay.

Kerchevik – a local name given to specific gypsum structures formed as a result of gypsum crystallization on the stems of saxaul and other plants growing on the shores of depressions and bays.

Kesh (Turkish) – a populated area, habitation, settlement.

Khiva campaigns of 1839, 1853, and 1873 – military campaigns of the Russian troops organized to conquer the Khiva Khanate, to consolidate the Russian Empire's domination in Central Asia, and to weaken the colonial positions of Britain. The first 1717 military expedition to Khiva led by A. Bekovich-Cherkassky ended in a failure. In the 18th–19th centuries, the people of Khiva often attacked Russian trade caravans and took the Russians to slavery, thus incurring great damage to the trade relationships between Russia and the Central Asian countries. In the early 19th century, after numerous raids of small Cossack units on Khiva, the Tsar government organized a large force expedition. On November 14, 1839, the expeditionary unit, numbering 5 thousand men with 20 cannons, went from Orenburg to Khiva led by Aide-de-Camp General *V.A. Perovsky* (see). Having lost nearly half of the troops due to diseases and a greater part of the camels due to exhaustion during a passage from the Emba fortification to Chushkul in the spring of 1840, Perovsky had to return to the Emba fortress and in June to Orenburg.

The second expedition was in May-June 1853 to the Kokand Khanate, which attacked Kazakh tribes for being allied to Russia. On June 27, 1853, a unit of 5 thousand men under Perovsky's command reached the mainstay of Kokand on the Syrdarya River, the Ak-Mechet fort, which was attacked for 5 days and seized on July 1. The Russian side lost 175 people. The fortress was renamed Perovsk and became the forward base for Russian advances in Central Asia. For seizure of Ak-Mechet, Perovsky was awarded the title of count. In December 1853, Kokand troops (12 thousand men) tried to win back Ak-Mechet, but its garrison, commanded by Colonel Ogarev (1 thousand men), held off all attacks and inflicted a complete defeat upon the enemy.

In spring 1873, a new march on Khiva was organized comprising the Turkestan unit (over 5 thousand men), the Orenburg (3.4 thousand), the Mangyshlak (about 2 thousand) and the Krasnovodsk (2.2 thousand men) unit as well as the Aral military fleet (2 ships and 3 barges). In all, over 12 thousand men with 56 cannons were led on the march by Aide-de-Camp General K.P. Kaufman, commander of the Turkestan Military District. The Orenburg and Mangyshlak units appeared near Kungrad, and on May 26 came up to Khiva from the north, while the Turkestan unit came from the southeast. The Krasnovodsk unit reached the Igdy waterbody on April 17, but due to the absence of water had to return to Krasnovodsk in early May. The Aral military fleet failed to enter into the Amudarya because of low water. On May 27–28, the troops were engaged in combat near Khiva, and on May 28 were victorious. As a result of these military expeditions, in 1873 the Khiva Khan became a vassal of Russia, and the Khiva lands on the right bank of the Amudarya were annexed to the Russia Empire.

Khodjeili – the second largest city by population in Karakalpakstan. It is located opposite Nukus, 6 km from it, on the left bank of the Amudarya. The railroad connected K. with Dashoguz (Chardjou), crossing through the city. Before construction of a bridge over the Amudarya, there was regular ferry service to the capital.

Kishi-Karatyup Bay* – located 20 km to the south-south-west of the *Ulken-Karatyup Bay* (see). It cuts into the mainland for 4 km and extends meridianally for nearly 14 km. The shores are low and sandy. The bay is shallow, and its depth vary from 2 to 3 m. The greatest depths are registered in its central part.

Kokaral Cape* – the southeastern tip of *Kokaral Island* (see), it is the western inlet cape in the Berg Strait leading from the Small to the Large Sea. K.C. is low and sandy, and its shores are overgrown with cane. It is encircled with a shoal less than 2 m deep.

Kokaral Dam – a dike with water outlet structures. The idea of its construction was shaped in 1992. It was constructed in August 2005 in the narrowest part of the *Berg Strait* (see) and connects the *Kokaral Cape* (see) with the mainland near the Syrdarya mouth. Its construction (length: 12 km and height: 8 m) made it possible to stabilize the water level in the Small Aral at 39–42 m. Investigations

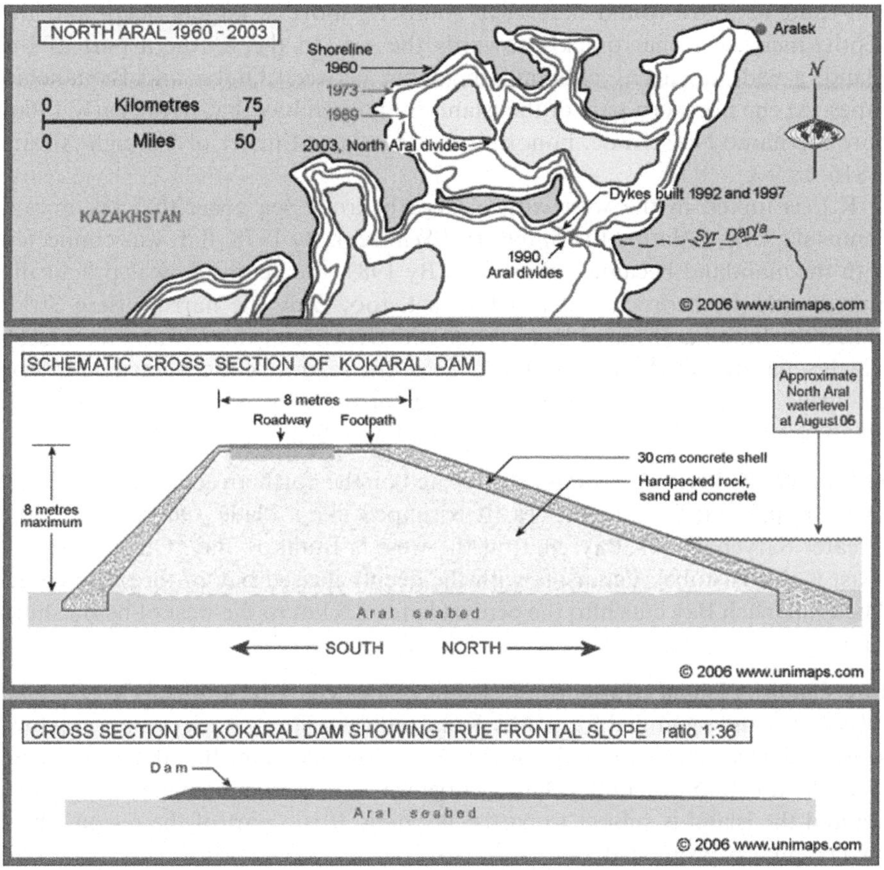

Fig. 34 Schematic view of Kokaral dam (www.unimaps.com/aral-north/aralschematic.gif)

are planned to consider options for attainment of the design water level of 46 m. Then water will cover a considerable part of the dried Sarychaganak Bay and come close to Aralsk City. The water salinity will be 4–17 g/l, and the discharge into the Arnasai Depression will decrease from 3 to 1 cu. km.

Kokaral Island* (Kazakh – *Kokaral tubegi*, "Kugara – "green island") – one of the largest islands in the Northern Aral. It forms a natural border between the Small and Large Seas. It is separated from the mainland by the Auzy-Kokaral Strait. K.I. runs along the parallel for more than 50 km, and its area is 273 sq. km at an altitude of up to 163 m (Dart City). The western and northern shores are low and sandy. Moving from the north to the south, the island becomes more and more elevated; the southern shore is mostly high and steep, reaching some 100 m. On the eastern shore, sand dunes 8-10 m high come very close to the water's edge. The eastern shore is shallow, incised with small bays. Shoals and sand bars are found here. The southern shore is largely deep, and the depths increase rather quickly towards the sea. In the southern part of the island, a wide open bay cuts into the shore between Ondyr and Baluanchin capes. At the northern part of the island is an open bay that in the early 1900s bore the name Nesselrode, honoring the Russian Minister of Foreign Affairs (1816–1856).

K.I. is linked in the west with a large bedrock sea coast that becomes a peninsula, thus isolating the *Small Aral Sea* (see). By 1978, K.I. was connected with the mainland became a peninsula. By 1989, the peninsula was practically linked with the bedrock coast in the east, too. Only the narrow Berg Strait linked the Small and Large Seas. After construction of the *Kokaral dam* (see), which separated the Small and Large Seas, the regime of the water bodies was changed. Regardless of repeated breaches of the dam, the water level in the Small Sea becomes higher than that in the Large Sea.

Kokturnak, Koktyrnak Peninsula* – located on the northern coast of A.S. (Small Aral), K juts out far into the sea. It is shaped like a blade. On the east is the Greater Sarychaganak Bay, and on the west is Butakov Bay. On the western coast is the Tastubek Peninsula with the deeply incised bay of the same name. The Zhalanash Bay cuts into the peninsula for 7.5 km to the west of he Baishura Cape.

Komsomolsky Island* (formerly Naslednik) – located to the south of *Vozrozhdenia Island* (see) and separated from it by Komsomolsky Strait. It is covered with low, sandy hillocks overgrown with shrubs. A shallow lake stretches along the island from the north to the south. During surge waves, it becomes connected with the sea and the island is subject to partial flooding. It is a part of the region where navigation was prohibited.

Komsomolsky Strait* – separates *Komsomolsky Island* (see) from *Vozrozhdenia Island* (see). The relief is very complicated here.

Komsuat Bay* – found to the north of A.S. between the Komsuat Cape and the mainland. The Shomyshkol fish farm is located in its northern part.

Konstantin Island* – a low, sandy island found 7 km to the south of the south-western tip of *Vozrozhdenia Island* (see). In some places it is overgrown with cane. Sparse shrubs grow over low hillocks. Shallow water areas up to 5 m deep extend to the north and south of K.I. The shores in the northern part of the island are shallow.

"Konstantin" schooner – a military double-cannon vessel built in Orenburg in 1848 under the supervision of Lieutenant *A.I. Butakov* (see). The schooner was 53 feet long with draft (without cargo) – 3 s feet. It was dismantled and in this form delivered to Raim where on July 20 it was set afloat in the Syrdarya River. The schooner was designed for the regular inventory of A.S. on which A.I. Butakov, Chief of the Inventory Expedition; A.I. Maksheev, warrant officer; K.Ye. Pospelov, land surveyor; A.A. Akishev; exiled poet T.G. Sehvchenko; a paramedic; and 20 crew members sailed.

Korzhindy Peninsula* – juts out not far into the sea between *Chernyshev Bay* (see) and *Kumsuat Bay* (see). This is a hillocky plateau ending at a cliff near the water edge. The shores of the peninsula are deep, the seabed steeply sloping southwards; the depth grows quickly, reaching 33 m at 4 km from the shore.

Kosaral, Kos-Aral Island* – located in the eastern part of A.S., to the northwest of *Bozkol Bay* (see), directly to the southeast of Akbasty Island and separated from the latter by a narrow strait. The island is surrounded by shallow waters at depths up to 1.5 m. A fort built in 1848 with a garrison of 50–60 men (abolished in 1854) and a fishing team of 20 fishermen were located on K.I. The sturgeons (Aral bastard sturgeon) and catfish were mostly caught here and after salting were transported to Orebnrug by livestock.

Kulandy Peninsula* (Kazakh – Kulandy tabegi) – found in the northern part of A.S. Elevated in its central part, K.P. has low shores, sometimes with a wide beach zone. The southern shore of the peninsula is not elevated, but steep, sloping down to the western low and sandy shore that gradually rises into cliffs at the central part of Chernyshev Bay. Cliff heights average 28 m, reaching 63 m near the Karzundy stow.

Kultuk – (1) bay of a sea or lake, mostly narrow, shallow, closed; (2) south-western wind in A.S.

Kumsuat Bay* – located in the north-west of A.S. Limited on the east by the *Korzhendy Peninsula* (see), and on the north and west by the mainland coast. The eastern and northern shores of the bay are low and sandy. The upland stretching along the shore of the Korzhendy Peninsula goes somewhat inland at the top of the bay, again approaching the sea at the northwestern shore of K.B., where it ends in a cliff that draws close to the coastline. At the entrance into the bay the depth is over 20 m.

Kungrad – an urban-type settlement, the center of the Kungradsky District of Karakalpakstan, Uzbek Republic, located in the Amudarya delta on its left bank, 8 km from the Ustyurt Plateau, and 6 km from the railroad station "Kungrad" (a terminal on the Dashoguz–Kungrad line). The name "Kungrad" originates from the name of one of the Turk peoples. For many centuries, K. belonged to Khorezm. More than once it became the capital of many states on the Circum-Aral territory. Cotton-ginning and cane-processing plants operated here; fishing and fish-processing were also done here. Main gas pipelines from Asia to Europe go via K. A gas pumping station is also located here.

Kungradsky main drain (KKS)* – the main drain of the land reclamation system on the left bank of the Amudarya delta that diverts water to *Sudochie Lake* (see). This drainage canal goes from the confluence of the *main left-bank drain (GLK)* (see) and the right branch of KKS. Its carrying capacity is

50 cu. m/s. The main drain has an insignificant slope (0.0001). It was built in unconsolidated loose rocks (sand, sandy loam) that after saturation with water become "floating," particularly at its terminal. Because of this effect, the width of the main drain is permanently increasing, while its depth is decreasing, which sharply deteriorates its drainage capacity, a reduction that also aggravated by water backup at its outfall in Sudochie Lake. The right branch of KKS runs along the railroad Khodjeily–Kungrad and at the 50th km joins KKS – KKS outfall. The water flow in the main drain varies from 0.6 to 12.0 cu. m/s.

Kunyadarya – see Daryalyk.

Kupak, Kupa – floating islands made of remnants of roots and stems of cane, mixed with silt and loam and thickets of cane.

Kuraily Island* – located near the eastern coast of A.S. It lies 6 km to the south-southwest of Zhingyldy Tubek Island. It is low and sandy, and overgrown with reed and cane; during surge waves, some parts of the island became inundated. There are many shoals around the island.

Kushkanatau – an anticlinal upland stretching nearly 80 km latitudinally to the west of the *Beltau* (see). Its absolute altitude is 138 m, while the elevation over the Amudarya delta is about 50–60 m. The slopes of the upland are flat due to deposition of parent rocks, while the *sai* (see) eroding it often form steep walls. The incise depth reaches 35–40 m. A water divide and slopes of K. are covered with eolian sands forming a ridge-hillocky relief. At the northern foot of the upland, solonchaks and small saline lakes are formed in inter-ridge depressions.

Kuvandarya – one of the ancient channels of the Syrdarya on the eastern coast of the Aral. On modern maps of the Aral, the K. delta is depicted as vast, deeply incising into the mainland Bozkol Bay. Here the lakes of the Aksai-Kuvandarya system are located. Its length from the Aksai channel mouth to A.S. was over 80 km. Before construction of a reservoir and irrigation development, these lakes had an area of 303 sq. km. By the early 21st century, all the lakes had dried out and many of their former beds were used for agricultural.

Kuzzhetpes Island* – located in the Eastern Aral Sea and separated from *Kosaral Island* (see) by a strait 2 km wide. Small dunes and barkhans covered with sparse shrubs are found on the island. With surge waves, the greater part of the island is inundated. Fisheries are developing in the bay that cuts into the southern end of the island.

Kyzylbai Island* – found in the eastern part of A.S., 5.5 km to the southwest of *Kaskakulan Island* (see). The island is low and sandy, and cane thickets grow along the shore.

Kyzyldjar – the lowest upland. Its absolute elevation at the highest point (absolute altitude 117 m) is no more than 50 m. The eolian cover with a ridge-hillocky relief dominates its surface. The upland is broken by submeridional *sai* (see). The relief in general is soft, with only slopes of some sai eroding;

the upland is steep. The breakdown depth does not exceed 20–22 m, often less. Solonchaks are developed at the foot of the slopes.

Kyzylkums (Turk – red sands) – a desert in the interfluve of the Amudarya and Syrdarya in Uzbekistan and Kazakhstan. It is confined on the south by the Nuratau ridge and the Zarafshan River valley, in the southwest by the Amudarya valley, and in the northeast by the Syrdarya valley. The area is approximately 300 thou sq. km. This is a plane sloping to the northwest, with elevations from 53 m in the northwest to 300 m in the south-east. It includes some closed depressions and isolated, heavily broken mountains: in the center are Bukantau, Aktau, and others; and in the northwest is Sultan-Uizdag ridge, which is composed of heavily disturbed metamorphized Paleozoic shales, hornfels, limestones, marble and granites. Near the foot of the mountains, springs are common. The greater part of the K. territory is covered by semi-overgrown meriodinally-oriented ridge sands with a relative height of the ridges being from 3 to 30 m, and the maximum height being 75 m. Flat spaces of K. are composed of Tirtiary clays and sands, while the northwestern spaces are of loamy and sandy-loamy fluvial deposits (dry ancient channels of the Syrdarya). The northwestern part of K. abounds in *takyrs* (see). The climate here is sharply continental. The summer is hot with an average temperature in June of +26–29°C and in January from 0 to –9°C. Precipitation is 100–200 mm a year, mostly in winter and spring. The soils are gray-brown, sands prevail, and stony soils are also found. Solonetz, solonchaks. Among the grass vegetation the ephemers dominate. For sands, the most typical are *saxaul* (see), kandym, cherkez in the uplands and sagebrush and sagebrush-shrub thickets along the valleys of dry channels – black saxaul forests. The following mammals are found here: antelope, gazelle, long-clawed gopher, sand eel, barkhan and steppe cat, wolf, fox and others; Common birds are Houbara bustard, saxaul jay, desert warbler, and others. Many pheasants live along old channels. Snakes (saw-scaled viper and blunt-nosed viper and others), lizards, and steppe tortoises are often found. K. is used for grazing mostly of karakul sheep, camels, and horses. Small oases where irrigated farming is practiced are found in K. In the Kyzylkums, natural gas fields of world significance are found (Gazliysky field). As a result of prospecting works, artesian basins are found in K., the fresh waters of which are used for irrigation.

Kyzyl-Orda (Kazakh – Kyzyl-Orda; red village, red capital, red army) – the center of the Kyzyl-Orda Region in Kazakhstan, known formerly as Perovsk, Ak-Mechet. Fortress Al-Mechet was built in 1820 in the Kokand Khanate. On July 28, 1853, the fortress, defended by the Kokand troops of Yakub-bek, was taken in an assault by Russian troops commanded by General *V.A. Perovsky* (see). After this, it was renamed Fort-Perovsky. From 1867, it was a district city – Perovsk of the Syrdarya Region. From 1925, it was known as Kyzyl-Orda. In 1925–1927, it was the capital of the Kazakh Autonomous Republic in RSFSR. Population: over 300 thousands (largely Kazakhs). Among the population are also Russians, Germans, Koreans, Chechens,

Uzbeks, Ukrainians, Buelorussians, Ingushes, and Greeks. In the early 1990s, the non-Kazakh population migrated mostly to Russia, Germany, Greece, and Ukraine. In 1937–1952, Kyzyl-Orda was a place of exile for repressed people. The major industries developed here are cellulose-cardboard, building materials, house-building plants, footwear, garment-knitting, food enterprises, as well as oil production. A pedagogical institute, polytechnics, and medicine and female pedagogical colleges are also available here. The Kazakh drama theatre performs here and the Regional historical museum has expositions. The city is surrounded by rice and melon fields.

Kyzyl-Orda region – part of the Republic of Kazakhstan. It was formed on January 15, 1938, and has an area of 232 thou sq. km. Its center is *Kyzyl-Orda* (see), and it borders with the Republic of Karakalpakstan on the south and southeast with the Chimkent Region on the east, and with the *Aktyubinsk* (see) and Karaganda Regions on the north. K.O.R. is located to the north of A.S. on both banks of the lower reaches of *the Syrdarya* (see). The greater part of the region is located on the *Turanian lowland* (see), a hillocky plane (from 50 to 200 m) sloping towards A.S. The northwestern part is covered by ridge-hillocky sands (the Circum-Aral Karakums), and the south and southwestern part by the hillocky-ridge sands of the Kyzylkums cut with ancient dry channels of Zhanadarya and Kuvandarya. In the southeast, the Karatau Mountains (1000 m) extend into the region.

The climate is sharply continental and arid with a long, hot, and dry summer and a short, low-snow winter. The average temperature in July is +26–27°C, while in January it is –10°C. The non-frost period lasts for 168–180 days. Annual atmospheric precipitation varies from 90 mm (near the A.S. coast) to 180 mm (in the Karatau piedmont area).

The vegetation period (with an air average temperature over +5°C) is 200–226 days.

The main waterway is the Syrdarya, which runs over the central part of the region from the southeast to the northwest for about 1 thou km. It has a meandering channel, numerous channels and arms, and a vast waterlogged delta. In 1956, the Kyzyl-Orda dam was constructed on the Syrdarya to prevent flooding of the banks. In 1958, river waters were directed to the Zhanadarya channel for irrigation of fields and pasturelands. There are many salt lakes in the region. Before 1960, there were 2583 of them with a total water area of 1164 sq. km. Among the 16 largest lakes (with an area of more than 10 sq. km) are *Kamyslybas* (see), Arys, Zhaksy-Kylysh, Akshatau, Tuschibas, Tibee, and others. Fresh-water lakes, such as Birkazan, Klemzhaigan, and others, are also here. The southwestern part of the region is washed by A.S. (at present – the Small Aral), the coastline of which before drying out was broken by multiple bays and lagoons (*Sarychaganak* (see), *Butakova* (see), *Shevchenko* (see), *Bozkol* (see), and others).

The soil-vegetation cover in the region refers to the desert zone. The prevailing are alluvial-meadow soils, such as gray, gray-brown, brown, desert, takyr-like, sandy, and sandy-loam soils, in the Syrdarya valley. The soils are mostly salinized. The salt content in a 2 m layer varies from 33 to 325 t/ha. The humus content in a plough layer is from 0.5 to 2.5% with the average being 0.8–1.2%.

Vegetation of the region is represented by the sagebrush-fescue of the ephemeral and saltwort associations; hillocky sands are fixed with white saxaul, tamarisk, winter fat, and others. In the Syrdarya floodplain, riparian-fringe forests (oleaster, poplar, Asiatic poplar, willow) and extensive cane thickets grow. In the past, on *Barsakelmes Island* (see) in A.S., there was a nature preserve for valuable and rare animals, such as the saiga, gazelle, and koulan.

Deposits of various salts (mirabilite and table salt), mineral pigments (ocher), and building materials (quartz sand, limestone, clay, and building stone) are also found here.

The population of K.O.R. is 596.3 thousands. The region is divided into 7 districts. Nearly all settlements are located in the Syrdarya valley and along its tributaries as well as along the railroad. While the Kazakh population prevails (over 80%), there are also Russians and Koreans. The large cities are *Aralsk* (see), *Kazalinsk* (see), Baikonur, and Djusaly.

In industry, the leading positions are taken by the food production, then light machine-building, metal-working, building material, and chemical industries. Apart from traditional industries, such as cellulose-paper, salt mining, cereal, well-developed is also power engineering based on Karaganda coals. The largest enterprise in K.O.R. is the *"Aralsulfat" Plant* (see), which produces sulfates and table salt. Before the drying of the sea, fisheries were of the country-wide importance: six fishery plants were located on the islands and peninsulas of A.S.

Agriculture is the basis of economics. Irrigated farming is practiced here. Up to 5 cu. km of water are used for irrigation every year, out of which up to 2 cu. km is for rice growing. The greater part of lands is used for grazing. The main agricultural crop is rice (over 80%), which is cultivated on an area of about 70 thou ha. Other cultivated crops are forage crops, wheat, cotton, millet, corn, and sunflower. Much area is given to melon crop growing. All cultivated lands are concentrated in the Syrdarya valley.

In the desert areas adjoining the valley cattle breeding has been developed, including karakul and fine-fleece sheep, meat-milk cattle, horse and camel. The lower reaches of the Sarysu, Circum-Aral Karakums, and Aryskum sands are used as summer pastures. In the wintertime, the cattle is moved south to the Kyzylkums.

The railroad Orenburg–Tashkent (736 km long) crosses the region along the Syrdarya valley from the northwest to the southeast. An automobile highway runs parallel to it, connecting all large settlements of the region. Before the 1960s, the share of the sea transport over A.S. (regular voyages Aralsk-Muinak, Aralsk – port Taldyk) was rather large in cargo traffic. Today, with the gradual restoration of the Small Aral Sea, local fisheries are being organized.

L

Land reclamation and water management (formerly "Hydraulic Construction and Land Reclamation") – the monthly scientific-production journal of the USSR Ministry of Agriculture (later, the theoretical and scientific-practical journal of the USSR Ministries of Agriculture and of Water Management). It was founded in April 1949. Its first editor-in-chief was K.K. Shubladze, then after 1952 A.I. Shklyarevsky, then from 1971 Ye.A. Nesterov. The journal highlighted the problems of land reclamation and water management in the Central Asian countries. Some issues were devoted to the Aral problems (No. 1, 2002).

Large (Aral) Sea* – this is the name of the part of A.S. located to the south of the largest *Kokaral Island* (see), formed as a result of the A.S. drying and its division into the Large and Small A.S. By hydrological characteristics, its depths, bottom relief, and water dynamics can be divided into 2 parts – eastern and western – separated from each other by an underwater ridge that stretches from the Tokmak-Aty Island via the *Lazareva Island* (see), the Bellinsgauzen Bank, Konstantin Island, Vozrozhdenia Island and extending finally out to the western part of the Kulandy Peninsula. This underwater ridge, which by 1990 has emerged from the receding water, is known as the tectonic *Arkhangelsky Rampart* (see).

To the west of this underwater ridge, an area of small depths is found that extends strip-like along the *Ustyurt chink* (see). The maximum depth (about 69 m) was revealed by *A.Butakov* (see) near the western shore in a narrow trough between the Aktumsyk Cape (in the south) and the Baigubekmuryn (in the north). In general, depths over 60 m composes only 0.5% of the whole sea area. An area with depths 30–60 m composes about 4% of the water area and extends as a narrow strip 22 km wide for 185 km along the western coast. The central part of the sea represents a rather flat plateau with depths varying from 20 to 28 m (1960).

The eastern part of the sea (to the east of the underwater ridge) is composed of a cup-like depression with depths less than 25 m. Near the eastern and southern coasts, a shallow area is found with depths up to 10 m that gradually increase towards the sea center (1960).

I.S. Zonn et al., *The Aral Sea Encyclopedia*, DOI 10.1007/978-3-540-85088-5_13,

Lazareva Island* – located 50 km to the east-southeast of the Aktumsyk Cape on an underwater upland with surrounding depths of up to 10 m. The island was low and sandy, an in some places flooded with water. A bar and several shoals stretched from its southern end. In 1961, a hydrometeorological station was opened here for monitoring the water level fluctuations of the Large Aral Sea. From 1983, "Lazarev Island" was one of the two stations (the other was Barsakelmes) where observations were ongoing. Deposit of shell limestone were found on the Island.

Letters Ad Patres, to the grandfather – the realistic novel of the Karakalpak people's writer, Hero of Uzbekistan, Tulepbergen Kaipbergenov, telling about the hard life of the people who, not through their fault, happened to be in a tragically hopeless situation in the Circum-Aral area, an ecologically hazardous region. The book was written in 1992 in the Karakalpak language. It was translated into Russian and published in 2004 by "Mir" Publishers in Moscow.

Licorice (***Glycyrrhiza L.***) – the salt-resistant plant, xeromesophyte, facultative halophyte. It can grow in saline environment with groundwaters occurring at a depth of 0.5–3 m. It includes 13 varieties, 7 of which grow in the territory of the Central Asian countries and in the south of Russia. Common licorice (*G. glabra L.*) and Ural licorice (*G. uralensis Fisch.*), however, are of commercial value as medicine, forage, and bioreclamation plants. The common and Ural licorice are perennial herbaceous plants that may reach 100–150 cm in height, with annually dying shoots that sprout from the main root and horizontal and vertical rootstocks. L. has vegetative and generative shoots. The stems are usually straight, ordinary and branching, not densely covered with short hair. It grows in different places, mostly forming vast thickets in floodplains and valleys of the Amudarya and Syrdarya. Common licorice yields 100–124 quintal/ha of hay and 84–120 quintal/ha of root mass. Due to its extensive root system, L. is capable of controlling deflation and erosion processes. It is very efficient as a land reclamation plant. The licorice root is included into the pharmacopeia of more than 30 countries, and by the harvesting volume, it is first in the world among medicinal plans thanks to the glycyrrhizic acid contained in it. Licorice-based preparations are widely used in treatments of allergic diseases, bronchial asthma, common colds, and others. The products of licorice root processing are applied in the food industry in baking, making sweets, halva, cakes, non-alcoholic liquors, beer, kvass, and others. In cosmetics, L. is the main element of shampoos and conditioners, creams, and soap. Substances extracted from the licorice root are used in production of ink, China ink, foaming fluids for fire extinguishers, solutions used in the drilling of gas wells, and others. In the Amudarya and Syrdarya deltas, common licorice roots are harvested at commercial scales.

Lobate type of the coast – see Coasts of the Aral Sea.

M

Main drain waters (KDV) – excessive ground waters formed as a result of irrigation or leaching of saline soils and diverted via drains. Considerable amounts of KDV are formed over the whole irrigated area in the Circum-Aral area. They are disposed into the Sarykamysh Depression and into water bodies in the area.

Main Turkmen canal Amudarya–Krasnovodsk – one of the "great construction projects of communism." On Stalin's initiative in 1950, the USSR Council of Ministers passed a decision "On the construction of the main Turkmen canal, Amudarya–Krasnovodsk, for the irrigation and water supply of lands in the south of the Circum-Caspian plain in Western Turkmenia, the lower reaches of the Amudarya, and western Karakums." Its construction was included into the directives of the XIXth party congress for the fifth five-year plan of the USSR's development for 1951–1955. It was planned to be completed in 1957. The canal length is 1100 km. The canal was to run from the lower reaches of the Amudarya at a water intake from a reservoir formed by the *Takhiatash headworks* (see) to the southwest, bypassing the Sarykamysh depression as far as the Charyshly pit and from there along the Uzboy natural riverbed. Dams with hydropower plants were to be constructed on it. A derivative canal was to branch from the upper dam and go southwards to Kazandjiku or to Kizyl-Arvatu and farther on along the western piedmonts of the Kopet-Dag up to the Atrek River. From the Lower Uzboy reservoir, the continuation of the main canal bed would go through the hollow separating the Greater and Lesser Balkhan ridges and strike via the oilfields as far as the Krasnovodsk Bay. At the tail end, the sluice was to be built to prevent disposal of the Amudarya waters into the Caspian. It was planned to irrigate 1.3 mln ha of lands to widen the cotton wedge in the region of the modern Amudarya delta (Karakalpakia), in the region of dry beds in the Sarykamysh delta, Tashauz Region (Turkmenistan), etc.; to supply water to 7 mln ha of pasturelands in the Karakums; and to create a basis for development of power generation and navigation. The canal carrying capacity is 17 cu. km/year (i.e. approximately one-third of the Amudarya's annual flow). In 1953, due to a changed political and economic situation in the Soviet Union, the project was closed. Later on, the route of the Karakum canal (Karakum-river) went further to the south.

I.S. Zonn et al., *The Aral Sea Encyclopedia*, DOI 10.1007/978-3-540-85088-5_14,

Manas Island* – located near the eastern coast of A.S., 11 km to the south-east of the Agurme Peninsula. The island was low and sandy, and in some places the barkhans rose to 304 m. Vegetation was rather scarce. To the south of it there was *Zhalpak Island* (see).

Map of Aral Sea dynamics – prepared by the Kazakh Aerogeodetic Institute in 1957 at a scale of 1:1,000,000. Interestingly, Sudochie Lake was shown on the map as the A.S. Bay. Obviously, due to the water level rise in A.S. that occurred in the second half of the 1950s, Sudochie Lake was linked with the sea (in the 1954 World Atlas the lake is separated from A.S.).

Map of desertification dynamic in the Circum-Aral Area – prepared in 1991 by A.V. Ptichnikov at a scale of 1:1,000,000. A landscape approach was taken in preparation of this map. A detailed analysis of the causes that spurred the modern dynamics of landscapes in the Circum-Aral Area enabled a clear-cut classification of the desertification processes into natural-anthropogenic and purely anthropogenic. The first category implies natural processes that are not initiated directly by, as a result, for example, of technogenous or pasture digression, but indirectly as a result of water intake in the upper reaches of a river, for instance.

Comparing the *"Desertification Map of the Southern Aral Region"* (see), however, it was found that instead of the application of labor-intensive particular indicators for assessment of desertification classes, as a complex indicator, landscape-genetic rows illustrating the dynamics of the natural-territorial complexes and the dynamics of exogenous processes initiated by desertification are quite sufficient. As a result, the present state of desertification may be defined by domination of certain stages of the landscape-genetic rows in the studied area, while the intensity (pace) of key landscape-forming exogenous processes (desertification processes) is determined, by Ptichnikov, by sharpness of borders of the areas where these processes were underway.

Mapping of the Aral Sea – the first more or less distinctive information about the territory of Central Asia was provided by the Greek scholars of the 6–5th centuries B.C. Herodotus (485–425 B.C.), in his "History," described the territory to the east of the Caspian Sea as "a flat, boundless terrain.". He also mentioned the Araks River (Amudarya). It allegedly flowed from the Matiens country as the Gind River, its estuary furcating into 40 arms, one of which flowed over an open terrain to the Caspian Sea, while some of the others were lost in the marshes and lagoons. In the stories of the marches of Alexander of Macedon marches, a new name, the Ox River, appeared. Later on Strabo, Arian, and Curtius, historians who lived in the period from the 1st century B.C. to the 1st century A.D., mentioned such rivers as Yaksart (Syrdarya), Polytimete (Zarafshan), Enarde (Murghab), and others. The Amudarya was called the Araks or the Ox in their treatises.

The first cartographic material about Central Asia was prepared by Claudius Ptolemy (2nd century A.D.), who pooled together in his works all of the rather extensive information about this territory available at the time. In his fundamental treatise, "Geography," he described the whole territory of Central Asia. On the map, he marked the Oxus that flowed into the Caspian Sea, the Oxian mountains, the Oxian lake, and other features. Interpretation of Ptolemy's maps showed that the Oxus River was Amudarya and that the Oxian lake was Sarykamysh or Aral, and that the Oxian mountains were the Ustyurt. The cartographic materials of Ptolemy had a long influence over the geographical notions of West-European scientists about Central Asia.

Of certain interest are the maps of Ibn Khaukal and Al-Istakhri, Arab travelers of the 10th century. The maps of Ibn Khaukal showed Central Asia with the Aral (Khorezmian) Sea, Amudarya (Djeikhun), Syrdarya (Sukhun or Shash), with both rivers flowing into A.S. The map of Al-Istakhri is superior to the map of Ibn Khaukal both by size and content. It shows A.S. the and Syrdarya and Amudarya Rivers flowing into it with their tributaries.

The valuable cartographic source of the 15th century is the map of Fra Mauro from Italy that was prepared in 1459. This map represented Central Asia more amply, but with some inaccuracies. Thus, the Yaksart (Syrdarya) River was depicted to the south of the Oxus (Amudarya) River. Both rivers took their origin from the Inzikol Lake and crossed the whole territory of Central Asia into the Caspian Sea.

In many world maps prepared by Europeans in the 16–17th centuries, Central Asia was not shown at all or was presented rather schematically.

Of special significance is the map made by English Ambassador *Jenkinson* (see) in the 16th century. In 1558, he went out with trade purposes to Astrakhan, crossed the Caspian Sea, landed on the Mangyshlak Island, crossed the Ustyurt, and reached Khiva and Bukhara. His map was published in 1562 and was reprinted several times. It is interesting because the map resulted only from Jenkinson's own observations and local interviews. It marked such Central Asian cities that could no be found in earlier maps. Jenkinson was also the first who defined geographical latitudes of some points in Central Asia (Mangyshlak Peninsula, Kunya-Urgench, and Bukhara). Of special significance was depiction of the deeply incising bay of the Caspian Sea and the Ogus River flowing into it; however, the map still abounded in vague, indistinct designations.

In the "*Book to the Great Drawing*" (1627), a whole chapter was devoted to Central Asia. It was an orohydrographic sketch naming such features as the Khvalim and Blue (Caspian and Aral) Seas, the Syr and Amedarya (Syrdarya and Amudarya) Rivers, and the Circum-Aral deserts Arakums, Karakums, and Barsukkums.

In 1687 on the basis of the detailed drawing of Siberia, the outstanding Russian scientist who made an enormous contribution into the development of cartography, S.U. Remezov, prepared the "Drawing . . . (Map)" on which the Blue (Aral) Sea was shown extended from the southeast to northwest, and one large river, called Syr (Syrdarya), flowed into in on the east. In the stretch between the Khvalynsky (Caspian) and Blue (Aral) Seas, the arm Urzan flowed from the latter (it was suggested that this was no other than the Uzboy).

In 1704, the Map of Russia prepared by N. Vitsen (published in Amsterdam in 1687) was reprinted. On this map for the first time in the West-European cartography, the Aral (Blue) Sea was shown in the territory of Central Asia ("Tataria"). But it was rather small in size.

The "*Drawing of a water way*" that was found in the materials of Peter I showed the whole territory of Central Asia with the Aral Sea, Syrdarya, Amudarya, and Zarafshan Rivers, the cities on them, and a irrigation network consisting of 9 canals in the Amudarya delta. The Aral was called the Special Sea, a name that had not been used in previous maps.

In 1715–1716, a map of Central Asia was prepared in Russia that showed the Caspian and Aral Seas. The Aral was presented schematically. The Amudarya, with the Zarafshan shown as its tributary, and the Syrdarya Rivers flowed into it. The cities on these rivers were also marked. The Uzboy river channel was shown with the caption, "former mouth of the Amudarya River."

During Peter I march to Persia in 1722, a "Map describing the Caspian Sea and the Uzbek country with its provinces, with the accurately shown route of the Tsar travel, and other noteworthy things" was made by hand. The map showed the whole territory of Central Asia. The Aral in the map has a round shape, and there is an inscription: "*Oralsky Lake that loses its waters, while the shores are filled with cane, the water is fresh, but in the middle of the lake the water is saline and bitter.*" The Syrdarya and Amudarya flowed into it. The first one has some tributaries, including the Farghan River. In the Syrdarya lower reaches on its right bank, there is a large Farabsky Lake linked via channels with the river. The Amudarya was connected with the sea via two arms or channels that were drawn very carefully. The northern arm took its origin from the confluence of the Amudarya into A.S. and went as far as the northern part of the Balkhansky Bay; perhaps this was the old bed of the Amudarya (Uzboy). Near the origin of this arm in the Amudarya delta a bridge was shown and there was a caption: "*Ancient bridge Khivaki.*" Another caption is near the lower (dry) section of the arm near the Balkhansky Bay saying: "*In this place, the river had been already dry for more than 100 years*".

A.S. was shown much smaller that it was in reality, and its southeastern part had many islands. The Ordarya River flowed into the sea from the northeast, and the Sydarya with tributaries flowed from the east; the large river Amudarya flowed into the sea from the south-southeast, and in its lower reaches (Khiva Khanate area) many canals were depicted. The map depicted three large, dry channels that flowed out from the southwestern shore of A.S., joined, and ran to Balkhansky Bay.

In 1723, G. De L'Isle published the "*Map of the countries lying near the Caspian Sea*" ("Carte des Pays voisins de la Mer Caspienne") in which A.S. was shown under the name "Glaukone Arapsky." In 1729, naval officer Mark Dubrovin, who was sent to Central Asia, visited A.S., Uzboy, and Khiva Khanate. He made a map that was published in 1731 by Ivan Kirillov. In 1730, Stralenberg (Tabbert) from Sweden, who was in captivity in Siberia, published a map where he gave a detailed presentation of the territory of Central Asia ("Tataria"). A.S. was depicted on it at 44–45° N, which accurately reflected its location. Hydrography of the region was also shown in it: the Amudarya (under three names: Amu, Gikhun, Oxus) flowing into A.S. with its three channels (delta). But he quite erroneously showed the non-existent left tributaries of the Amudarya that connected it with the Syrdarya.

In 1731, M. Tevkelev, a translator of the Foreign Affairs Board, and two officers who were land surveyors, carried out the first surveys of A.S. shores. In the map of the Russian Empire of I. Kirillov that was published in 1734, the northern shores of A.S. were delineated most likely on the basis of the Tekvelev surveys. A.S. was shown much smaller than its actual size. In the same year, I. Kirillov, the outstanding Russian cartographer and Senate Chief Secretary, published the first Atlas of the Russian Empire, comprising 14 specific maps and 1 general map in which he gave a verified depiction of Central Asia. Thus, the position and configuration of A.S. was changed significantly: the area of A.S. became more and the sea was shifted more to the west and stretched longitudinally.

In 1741, on the basis of the results of surveys conducted in the Syrdarya lower reaches by Lieutenant D. Gladyshev and land surveyor I. Muravin, the latter prepared the map, "*New land map of a tract from Orenburg via Kyrghyz, Karakalpak and Aral holdings to Khiva and a part of the Aral Sea with the rivers flowing into it, a part of the Syr-Darya, Kuvan-Darya, Ulu-Darya*" (Ulu-Darya is Amudarya). In 1850, this map was published by Ya. Khanykov, and the data was used in the "All-Russia Atlas" published by the Academy of Sciences in 1745 under the guidance of academician G. De L'Isle. In 1744, a handwritten Atlas of the Orenburg Province comprising 13 maps, two of which showed the territory of Central Asia was prepared. The first map, being most general, was called "*General land map of the whole Orenburg line, including both a part of the Kazan and Simbirsk Provinces, and other Asian holdings as well as lands belonging to the Orenburg Administration.*" The Aral was presented in it rather schematically. More detailed was the other map – "*Land map of the Khiva and Aral holdings, including a part of the Caspian Sea and the Aral Lake.*" It showed the Uzboy channel with the following caption: "*here the ancient river was that flowed from the Caspian Sea into the Aral Lake.*"

In 1755 in Orenburg, land surveyor I. Krasilnikov prepared the "*Map representing the greater part of the Caspian and Aral Seas with nearby lands.*"

Of great interest is the map that was also made in Orenburg in 1780 – "*General map of the Orenburg Province and nearby areas.*" It presented in detail (according to Muravin), the eastern shore of A.S. with the lakes, "sandy hills," villages, and spring channels. On the map, the Syrdarya ended not far from Turkestan City.

In 1759, French scientist D'Anville prepared the map, "*Asian part of the Russian Empire*" ("*Partie de l'Empire de Russie comprise en Asie*"). On this map, A.S. was shown narrow, stretching from the north to the south. To the north of it there was a caption "Karaklpaki," to the west the Solenoye Lake was shown.

In the little-known "*Map of the Kyrghyzkaisatsky Steppe, Bukhara, and a part of Russia*" prepared by General Ya. Bowver, allegedly in 1775, at the top right corner a part of A.S. and great lake "Sarykamyshnyn arali" were shown.

In 1776, the Geographical Department of the Academy of Sciences initiated actions on revival of cartographic activities. In the same year, I.F. Trusscot and I.F. Shmidt made a new General Map of the Orenburg Province that represented schematically the Aral Sea with a part of the Syrdarya, "the sands called "Barsuki," etc. And they used the data of Krasilnikov.

The full territory of Central Asia might be found in the "Map of Asia" carved in Petersburg in 1795. The contour of A.S. and river channels was presented schematically. The Syrdarya with tributaries was depicted correctly; however, in the lower reaches of the Amudarya, non-existing arms were shown and the contours of the A.S. western shore were distorted.

In January 1801, Emperor Pavel I prepared for a march to India to push out the British and bring the country and its trade under the rule of Saint Petersburg. He openly said in his instruction to the chiefs of the expedition that "My maps extend only as far as Khiva and the Oxus River. Farther on from these places you will have to collect information..." The march of 22,000 Cossacks in severe winter conditions only covered 600 km in a month, ending at the A.S. coast and nearly ending in tragedy. Only the death of Pavel stopped this wild idea.

In 1816, the "*Map of a part of Central Asia containing the lands of Kyrghyzkaisakov, Turkhments, and Bukharians*" was prepared. Its geographical content was very contradictory. Thus, the western shore of A.S. was shown incorrectly.

In 1825, the expedition of *Colonel F.F. Berg* (see) visited the western shore of the Aral. Here some astronomical observations and reconnaissance surveys were carried out. The obtained data were the basis for the map of A.S. published in 1832.

Graphically, the whole complex of works on preparation of the map of Central Asia from 1830 to 1856 was vividly presented on a special map (scale: 100 verst per inch) called the "*Summary map of topographical field surveys in the Orenburg area from 1830 to 1856.*" It showed the borders of the regions of Central Asia covered by surveys (two- and five-verst scale). Among the surveyed regions were Mangyshlak and Buzachi Peninsulas, the northern parts of Ustyurt, A.S., some Syrdarya stretches from the Aral to a confluence of the Kuvandarya River, and as far as the Djulek fortification, as well as paths to Khiva and Bukhara. So by 1856, approximately 40 thou sq. verst of the territory of Central Asia was covered by the partially automated surveys at two- and five-verst scale.

In 1842, Colonel G.N. Danilevsky traveled to Khiva along the western coast of A.S. He carried out surveys of the whole Khiva Khanate and prepared detailed traverse and general maps. F. Baziner, the natural scientist who accompanied Danilevsky in this travel, prepared a detailed map of A.S., using the results of expeditions of Nikiforov and Danilevsky and also N. Khanykov. In 1848, this map was published in German as "*Skizze des Aral-Sees und Amu delta*" (Plan of the Aral Sea and Amu-Darya delta).

For two months in 1848, the expedition of *A. Butakov* (see) carried out surveys of the A.S. coast, and as a result the first accurate map of the Aral appeared – "*Map of the Aral Sea prepared on the basis of surveys and observations during a voyage over this sea on the schooners 'Nikolai' and 'Konstantin' attached to the Army's Topographer Corps by warrant officers Golov and Akishev during summer 1848.*"

In 1849, A.I. Butakov and K.Ye. Pospelov completed works determining astronomical points, surveying the Aral shores, and measuring depths and making general descriptions. On the basis of these materials, in 1850 the Geographical Department of the Marine Ministry published "*Map of the Aral Sea.*" It was so accurate that the seafarers used it up to the 1950s.

In 1854, the greatest number of surveys was carried out on the right bank of the Syrdarya's middle and lower reaches. The results were included in the "*Map of Syr-Darya lower reaches from the Perovsky Fort to the Aral Sea prepared by the materials of the recent surveys of Captain-Lieutenant Ivashintsov, 1854.*" N.L. Ivashintsov published it on the basis of the maps that he found in the archive of the General Headquarters of the Orenburg Corps: the map that covered the region from Raim to Kosh-Kurgan (it was prepared in 1842), and the map that included the territory from the Mailibash stow to the Syrdarya mouth (prepared after 1847).

In 1859, A.I. Butakov, using materials from all of the past expeditions to A.S., prepared the handwritten "*Flat map of the southern coast of the Aral Sea and arms of the Amu-Darya River made by Captain 1st Rank Alexey Butakov on the basis of his astronomical observations and surveys carried out under his guidance in 1848, 1849, 1858 and 1859...*". The map presented the carefully delineated southern coastline of the A.S. and showed the coastal shallows and islands as well as a large bay, Aibugir Lake.

In 1899–1902 and later in 1906, the Aral was investigated by *L.S. Berg* (see), the outstanding Russian geographer. In 1908, he published his book, "*Aral Sea: Experience of Physiographical Monograph"* (see). As an enclosure, the book had two maps of the A.S., one of depths and one of the specific weight of water on the surface. They were prepared from the data of Butakov and Pospelov (1848 and 1849, respectively), a 20-verst map of the Military-Topographical Department of the Headquarters, a 10-verst map of Khiva published by the Turkestan Military-Topographical Department, and the results of measurements made by L.S. Berg himself and other sources.

Mezhdurechensky reservoir – constructed on the left bank of the Amudarya in its delta area. The western and northern parts of the dam extended along the Kipuakdarya channel, while the eastern part turned around the Akdarya River channel. Near Kyzylzhar settlement, the water intake of the "Glavmyaso" canal was constructed in the dam, and near Porlatau settlement was the intake of the "Marinkinuzyak" canal. During the summer high-water period, the water surface area of M.R. might be 18–19 km long and 15–16 km wide. In 2002, construction of a side weir in the downstream part of the eastern dam was completed. A length of the weir reached 1800 m with a design discharge of 3600 cu. m/s. The water salinity in the reservoir was 0.6–1.0 g/l. The water was categorized in the sulfate-chloride class, sodium and magnesium group by the salinity level, which means it is fresh enough to be used for domestic-drinking and industrial water supply as well as for all kinds of fishery activities. The newly formed ichthyofauna of the reservoir was characterized by high fish productivity. After construction of M.R. and the halting of the flow along the Amudarya's old channel, the estuary seashore was changed.

Meteorological stations of the Aral Sea – in the early 1960s there were 11 meteorological stations around A.S. and in its water area. Provisionally, they may be divided into 3 categories – marine, coastal, and mainland. Marine stations included those located on Lazareva and Barsakelmes Islands; coastal included those on A.S. and on Uyala, and mainland included those in Saksaulsk, Monsyr, Djusaly, Kyzyl-Orda, Kazalinsk, Chirik-Rabat, and Karak. After the drying of the A.S., most of these stations were closed.

Mityk Island* – located in the western part of A.S., eastward of the *Karatyuk Peninsula* (see), 2 km to the south of *Kendyrli Island* (see). From Biyurgundy Island, it was separated by a narrow strait. The island was low and sandy. Its coast was slightly sloping to the east. It was covered with the sparse vegetation.

Motornaya Tropa strait* – connected Muinak Bay in the east with A.S., passing over territory overgrown with cane. Depending on the A.S. water level, its depth fluctuated from 1.2 to 2 m, allowing only small river vessels sailing on the route from the Amudarya bar (mouth) to the port at Uch-Sai could use it.

Muinak (Muinok) – the center of the Muinak Region in the north of the Kapakalpakstan Republic, the former capital of the Aral fishermen. In the early 1960s, it was located on the Tokmak-Ata small island, which became *Muinak Peninsula* (see), separated from the coast by a narrow and shallow strait. The population is 14 thousand. M. boasted of one of the USSR's major fish canning plants. Apart from this plant, one could find a fishery plant, fishing stations, and a research fishery station. The Amudarya muskrat farm was also located here and it reared not only muskrats, but silver foxes, too. The fishery and passenger vessels did not call at M. proper but at port Uch-Sai, located not far from the city. In 1968, the A.S. waters receded from M., and today a distance of 200 km separates the city and the sea.

Fig. 35 A monument in Muinak city. Photo by Dmitry Soloviev, June 2008

Muinak Peninsula – the former Tokmak-Ata Island, later called Muinak Island, which with time became a peninsula. It was considered the largest island in the Southern Aral Sea. It was located in the southwestern part of A.S. On the southwest, it was washed by *Muinak Bay* (see), and on the east byRybachie Bay. Its area was 127 sq. km. It was composed of sandstone and clay. On its elevated southeastern part, there were cliffs up to 25 m high. In the eastern and northeastern parts, M.P. gradually dipped to the water's edge. Its southeastern end was low and inundated by surge waves. A narrow sandy bar 25 km long went out from the mid-northern part of the peninsula. It ran to the northwest and west and ended with the *Tigrovy Khvost Cape* (see). Up until the 1960s, the northeastern coast of M.P. accommodated rest houses, sanatoriums, and children rest camps. From 1978, they were all gradually closed.

Muinak International Research Station (MIRS) – was created on the initiative and with financial support of UNESCO and the German Ministry of Science and Technology on the basis of a research station of the Integrated Institute of Natural Sciences belonging to the Karakalpak Branch of the Uzbek Academy of Sciences. This station is located in the north of the Amudarya delta in the Muinak Region of the Kapakalpakstan Republic, 7 km from Muinak, on the left bank of the "Glavmyaso" canal. The first investigations under the Program "UNESCO – Aral Sea" were organized and conducted at this station in spring 1993.

Muinak Bay* – located to the east of the Smaller Kusatau Cape, formed the eastern part of Adjibai Bay (1960). The southern mainland coast is low, waterlogged, and overgrown with reeds, while the northern coast was formed by the southern shore of Muinak Island. It was an elevated coast, having in some places sandy beaches, but largely overgrown with reed. The coastline was broken, but slightly, and its configuration was vague and periodically changed. The bay was shallow at 2 to 3 m. It was protected from winds from all directions excepting the west, but even with the western winds waves did not develop here because of its shallow depth.

The source of water supply of the reservoir was the Shege-Muinak canal ("Glavmyaso"). The reservoir extended along the dam as a water band with the central pool of the round form. The length of the near-dam section of the reservoir reached 6.5 km, while the width was 150 to 200 m; the length of the central pool was 2.0 km, while the width was 1.0 to 1.2 km. Water depth near the dam was 2.5–3.0 m, while in the open pool it was no more than 0.7–0.9 m. The northern part of the bay up to Muinak City formed numerous small overflows, while the southern part remained dry. The water level in the reservoir was 50.9 m.

The water salinity in the reservoir was 3.5–3.8 g/l. The water was classified in the sulfite class, the sodium group; the moderately saline waters were suitable for all kinds of fishery activities, but due to the disastrous scales of fishing the reservoir lost its fisheries significance. No waterfowl were found in the bay.

The coastal zone as well as the newly inundated territories and the southern part of the bay were covered with cane (*Phragmites australis*) and to a lesser extent cattail (*Typha angustifolia*) thickets. The cane that grew in the near-dam part of the reservoir were partially eaten by the Asiatic hopper.

Muinak fish canning plant – construction began in 1933 as a meat processing plant on the basis of the fishery in A.S. and was commissioned in 1941 as a fish-meat plant. The first products of the plant – canned beef and tortoise meat – was directed to the army. Up until 1956, M.F.P. used both kinds of raw materials. Later, the plant represented an association of 6 enterprises: 5 fish plants and one canning plant. The fish plants were located in Muinak, Urga, Uchsai, Kazakhdarya, and Porlatau. The plants manufactured various fish products: smoked, dried, saline, fresh, frozen, and mostly canned products. In 1958, the plant produced 21.5 mln cans which were sold in various regions of the USSR and abroad. As a result of the Aral drying and the lack of fish, the plant operated only at 30% of its capacity using raw materials supplied mostly from the Atlantic Ocean. After the late 1970s, the fish-processing plants stopped functioning. The total annual losses of the fishery in the Amudarya delta were evaluated at approximately 20 mln rbls (in 1980s prices). The plant provided employment for 1000 people working in 3 shifts. Now only 300 people work at the plant in 1 shift and only during one week in a month. In 1991, the plant produced 17 mln conventional cans. Until 2000, sprat was supplied here from the Caspian, then from internal water bodies.

Muskrat *(Ondatra zibethica)* – commercial mammal of the rodents family. Its body length is about 30 cm, and its weight is about 1 kg. The fur is dense, silky, with coloring varied from red to silver. It came from North America, and lives

Fig. 36 Muskrat (http://www.outdoors.ru/hunter/img/ondatra.jpg)

along banks of lakes, rivers, ponds, and in marshlands. M. was brought to the Circum-Aral area in autumn 1943 and spring 1944, and by the early 1960s it populated the whole delta of the Amudarya. M. quickly acclimatized and propagated. From 1946, M. hunting was already of commercial significance. Every year the Amudarya M. farms delivered more than a million in valuable pelts. The main requirement for successful M. farming is availability of cane and cattail thickets which both protect and feed. The water body depth should be no less than 1.5 m. Density of the M. population was 4–5 families per ha, and catching of 20–25 animals in a year do not affect the cane thickets. With deterioration of the environmental conditions, the population and area of M. started decreasing. Thus, in 1950–1960 up to 1 mln heads or 15–20% of the animal fur procured in the USSR came from M., while in 1968 it was only 9 thou pelts. Today this business has practically disappeared.

N

Natural and anthropogenic soil salinization in the Aral Sea basin (geography, genesis, evolution) – the monograph prepared by a group of Russian authors (Ye.I. Pankova, I.P. Aidarov, I.A. Yamnov, A.F. Novikova, N.S. Blagovolin). It was published in Moscow in 1996. It provided an objective analysis of soil salinization of natural and irrigated ecosystems in the A.S. basin, considered the causes and specific features of oases in regard to peculiarities of the natural and economic conditions, and outlined the vector and intensity of the modern salt buildup processes and provided recommendations for their regulation. Validated a new concept of irrigation development in the region.

Natural resources of Uzbekistan – a collection of maps at scale 1:500,000 prepared in 1980 by the State Center "Priroda" and its Uzbek branch on the basis of satellite photographs. These maps covered nearly the whole basin of the Amudarya and a part of the Syrdarya basin as well as the whole territory of the Southern and Western Circum-Aral Area (within the borders of the Karakalpakstan Republic). This collection included descriptions of the natural conditions of separate components and landscapes, specifically their resources, condition, and utilization.

I.S. Zonn et al., *The Aral Sea Encyclopedia*, DOI 10.1007/978-3-540-85088-5_15,

O

Obodina – a small bay not far into the mainland in the Aral and Caspian Seas.

Obrucheva Island, bank* – located to the west of the Akpetkinsky Archipelago, 15 km to the southwest of *Tolmacheva Island* (see). The depths around the bank were 5–10 m. It was formed as a result of island erosion and composed of sand.

On the environmental condition of the circum-Aral territory in the Turkmen SSR and ways to improve it – Resolution of the Supreme Council of the Turkmen SSR on November 17, 1990 that defined the territories of the Tashauz Region and the Darganatinsky District of the Chardjou Region to be a zone of environmental disaster and defined a special regime of economic activities in it.

On measures for accelerated economic and social development of the Karakalpak ASSR – Resolution of the CPSU Central Committee and the USSR Council of Ministers on March 17, 1986 that outlined the main economic and social actions to be taken for accelerated development of the economy of this territory with regard to the Aral Sea level dropping.

On recognizing the territory of the Karakalpakstan Republic a zone of environmental disaster – Resolution adopted by the 8th Meeting of the Supreme Council of the Karakalpakstan Republic on April 11, 1992. In this context, the meeting appealed to the UN and to the world community with a request to recognize the resolution in order to overcome the crisis.

On the termination of works on partial flow transfer of northern and Siberian Rivers – Resolution of the CPSU Central Committee and the USSR Council of Ministers on August 14, 1986. Regardless of the termination of works, this Resolution also assigned "to the USSR State Committee for Science and Technology, the USSR Academy of Sciences, and the USSR Agricultural Academy a mandate to continue with the scientific study of the problems related to the regional re-distribution of water resources on the basis of comprehensive economic and environmental studies, application of advanced economic-mathematical methods, construction of technological facilities, and in-depth analysis of the Russian and foreign experience in this field."

I.S. Zonn et al., *The Aral Sea Encyclopedia*, DOI 10.1007/978-3-540-85088-5_16,

Ondyr Cape* – located 9 km to the west-south-west of the Kokaral Cape. The Ondyr Cape is high, abrupt, and steep.

Orussengir Island* – located 9 km to the west-southwest of the Akpetki Island. It was low and sandy, and in some places eroded by water. The depths around O. were up to 2 m.

Oryskamys Island* – located in the eastern A.S., 7.5 km southward of *Kaskakulan Island* (see). The island is low and sandy.

Ox – the ancient name of the Amudarya River.

Oxian Bog – the ancient name of the Aral Sea.

Ozen (Turkish) – a river, small river, river arm, or channel, often marked on the maps as *uzyak* (see).

P

Partial transfer of the Siberian Rivers flow to the Aral Sea basin – among the first developers of this project were such well-known engineers and scientists as Ya. Demchenko (1871), the Monastyrievy brothers (1907–1923), B. Bukinich (1920–1930), N. Botvinkin (1924–1934), A. Miler-Schulga (1934–1950), I. Gerardy (1970), A. Voznesensky (1972), and G. Voropaev (1978). The need of such partial transfer of the Siberian river flow to the Aral Sea basin was substantiated by the USSR's leading research and design-survey institutions.

In 1970, the USSR Government passed the Resolution, "On perspectives of land reclamation and development for 1971–1985, flow regulation and redistribution." It was called for to stimulate activities of the design-survey and academic institutes. In 1976, the following statement was included into the "Main Directions of the USSR Economy Development" approved by the 25th CPSU Congress: "Carry out research and on its basis develop issues related to the partial flow transfer of the Siberian rivers to Central Asia, Kazakhstan, and the Volga basin," which required an integrated approach to addressing all economic, environmental, and other problems. In the same year, the USSR State Committee for Science and Technology (SCST) approved a comprehensive scientific-technical program for 5 years (1976–1980), the head operator of which was appointed by the Institute of Water Problems (IWP) of the USSR Academy of Sciences. In April 1978, the USSR State Committee for Planning adopted the resolution "Basic provisions of the feasibility report on partial flow transfer of Siberian rivers to Central Asia and Kazakhstan." A special institute on flow transfer of Siberian and northern rivers to the south was established in Moscow on the basis of "Sojuzgiprovodkhoz."

The 26th CPSU Congress approved the preparatory works on transfer of Siberian river flow to the Volga basin and also ongoing scientific and project developments on partial flow transfers of Siberian rivers to Central Asia and Kazakhstan. USSR SCST approved development of the new scientific problem, "Elaboration of the scientific-technical basis for territorial redistribution of water resources" for 1980–1985. In 1980, USSR SCST received the "Draft feasibility report on partial transfer of the Siberian river flow to Central Asia and Kazakhstan." The parameters of the canal for transfer of Siberian river flow were as follows: length – 2230 km, carrying capacity – up to 1300 cu. m/s,

I.S. Zonn et al., *The Aral Sea Encyclopedia*, DOI 10.1007/978-3-540-85088-5_17,

water lifting by several pumping stations to 113 m, installed capacity of pumping stations – 2300 MW, power consumption – 8 to 8.2 bill kWh.

At the same time, the public, which called this project a "river turn-back," raised their voice against it. Among the first who disapproved of this project were writer S.P. Zalygin, academicians A.L. Yanshin and L.S. Pontryagin, ecologist M.Ya. Lemeshev, and many others. They won this struggle, and on August 20, 1986, the *Resolution of CPSU Central Committee and the USSR Council of Ministers* (see) was adopted that stopped the works on the transfer project.

In recent years, this issue was again put to the fore. In early 2002, Mayor of Moscow Yu.M. Luzhkov submitted a proposal to the RF President stating that it was time to return to this issue once more. According to the Luzhkov project, 27.2 cu. km of water a year will be taken from the Ob River (the total Ob flow is 316 cu. km) from the "Belogorie" site near Khanty-Mansiysk. The canal length will be 2550 km, and its carrying capacity will be 1150 cu. m/s (an open canal 200 m wide and 16 m deep). From the water intake site, the canal route will pass over the right bank of the Tobol River, and coming over the water divide along the Turgai valley, will go to the Syrdarya near Djusary City, where it will cross the interfluve between the Syrdarya and Amudarya and join the Amudarya at a site between the Tuyamuyun and Takhiatash waterworks. To overpass the Turgai water divide, the water will be lifted by 7 pumping plants to a total height of 110 m. The installed capacity of the pumping plants will be 2.7 mln kW, and the annual power consumption will be 10.2 bill kW.

Several water reservoirs – the Tobol headworks on the Irtysh River, the Tengiz and Kairasor reservoirs, and a reservoir before the canal inlet into the Amudarya – will be constructed.

The water supplied along this canal will be distributed as follows: to the Russian districts in the initial stretch of the canal – 4.9 cu. km; to Northern Kazakhstan – 3.4 cu. km, and to recharge of the Syrdarya and Amudarya – 16.3 cu. km (including to Uzbekistan – 10 cu. km).

The total would 24.3 cu. m (net), including expected losses of 2.9 cu. km (12%). The area of lands that may be additionally irrigated with this water will be 1.5 mln ha in Russia and 2.0 mln ha in Central Asia.

The President of Kazakhstan, N. Nazarbaev, discussed the problem of water transfer to Central Asia in September 2006 and in June 2007 at the Saint-Petersburg Economic Forum.

Paskevich Bay – see Shevchenko Bay.

Past, present, and future of the Aral Sea – Monograph prepared by two well-known Uzbek scientists, Z.M. Akramov and A.A. Rafikov, and published in 1989. This monograph describes the hydrological condition of A.S. before 1961, its environmental and socioeconomic significance, the origin of the Aral Depression and the age of the Aral, as well as natural peculiarities and resources of the Circum-Aral Area. The monograph states the causes of the level drop, the development of the desertification processes, the socioeconomic and environmental consequences of the sea's drying, the modern state of the natural environment, and the dynamics of natural processes on the dried seabed of the Aral, including the drift of salts and salt dust in the environment. It provides a forecast of environmental changes in the Circum-Aral Area and the dried seabed until 2010. In conclusion, it assesses the future state of the Aral after the dropping of its level by 29 m abs.

Perovsky Bay – see Butakov Bay.

Perovsky campaigns – see Khiva Campaigns.

Perovsky, Vasily Alekseevich (1795–1857) – a Lieutenant General, Count Perovsky took part in the 1812 War and in the Russian-Turkish War of 1829–1831. He strongly advocated expansion to Central Asia. In 1833, he was appointed the military governor of Orenburg. In the same year, P. appealed to the Director of the Asiatic Department of the Foreign Ministry, K.K. Rodofinikin, with a message insisting on more energetic actions in Central Asia, partially in reaction to the active infiltration of the British in this region. In February 1839, his efforts were successful: his plan of a great march to Khiva received approval, and he was given command of over 5000 men. (Among the participants of this Khiva campaign was Dal Vladimir Ivanovich, later the author of the classical work, the "Dictionary of the Live Russian Language.") They were provided with collapsible boats for crossing A.S. and canvas pontoons for crossing rivers. This march, which occurred during severe winter through the semi-desert, ended in a failure, but it had an enormous international response because Central Asia became a further site in the Russian-British conflict. In 1842, P. was removed from the position of military governor of Orenburg, and in 1851 was appointed to this position once again. In 1853, marching from Aralsk, P. conquered Ak-Mechet (now Kyzyl-Orda). For capture of the Kokand fortresses, P. was awarded the Emperor's commendation of acknowledgement and the Ak-Mechet fortress was renamed Fort Perovsky in his honor. One of the A.S. bays was also given the name of P., though it was later renamed Butakov Bay. During Perovsky's lifetime, one of the vessels of the Aral fleet that was actively involved in military actions navigating over A.S. and Amudarya as far as Kungrad also bore his name.

Physical oceanography of the dying Aral Sea – Monograph of Doctor of Geography P.O. Zavialov (P.P. Shirshov Institute of Oceanology, Russian Academy of Sciences) published in 2005 by Springer Publishers. It is the result of field expeditions to the Aral in 2002–2004 by IO RAS through a grants of RFBR, the Russian Ministry for Science and Technology, NATO, the USA National Geographic Society, and the Uzbekistan Academy of Sciences. The monograph is divided into five chapters that give a historical overview of A.S. research, a brief description of A.S. hydrology in the period of its drying, the present-day hydrological and hydrochemical state of A.S., and outlooks on what to expect on the Aral crisis from a global perspective (A.S. analogs). This work was prepared using extensive scientific sources published in the former

USSR with the addition of new results from the expeditions. The latter include a description of a 3D model of the thermohaline fields of the Large Aral, the mass exchange between the eastern and western basins of the Large Aral, new information about the circulation of the modern Aral on the basis of direct measurements, modeling and remote sensing, H_2S content, and digital evaluations of likely scenarios of future development in regard to river flows and evaporation rates.

Fig. 37 P.P. Shirshov Institute of Oceanology expedition to the Aral sea in October–November 2002. P.O. Zavialov is third from the right, A.G. Kostianoy is rightmost

Pilot's sailing directions – the first short sailing directions of A.S. and the Amudarya delta prepared by D.P. Malinin on the basis of investigations carried out in 1920–1921. In 1921, it was published as a manuscript by the Chief Hydrographic Department. The first hydrographic investigations of A.S. were carried out in 1848-1849 under the guidance of *A.I. Butakov* (see), as a result, the navigation map appeared. After this, hydrographic investigations were carried out only from time to time, and only in 1963, on the basis of the detailed hydrographic investigations conducted by the Hydrographic Service of the Caspian Fleet in 1958–1960, did the Navy's Department of Hydrographic Services publish a new original guidance for navigation over A.S. – "Pilot's sailing directions."

Presarykamysh delta of the Amudarya River – a sandy-loamy alluvial plain evenly sloping to the west and northwest and limited on the east by the modern delta of the Amudarya. Its surface elevations was 80–90 m abs. It was located east of the Sarykamysh Depression within the Daryalyk-Daudan sag. The plain was formed by the Amudarya during the Late Pleistocene and Holocene. Its takyr surface was cut by abundant channels and the traces of their meanders and oxbows. The depth of the most recent and significant of them, such as Daryalyk, Daudan, and Kangadarya, varied from several meters to 40–60 m. The plain was composed of a series of interbedding sandy and clay sedimentary layers featuring different thickness because it occurred on the eroded surface of the parent rocks (it was measured, on the average, by several dozens of meters). Some table uplands surrounded by chinks up to 50 m high rose over the plain's surface. The largest of them were Tarymkaya (35 km long and 4–15 km wide), Bukentau (20 km and 5–6 km, respectively), Kangagyr (23 km and 13–15 km), Tuzgyr (20 km and 12 km), and others. They were mostly concentrated in the western part of the Presarykamysh delta and represented residual mountains separated from the Ustyurt in the north and from the Zaunguz Karakums in the south.

Massifs of barkhan sands were found largely along the ancient channels and in the west along the Sarykamysh Depression. They were formed as a result of drifting in the wind of riverbed alluvium and lacustrine deposits. For many centuries, the eastern and central part of this delta was used for irrigated farming, and its surface here was overlain with recent deltaic and irrigation sediments, though the relief was transformed by numerous irrigation and drainage canals and other structures. Lands of ancient irrigation were widely represented here.

Present status of landscapes in the Aral Sea region – map prepared by Ye.V. Milanova, Ye.V. Glushko, and A.V. Ptichnikov (Geographic Faculty of Moscow State University), and included as a case study into the UNEP "World Atlas of Desertification," published in 1992. The Amudarya delta was selected for the case study. By the time of this map's preparation, the water level in A.S. had dropped by 14.2 m (1957–1989). The map was prepared at scale of 1:1,000,000 on the basis of field investigations and space photos made from the orbital station, "Salyut" (resolution 70 m), and from the satellite "Meteor" (resolution 10–30 m). The coastline and contours of deltaic lakes were taken from photos made in 1969. The present status of landscapes is characterized by four principal categories: modal, derivative (secondary), anthropogenically-changed, and technogenic complexes. Map analysis has enabled the identification of changes in the landscapes over the course of their utilization and the tracing of their evolution as affected by desertification.

Problem of the Aral Sea – the generalized notion of the sea's historical development. The problem was shaped in the early 20th century when the seabed was still full and there was even a tendency for the water level to rise. At that time, many well-known researchers, for example *A.I. Voeikov* (see) and F.P. Morgunenkov, thought that with time, due to the extension of irrigated areas in Central Asia, the sea level would, of course, drop because of the flow reduction in the Amudarya and Syrdarya mouths; however, in those years there was no reasoning for the likely consequences of such a drop.

In the recent five decades, the Aral Sea has changed enormously. The water level has dropped by 24 m, its area has shrunk to 16 thou sq. km, and its volume has reduced to 75 cu. km. The area of the dried seabed is now more than 3 mln ha. The coastline has become less jagged. Many islands have become connected to the shore, turned into peninsulas or even mainland. Saline sand dust from the barren 3 mln ha of the dried seabed has affected up to 15 thou ha of pastures in the Primorskaya zone every year. In late 1980, the Aral separated into the Large and Small seas. At present, the Large Sea is practically divided into two parts, the western and the eastern.

Sea water salinity has increased up to 100–150‰ in the Large Sea. The climate has changed sharply, with annual temperature variations becoming more significant. The summer has become hotter, while the winter has become colder. In the Amudarya and Syrdarya lower reaches, hundreds of thousands of hectares of pasturelands that were once characterized by great species diversity, tugai, and cane thickets have dried out. Many valuable plant and animal species (in particular birds and fish) have disappeared. Many bays and lakes around the sea have intensively dried out, and solonchaks have formed not only on the seabed as it became exposed after the retreat of the water but around the sea, too. Bottom sand deposits, dust, and salt have been transferred with wind and storms. Every year, their total drift reaches 75 mln tons across a zone of 1000 and more kilometers. The zone of dust-salt storms is 40 km wide and 400 km long. The dust and salt first drifted with the winds and fell with the atmospheric precipitations in the Circum-Aral Area where 3.5 mln people live at present, causing grave diseases in the local population which include hepatitis, blood diseases, abnormal development and births, and stomach diseases. P.A.S. refers to the environmental and social disaster of the Aral Sea.

Problem of the Aral Sea – collection of articles of the Institute of Geography of the USSR Academy of Sciences edited by *S. Yu. Geller* (see). It was published in 1969 and analyzes the likely consequences of the A.S. level drop, which was then expected due to the increased withdrawal of waters from the Amudarya and Syrdarya for the development of irrigated farming. The articles consider the long-term (decadal) forecasts of the Aral level drop, and its fluctuations in retrospect, the role of A.S. in the water cycle in Central Asia (K. Kuvshinova); the anticipated changes in the land stock in the Amudarya delta due to the reduced inflow of river waters and the water level drop (V. Kostyuchenko et al.); the effect of the reduced river flow on modern sedimentation in the river delta and the input of fertilizers into the delta with suspended matter (A. Klyukanova); the conditions of and possibilities for utilization of the exposed area of the A. seabed (S. Geller); the evaporation from cane thickets in the Amudarya delta that played an important role in the Aral water balance (Ye. Minaeva); and a verification of the morphometric characteristics of the Aral based on the results of new investigations (R. Nikolaeva).

Problems of desert development – the international scientific-practical journal published since 1967 in Ashgabat (Turkmenistan) jointly by the Institute of Deserts, Fauna, and Flora and IFAS and the Regional Center of ESCAP/UNEP. Periodicity is 4 issues a year. It discusses the results of investigations and experiences of desert development both in the Central Asian countries and abroad. The editorial board of the journal includes representatives of Russia, USA, China, Saudi Arabia, Israel, Uzbekistan, Kyrghyzstan, Kazakhstan, Tajikistan, and Azerbaijan. In 1999, a new heading appeared in the journal, *"Aral and its problems"* (see), under which original scientific papers and practical recommendations on the Aral issues are published. Some issues of the journal have been specifically devoted to the Aral issues (No. 2, 1979; No. 3–4, 1988; No. 6, 1991; No. 3, 1993).

Fig. 38 Journal "Problems of Desert Development"

Problems of the Aral Sea and Amudarya delta – collection of papers presented at the joint meeting of the Presidium of the Uzbek Academy of Sciences and the Scientific-Practical Conference, "Problems of the Aral Sea and Amudarya Delta" that was held in *Nukus* (see) on December 2–3, 1980. It was published in Tashkent in 1984 (at that time marked: "For internal use only"). The visiting meeting of the Presidium of the Uzbek AS (the first meeting was held in Nukus in 1945) addressed the most urgent problems of the Karakalpak Autonomous Republic, mostly employment and the problems of A.S. and the Amudarya delta. Many leading scientists and specialists from 57 ministries, departments, and research organizations; Academies of Sciences of the USSR and union republics; and representatives of the leadership of the Central Asian republics and Kazakhstan took part in this conference. Urgent actions concerning the development of emplotment in the region, reduction of the negative consequences of the A.S. level drop, and desertification in the Amudarya and Syrdarya delta were considered and outlined; however, no radical efforts were taken.

Problems of the Aral Sea: the condition of the water area and dried seabed of the Aral Sea – collection of papers published in Almata in 1983.

"Program of concrete actions for improvement of the environmental and socioeconomic situation in the Aral Sea basin for the period of 2003–2010" ASBP-2 – in October 2002, the governments of the Central Asian countries initiated a new IFAS program. They approved the main directions and ordered to the IFAS Executive Committee, together with ICWC and ICSD, to elaborate a "Plan of actions for 2003–2010 on improvement of the environmental and socioeconomic situation in the Aral Sea basin." This work was supported by the Swiss Mission on the Aral Sea Problems, OSCE, and USAID. The Program was approved by the IFAS Board on August 28, 2003. It included the following priority directions: development of coordinated mechanisms for water resources management in the A.S. basin; restoration of waterworks and improvement of water and land resources management; improvement of environmental monitoring; program on combatting natural disasters; program on rendering assistance in addressing the social problems of the region; consolidation of the material-technical and legislative base of interstate organizations; development and implementation of regional and national programs on environment protection in the flow formation zones; development and implementation of regional and national programs on rational water consumption by farms in the Central Asian countries; development and implementation of international programs on hygienic-environmental improvements of settlements and natural ecosystems of the Aral region; development and implementation of international programs on restoration of environmental stability and bioproductivity; the concept of sustainable development of the A.S. basin; regional programs of action on desertification control; development of wetlands in the Amudarya and Syrdarya lower reaches; and rational management of saline drainage waters.

Project on management of water resources and the environment (PMWRE) – the largest regional project, sponsored by the Global Environmental Facility, the governments of the Netherlands and Sweden.

The project consisted of six components:

Component A: "Water Resource Management and Control of Soil and Water Salinity." The purpose of this component is to elaborate regional and national scenarios and strategies of sustainable management of water resources and their distribution with regard to the environment requirements in the basins of the Amudarya and Syrdarya rivers and to render assistance to the decision-makers in five countries on preparation of medium- and long-term agreements on water resource management.

Component B: "Population Awareness." The purpose of this component is to assist in the development of awareness in the general population in understanding the urgency of water saving and the development of more prudent attitudes toward water resources on the part of water users.

Component C: "Management of Dam and Reservoir Security." The purpose of this component is to evaluate the security of dams in the region, update the monitoring and warning systems on some dams on the basis of pilot projects, and prepare draft actions on urgent dam rehabilitation.

Component D: "Monitoring of Transborder Waters." The purpose of this component is to organize monitoring, with the assistance of independent institutions, of the qualitative and quantitative parameters of river flow at 37 transborder gauging stations.

Component E: "Restoration of Wetlands." The purpose of this component is to restore the wetlands of Sudochie Lake, which are the nestling place of endangered species of migratory birds.

Component F: "Project Management Support."

PMWRE was completed by October 31, 2003 as established by the guidelines of the Facilities Board.

Projects on the Aral Sea funded by INTAS (INTAS-CNRS-DFG call for proposals for research on the Aral Sea basin, 2000):

1. "Mathematical modelling of ecological processes in the Aral Sea basin aimed at economic water use and the prevention of salination of agricultural lands" coordinated by Prof. Jean-Claude Legros, Department of Chemical Physics in the Faculty of Applied Sciences (Free University of Brussels), Belgium.
2. "Economic assessment of joint and local measures for the reduction of socio-economic damage in the coastal zone of Aral Sea" coordinated by Dr. Helmut Weidel, Mountain Unlimited, Austria.
3. "Prospect for the development of natural-economic resources in the Kazakh Priaralie" coordinated by Prof. Michel Maignan, Institute of Geostatistics in the Earth Sciences Department (University of Lausanne), Switzerland.
4. "Satellite image processing techniques for effective management of land use and irrigation demand in the Aral basin" coordinated by Prof. Christopher Clayton, Department of Civil and Environmental Engineering (University of Southampton), UK.
5. "Improvement of drinking water supply in the Amu Darya basin by using groundwater resources of fresh water lenses" coordinated by Prof. Wolfgang Kinzelbach, Institute of Hydromechanics and Water Resources Management, Swiss Federal Institute of Technology Zurich (ETH Zurich), Switzerland.

6. "Restoration and Management Options for Aquatic and Tugai Ecosystems in the Northern Amu Darya Delta Region" coordinated by Prof. Helmut Lieth, Department of Biology and Department of Mathematics & Informatics, Institute of Environmental Systems Research (University of Osnabrueck), Germany.
7. "Holocene climatic variability and the evolution of human settlement in the Aral Sea Basin" coordinated by Dr. Hedi Oberhaensli, Projektbereich Beckenanalyse (GeoForschungsZentrum), Germany.
8. "Study of Groundwater Contribution to the Aral Sea Region Water Supply and Water Quality: Strategies for Reversibility and Pollution Control" coordinated by Prof. Corinna Schrum, Institute of Oceanography (University of Hamburg), Germany.
9. "Assessment of the status of children's health and the psychological status of dependent families under ecological stress in the Aral Sea Basin" coordinated by Dr. Lothar Erdinger, Environmental Laboratory, Dept for Hygiene and Medical Microbiology (University of Heidelberg), Germany.
10. "Study of the Role of Groundwater in Water Resources of the Aral Sea Region: Ecological Policy, Assessment, and Prediction" coordinated by Dr. Iain Muse, EMonument, Belgium.
11. "Monitoring of Aral Sea level variations and consequences on lacustrine and riverine ecosystems" coordinated by Dr. Jean-François Cretaux, Groupe de Recherche en Geodesie Spatiales, Centre National d'Etudes Spatiales (CNES), France.
12. "Development of methods to rehabilitate degraded riparian forest ecosystems in the Aral Sea Basin's river plains and deltas" coordinated by Mr. Succow, Institute of Botany, Landscape Ecology, and Nature Conservation, University of Greifswald, Germany.
13. "Plant adaptation to salt stress mediated by nitrogen metabolism, 14-3-3 proteins, and ion transporters" coordinated by Dr. Albertus De Boer, Faculty of Biology – Developmental Genetics (Vrije Universiteit Amsterdam), The Netherlands.
14. "The present state of Aral Sea Basin herpetofauna and implications for conservation of natural habitats and biodiversity" coordinated by Dr. Claude Miaud, Laboratory Biology of Alpine Populations (University of Savoie), Switzerland.
15. "The use of halophyte species diversity for the rehabilitation of salt-affected soils and the production of biologically active compounds in the Aral Sea region" coordinated by Prof. Stanley Lutts, Laboratory of Cytogenetics (UCL), Belgium.
16. "Mass transfer phenomena in membrane systems and ion exchangers: Theoretical and experimental research for improvement of electromembrane technology in implementing a new technique to produce quality drinking water in the Aral Sea Basin" coordinated by Prof. Gérald Pourcelly, Institut Européen des Membranes (University Montpellier II, CNRS), France.

17. "Development of integrated water management tools for the Tuyamuyun reservoir complex for the improvement of the drinking water supply and health in the disaster zone of the lower Amu-Darya (IWMT)" coordinated by Prof. Jochen Fröbrich, Fachgebiet Gewässergütemodellierung, Fachbereich Bauingenieur- und Vermessungswesen (Universität Hannover), Germany.
18. "Development of novel hydrogel materials and water desanilation methods to produce drinking water" coordinated by Prof. Heikki Tenhu, Laboratory of Polymer Chemistry in the Department of Chemistry (University of Helsinki), Finland.
19. "Determination of the quality of ground and surface waters suitable for drinking water production in the Khorezm Region, Republic of Uzbekistan" coordinated by Dr Yngvar Thomassen, Department of Occupational Hygiene (National Institute of Occupational Health), Norway.

Projects on Aral Sea preservation and restoration – with the dropping of the A.S. level (drying), beginning from the late 1980s, many proposals were formulated on its restoration and maintenance. Among these proposals were: transfer of waters from Sarykamysh Lake (25 cu. km); partial transfer of the Indus River flow to the Amudarya; augmentation of the river flow by artificially inducing precipitations in the mountain and piedmont areas of the Pamirs and Tien-Shan with a view to increase the river flow by 10–25 cu. km a year (in this context a special Resolution was adopted by the USSR Council of Ministers in 1989 "On the organization of works for the artificial increase of precipitations in Central Asian mountain regions with a view to augment water resources of the Amudarya and Syrdarya Rivers and the Aral Sea"); release of waters of Sarez Lake on the Murghab River (the Amudarya basin) (the lake was formed in 1911 when, during the 9-point earthquake, 3 cu. km of mountain rocks fell into the river, resulting in a depression that accumulated over 20 cu. km of water); construction of the Trans-Kyzylkum main drain from Termez to the Aral (1500 km long, supplying 5–6 cu. km of water); drilling of 10–12 thou wells 500–1500 m deep for withdrawal of 30–35 cu. km of ground waters; transfer of 90–100 cu. km of waters from the Caspian Sea via a canal 600–700 km long with a water lift of 100 m; partial transfer of the river flow from the Angara-Yenisey basin over Turgai; partial transfer of the flow of the Siberian rivers Ob and Irtysh (*Canal Sibaral*, see); release of waters from Issyk-Kul Lake via the Chu and Syrdarya Rivers; accumulation of approximately 6 cu. km of irrigation and drainage waters from the Circum-Aral Area (other than from Sarykamysh Lake); intensification of glacier melting by blackening surfaces (about 4–5 cu. km of water) to increase the Amudarya and Syrdarya flow. The idea of diverting drainage waters into the Aral was first formulated in the 1970s, and although the sea was in this case to play the role of a wastewater settling basin, there was some rationale in this proposal. In the 1970s, the volume of A.S. was much greater, thus the wastewaters would have been diluted to a greater extent, more so as their greater part came to the sea with the river flow, which was proved by the dynamics of the Aral pollution. Creation of controlled water bodies in the A.S. region were also proposed.

In the recent decade, projects based on the import of water resources from other regions acquired wide and often controversial publicity, and even now there is no assurance that these projects have been completely lain aside because from time to time they become the focus of attention once again. The redistribution of water resources did not assume stabilization or restoration of the A.S. level; on the contrary, importing additional amounts of water was planned to renew irrigation, without saline, to far lands not yet completely lost for agricultural use.

Assessing the situation in Central Asia critically, many specialists have come to the conclusion that an operative stabilization of the Aral (Large Aral) level even on present-day elevations is hardly possible, which is why various measures to prevent the complete drying of the sea have been proposed, including the creation in its place of a system of regulated water bodies (*wetlands*, see) where, due to a through-flow regime, a relatively low water salinity would be maintained and a system of non-regulated discharge water basins. There are also proposals envisaging creation of regulated water bodies in place of the Small Sea, Sarychaganak Bay, and the shallow central part of the Large Sea with disposal of saline waters to the western deep-water depression, or construction in the Amudarya lower reaches of the Djaltyrbass and Adjibai ecological reservoirs and others; however, none of the above proposals have been thoroughly elaborated.

R

Rand coast, coast of the Rand type – characterized by wide (from 1 to several kilometers) recently dried areas. In the lower part closer to the sea, the surface was densely covered by salt-loving plants (halophytes). With the dropping of the groundwater table, the vegetation cover dried, and with proximity to the sea, the new halophytes appeared (i.e. its border with the lowering of the sea water level moves towards the sea). The middle and upper parts of dried areas had no vegetation. Here, the sandy-aleurolite sediments were intensively affected by deflation and as a result chains of barkhans appeared on the dried area with heights varying from 1 to 3 m, and seldom over 5 m. The underwater shoreface of R.C. usually had insignificant slopes, and beaches were not formed here because waves not reaching the coast are attenuated in the shallow zone. R.C. was intensively developed in the southwest (Adjibai Bay), south (Tigrovy Khvost bar), southeast (between the Akala Cape and the Akpetkinsky Archipelago), and north of A.S. (the northern parts of the Kokaral Island).

Regime of Aral Sea waves and wind: a practical guide – prepared by the State Institute of Oceanography and edited by G.V. Rzheplinsky. It was published in Moscow in 1963. This guide was intended for practical application by dispatchers and captains of the Navy and fishery fleet of A.S. and also for forecast hydrologists in the Hydrometeorological Service. The guide includes a short description of synoptic processes causing stronger winds on the sea; duration of storms; maximum wind speeds; methodology for estimation of the elements of waves and plotting of maps; comparison of measured and estimated maps; specific features of wave regimes in various regions of A.S.; and wave maps of A.S. It also provides examples of application of maps, elements of waves, tables, and graphs.

Region prohibited for navigation* – the water area of A.S. adjoining the *Komsomolsky* (see), *Vozrozhdenia* (see), and *Konstantin* (see) Islands where navigation was prohibited. Limits of the region were indicated on the navigation maps of the Aral Sea.

Relics of the Aral Sea (of Latin *relictum* – remnants) – varieties and other taxa of animals and plants, and remnants of extinct fauna and flora. R. were usually

I.S. Zonn et al., *The Aral Sea Encyclopedia*, DOI 10.1007/978-3-540-85088-5_18,

distinguished by the geological age of the biotas from which they survived. The aquatic flora and fauna of the Aral basin also included relics from historical time.

Uzboy relics survived in the Uzboy Lakes after termination of the Amudarya flow into the Caspian (approximately 1573). For instance, in the Topjatan Lake, the Turkmen aqueous donkey (*Asselus messerianus*), Aral stickleback (*Pungitius aralensis*), and 6 other Amudarya fish species that for the other part of the Aral paleobasin were relics of earlier geological periods. In the saline lakes Yaskhan and Karatogelek Caspian fish (bullheads and aterina), shellfish and mollusks – the relics of the Novo-Caspian transgression – survived.

Oxian relics appeared during the drop of the Aral lake level in the 2nd–4th centuries. About 1600±100 years ago, the eastern (shallow) part of the Large Sea dried out and mirabilite distillation occurred in brines of the deep depression (at a depth of 31 m) where salinity levels were over 110%. Later on with the increase of river flows, the shallow (1–3 m deep) tugai, which were described in the Roman historical chronicles under the name "Oxian wetlands," appeared. The presence of fossil shells of *Acroloxus aff. lacustris* in the Oxian deposits was unique proof that in this period the lake was fresh-water. During the water level drop (during 250±200 years) and growing salinity, the relatively stenohaline and stenothermy Caspian salt-water species (e.g. *Ninnia jukovi* and possibly *Oxypyrgula spica*) that intruded into the Paleo-Aral along the Uzboy (6600±400 years ago) died out, while euryhaline and eurythermy species adapted for propagation in fresh waters.

Oxian relics were not endemic of the Aral. Some of them lived in Aral rivers. For example, live sertsevidki were found in the Sarykamysh Depression after rushing there with the Amudarya waters in 1857 and 1862.

Late Wurmian relics appeared in the Syrdarya basin from West-Siberian species that washed into the Aral depression with the waters released from the periglacial West-Siberian lake approximately 10 thousand years ago. The flow via the Turgai Strait was marked with a thick sandy bench with intercalation of rare lenses of fine-disperse material, coarse and small sand in the profile of bottom sediments in the central part of the Aral. The river crawfish (*Pontastacus kessleri*) was found in the drainless area of Irgiz-Turgai on the northern slope of the Aral Depression and in aryks and springs of the western slopes of the Karatau (the Syrdarya basin).

Availability in the piedmont areas of the Syrdarya of typical periglacial deposits indicated that there were no river flows, the glacier shield in the period of the maximum Wurmian glaciation (20–16 thousand years ago) covering practically the whole watershed area. Thus, it was quite obvious that during the Wurmian glaciation in the eastern part of the Aral basin only those hydrobionts could survive that presently live in the springs of high-mountain periglacial zones. The Chardarya dam was constructed in 1968 near the borders of the Wurmian glaciation, and its watershed area covered 80% of the Syrdarya basin. Therefore, the extinction of fish at the beginning of the Wurmian glaciation may be demonstrated by the disappearance of Syrdarya pseudo-bastard sturgeons during the hydraulic construction. In 1947, the dam of the Farkhad hydropower plant cut off their main reproductive areas from the fattening zone. The bastard sturgeon Fedchenko (*Acipenser fedtschenkoi*) was caught for the last time in 1953, while the bastard sturgeons (*A. brevirostris*) were found in the Karadarya in 1969. Construction of the Charvak (1970) and Andijan (1978) hydropower plants left small hope that the Syrdarya bastard sturgeons still existed in the non-regulated tributaries of the Syrdarya.

Wurmian relics were found in the upper part of the Amudarya basin from the period of the Wurmian glaciation. They most likely included the Samarkand khramulya (*Varicorhynus capoeta heratensis natio stendachneri*) that had ancestors among the fish of the mountain-Asian complex, the Zarafshan dace (*Leuciscus lehmani*), and the Bukhara roach (*Rutilus rutilus bucharensis*), whose ancestors lived in the Palearctic area. Biotypes of these endemics were confined solely to backwaters and lakes in the Amudarya piedmonts, which indicates low water availability in the last glacial cycle of the Holocene. In winter, the flow of the paleo-Amudarya might have reduced so that even carp fish that were not very demanding of the oxygen regime could not exist in the plain part.

Balakhandsky relics survived in the Amudarya paleobasin (including the modern Issyk-Kul and Balkhash) from the time of the hyperhaline Balakhansky Lake that existed 5.3–3.5 mln years ago. At that time, fresh-water mollusks (*Theodoxus aralensis* and *Dressena obtrusecarinata*), crawfish (*Turcogammarus aralensis*), and fish (*Pungititus aralensis*) survived only in the Aral paleobasin. *Victorella bergi Bryozoa* lived in the Aral basin, while its close fresh-water species (*V. continentalis*) lived in the Issyk-Kul Lake.

Postglacial relics were maintained in the Aral rivers from the time of the postglacial linking of the basins of the paelo-Amudarya and paleo-Indus; the time is awaiting further verification. In the Aral rivers were found Sino-Indian mollusks, crawfish, fish, and their parasites. Ichthyologists believe that the Turkestan catfish (*Glyptosternum osciani*) were co-specific of *G. reticulatum* from the Indus. It was confined to the cool (10–18°C) mountain rivers. Close comparative study of a pike-like asp (*Aspioluscus esocinus*) with *Paralaubuca harmandi* from Vietnam did not exclude the Sino-Indian roots of its ancestors; however, the first was a carnivore, while the second was a planktonophag. The pike-like asp propagated in the Aral rivers in winter (February–March, 5–10°C), while *Paralaubuca* was tropical fish.

Republic of Karakalpakstan ("kara" means black, "kalpak" means cap, "stan" means country) – an autonomous Republic located in the northwestern part of Uzbekistan. Its area is 166.6 thou sq. km, or 37% of the Uzbekistan territory, out of which 80% is occupied by deserts. This is the only region in Uzbekistan that is washed by A.S. (the coastline length is 420 km). It encompasses the northwestern part of the Kyzylkum Desert, the southeastern part of the Ustyurt Plateau, the southern part of A.S., and the Amudarya delta. The capital is

Nukus (see). The population is 1,559,000 people (2004), out of which 746.1 thou people are the urban population. The population density is 7.5 people/sq. km. The Karkalpaks were resettled here from the lower reaches of the Syrdarya in 1811 by the Khiva Khan. In 1873, the right-bank of K. was joined to Russia, then Soviet power was established in December 1917. After April 1918, it became a part of the Turkmen Autonomous Republic, with its left-bank belonging, as before, to the Khiva Khanate. In April 1920, it was joined to the Khorezm People's Soviet Republic, and in the course of national demarcation, K. was united. In February 1925, the Karakalpak Autonomous Area was formed as a part of the Kazakh Autonomous Republic (After July 20, 1930 directly to RSFSR). On March 20, 1932, it was given the status of the Karakalpak Autonomous Republic, and on December 5, 1936 was included into the Uzbek SSR. The Republic of Karakalpakstan was formed on December 21, 1991. The largest cities are Turtkul, Takhiatash, Khodjeily, and Muinak.

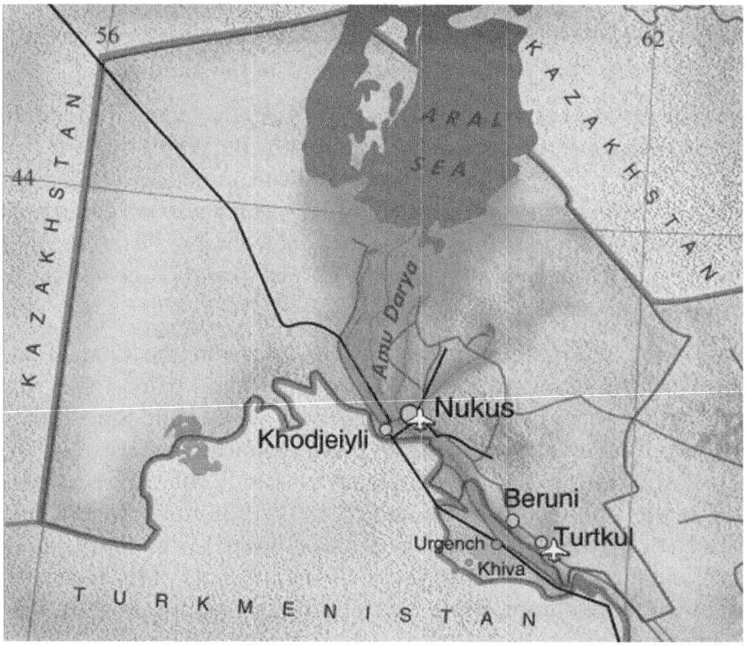

Fig. 39 Republic of Karakalpakstan (www.orexca.com/img/karakalpakstan/map/jpg)

The border of Karakalpakstan crosses the Turan Depression. On the southwest, it borders on the Karakums, on the northwest it borders the Ustyurt Plateau, and on the northeast it borders the Kyzylkums. The territory of Karakalpakstan also includes the southern part of A.S. on the dried bottom of which (and the lower reaches of the Amudarya River) a new solonchak desert, *Aralkum* (see), is forming.

Karakalpakstan is characterized by a sharply continental climate with a hot dry summer and a relatively cold winter. The average annual temperature is 32–37°C, with the absolute temperature in the irrigated zones being 75–80°C and 80–85°C in the desert zone. The winter is cold here with the average temperature in January being −11°C (north of Ustyurt) and −5°C (in Turtkul in the southern part). The absolute minimum temperatures vary from −28°C to −37°C (on the Ustyurt).

Among the most important mineral deposits and mineral raw materials here are natural gas, building materials, rare-earth metals, and mineral salts. The forecasted potential of oil and gas reserves in the most promising areas of the Ustyurt is estimated at 1.685 tril tons of liquid hydrocarbons. In the Ustyurt gas fields, 220 wells are in operation. Quite promising for future industrial-scale processing are porphyrites, facing stone, ceramic raw materials, and lime.

Of special significance are reserves of table salt. Explored resources in 10 deposits are enormous, with an estimated 11 bill and 40 mln tons, respectively, in Karakalpakstan's largest deposits of Barsakelmess and Karaumbetsky. Another precious natural resource is groundwater used for cattle grazing and as drinking water for the municipal-domestic water supply of cities and settlements.

Karakalpakstan is an agrarian republic. Agricultural lands cover an area of 10211.8 thou ha. They are concentrated in the coastal belt of the Amudarya, from the Tuyamuyun narrow to the river delta. Out of the whole arable land stock in Karakalpakstan (2 mln ha), about 500 thou ha are used for cultivation of agricultural crops, and 12.0 thou ha are used for growing of gardens, vines, melon crops, and other perennial plantings. The free lands suitable for irrigation make 1.2 mln ha. In Karakalpakstan, only irrigated farming is practiced; atmospheric precipitation meets only 7% of the plant need in water, and during the vegetation period less than 1%. The only source of irrigation water is the Amudarya. Before commissioning of the Takhiatash waterworks and later the Tuyamuyun and Nurek reservoirs, the water intake into irrigation canals depended on the natural regime of the river, mainly on the water level. After construction of the Takhiatash and Tuyamuyun waterworks, the possibility for efficient utilization of water resources in the Spring low-water period emerged. But in recent years water availability in the Amudarya has decreased sharply.

By the early 21st century, 96% of lands in K. were saline to varying degrees. The following factors contributed to such a condition of the irrigated lands: a sharp drop of the water level in A.S. and the follow-on drying out of the vast area of its bottom as a result of which salt drifts to irrigated lands began, becoming more intensive with each passing year and intensifying salt accumulations on agricultural landscapes; deterioration of the quality of river waters, which mineralization has nearly doubled (from 0.33% to 1.2%), as a result of disposal into the river (in its upper reaches) of highly saline drainage waters heavily polluted with various chemicals and agricultural wastes; great losses of water to seepage during water delivery along canals to irrigated fields as well as a failure to conduct leaching of saline lands in due time during agricultural operations; and poor efficiency of drainage networks due to their insufficient lengths and untimely cleanings, repairs, and rehabilitations.

Agriculture is the main source of the national income. Thus, the intensified processes of water and soil salinization has caused irreparable damage to agriculture in the region. The principal agricultural crops are cotton (in 2003, the yield of raw cotton was 88 thou tons) and rice (approximately 20% of the whole rice production in Uzbekistan). Cotton-ginning, butter, fish canning, as well as fur animal breeding, sericulture, and animal breeding (karakul sheep and large-horned cattle) are developed here.

The territory of R.K. is a kind of archeological nature preserve. Over 300 archoeological features are found here. In ancient times, this territory was called Khorezm.

Republic of Kazakhstan, Kazakhstan – state in the southwestern part of Central Asia at the very center of Eurasia. R.K. is washed on the west and northwest by the Caspian Sea, on the southwest is bordered by Turkmenistan, on the south by the Republic of Uzbekistan and the Republic of Kyrgyzstan, and on the southeast and east on the People's Republic of China. Its territory is 2724.9 thou sq. km. The capital is Astana. R.K. is the most urbanized of all other Central Asian republics. The largest cities are Almata (1.13 mln people), Karaganda (560 thou people), Shymkent (409 thou people), Djambul (320 thou people), Pavlodar (300 thou people), and Semipalatinsk (270 thou people). Administratively, it is divided into 14 regions and 3 urban *akimats* (Astana, Almata, Leninsk), 160 districts, and 2150 *aul* (rural) areas. The largest regions are Aktyubinsky, Almatinsky, East-Kazakhstansky, Karagandinsky, and Kustanaisky. The population is 15.5 mln people (2006). The national composition is as follows: Kazakhs 53%, Russians 30% (in 1989 6.1 mln Russians lived in R.K., and in 2000 only 4.5 mln were left), Ukrainians 3.7%, Germans 2.4%, Uzbeks, Tatars, Uigurs, Belorussians, Koreans, and others. The economically active population of R.K. is 7.2 mln people, out of which 6.2 mln people are involved in different branches of the economy. The official language is Kazakh. The greater part of the population is Moslem of the Sunni branch, with Russians professing Christianity and Germans being Lutheran. The currency is the Tenge.

As a result of the referendum of August 31, 1995, the people of R.K. adopted the Constitution that mandated a presidential system combined with a permanently acting professional Parliament consisting of two houses, the Senate and the Majilis. R.K. is a democratic, secular, unitary state. The leader of the state is President that heads the single executive power of the republic. In direct elections held on December 1, 1991, Nursultan Nazarbaev was elected President of R.K. for a term of 5 years. in further election on April 29, 1998, his term of office was extended. The highest organ of state power is the Parliament.

The topography of R.K. is very diverse. Low plains are replaced with uplands and small hills, and in the east and southeast, high ridges covered with permanent snow and glaciers rise up to 5 thou m high. In the west is the lowest place in the country, the Karagie Depression (132 m below sea level). The

Fig. 40 Republic of Kazakhstan (www.odyssei.com/images/maps/big/kazakhstan.jpg)

surface of R.K. is mostly represented by plains and lowlands. A considerable area is taken by the Circum-Caspian Depression that lies near the Caspian Sea at 28 m below sea level. Further northwards, the land rises to 50–60 m. In the northeast, the Circum-Caspian Depression is limited by the Ural Mountains and Mugodjars. To the east of the Mugodjars is the Turgai Plateau (250–300 m) which passes in the south to the Turanian Lowland occupied largely by the *Kyzylkums* (see) and *Circum-Aral Karakums* (see). Westward of the Aral Sea, the *Ustyurt Plateau* (see) rises (up to 300 m high), breaking down to all sides with steep benches ("*chinks*" (see)). The central part of the country is characterized by the small hummocky relief representing the remnants of the ancient mountain system with some mountain massifs, the Kyzylrai and Karkaraly. The southern part of the Kazakh small hummocky territory passes into the Betpak-Dala, a vast desert plateau (300–400 m high) with a large territory to the south occupied by the Muyunkum sands. The southern chains of the Altai, the Tarbagatai ridges, Saur, and Djungarsky Alatau, and the northern chains of the Tien-Shan and Zailyisky Alatau run in the east and southeast of R.K. along its southern borders.

The climate is sharply continental and arid. The southern areas are significantly affected by tropical air encroaching from Iran and the Caspian Sea. The average temperature in January is from –18°C in the north to –3°C in the south, while in July the temperatures are, respectively, 19°C and 28–30°C. In the north, precipitations reach 300 mm a year (mostly in the summer season), while the deserts receive less than 100 mm (falling mostly in spring), and the mountains receive up to 1600 mm. There are more than 2700 glaciers in R.K.; the total glaciation area is approximately 2000 sq. km.

Apart from the Irtysh River, other rivers in R.K. are in the basins of the Caspian and Aral Seas and Balkhash Lake. The Irtysh runs to the basin of the Arctic Ocean. The largest rivers are Irtysh, Ural, Syrdarya, Emba, Turgai, Nura, and Sarysu. In summer, many rivers dry out. The Ily, Karatal, Aksu, and Lepsa Rivers bring their waters into Balkhash Lake. The rivers are regulated by 180 reservoirs, the largest of which are Chardarinsky, Bukhtarminsky, and Kapchagaisky. There is also the Irtysh-Karaganda canal, which is 478 km long.

R.K. has more than 48 thousand lakes, located largely in deltas and floodplains. The major of them are Balkhash, Zaisan, and Alakol. After the drying of A.S. and its division into three water bodies – the Northern (Small), the Eastern, and the Western – only the smaller sea belongs completely to the republic and is being rehabilitated at present.

In R.K., chernozem soils in the north are gradually replaced to the south with dark-chestnut and chestnut soils. In the mid-1950s, during the development of the virgin lands here, nearly all of these soils were brought under cultivation. The southernmost territories are composed of gray soils, while the intermountain dry depressions are of brown soils.

The steppes of grass-cereal and fescue-feather grass vegetation (mostly ploughed) are replaced with wormwood-cereal semidesert and wormwood-thistle desert vegetation. The piedmont areas are covered with dry steppes, the middle mountain areas are covered with coniferous forests, and still higher are the sub-Alpine and Alpine meadows. The fauna is represented by multiple rodents, reptiles, and carnivorous animals (fox, wolf, badger, and others). In semideserts and deserts, the saiga and gazelle are found; such birds as bustard, gerfalcon, waterfowl (ducks, geese, swans, pelicans and others) are also in R.K. In the mountains, the argali are found, and in the Tien-Shan is the snow leopard.

In the Caspian Sea and in the Ural River, herring, sea roach, sturgeon, and starred sturgeon are fished, and seal hunting is also developed. In the Small Aral Sea, asp, bream, and common carp are caught.

In R.K., there are 8 nature preserves: Almata, Aksu-Djabagly, Barsakelmes, Naurzumsky and Kurgaldjinsky, Markakolsky, Western-Altaisky, and Ustyurtsky. In 1997, the Bayan-Aulsky natural preserve was established.

R.K. boasts a great diversity and rich resources of mineral deposits. Out of 105 elements of the Mendeleev's Table, 99 are found in R.K., with 70% of the reserves of each having been explored and over 60% used in production. In terms of explored reserves of lead, zinc, and bismuth, R.K. ranks first among the CIS member countries, and in reserves of copper, molybdenum, boxites, oil, phosphates and cadmium, it ranks second. Out of all CIS countries, Kazakhstan possesses the largest hydrocarbon resources in the shelf of the Caspian Sea and the adjoining land area. R.K. is 13th in the world in terms of prospected resources of oil and gas, and 26th in terms extraction level. At present, more than 180 oil and gas fields have been explored in R.K., while only 60 of them have been developed. Explored reserves of hydrocarbons include 2.2 bill tons of oil, 0.7 bill tons of gas condensate, and 2.7 tril. cu. m of natural gas. Potential resources of R.K. on the mainland and in the shelf are evaluated at 12 bill tons of oil, 1.6 bill tons of gas condensate, and 5.9 tril. cu. m of natural gas. The extraction of revealed oil resources is not easy, however, because they contain high concentrations of asphalt, resin, and paraffin components. Apart from major oil and gas fields, R.K. also has significant resources of phosphates, chromium, and uranium ores. In the north, there are rich deposits of iron ores, gold, and coal; also, bauxites and table salt are extracted. Deposits of titanium ores and asbestos are also found here. The eastern part is the main source of nonferrous ores, primary of which are polymetallic. In many instances they are confined to the fields of the Ore Altai. In the center of the country, deposits of iron and manganese ores, copper, and Kazakhstan's largest deposit of cocking coals (Karaganda basin) are concentrated. In the south, the CIS's largest deposit of phosphates in Karatau is found. In addition, the country possesses sufficient raw materials for development of a building material industry.

The ancestors of the Kazakhs settled the are in approximately the 1st century A.D. The Sax tribes populated a vast territory of the present-day Semirechye and Syrdarya basins. In the 6th–7th centuries, different early feudal states that united the Turkish-speaking nomadic tribes replaced one other in this area. In the early 13th century, the whole of Central Eurasia entered into the Mongolian state. In the 15th century, the White Orda consisting of the Uzbek-Kazakh tribes broke down, and the Kazakh Khanate was formed in its place, divided into three "zhuz." Due to non-conformity of the feudal superstructure to the patriarchal basis, however, it failed to make a strong state. The process of disintegration of the Kazakh Khanate that started soon after its formation reached its apex in the early 18th century when in 1731 the Junior and in 1740 the Middle "zhuz" voluntarily joined Russia. Approximately from that time on, the period of active penetration of the Russian Cossacks to the territory of modern K. began. In addition to the growing influence of the Yaik Cossacks, the so-called "linear" Cossacks started settling in the expanses from Omsk to Orenburg along the chain of the bitter-saline lakes ("bitter line"). In 1819, the tsarist government liquidated the khan ruling on the territories of both "zhuz" and introduced a new system of administrative control. By the mid-19th century, the whole territory of the Senior "zhuz" was joined to Russia, an in 1867, the "Interim Regulations on Steppe Territory Administration" was adopted. At the same time, from the 1880s–1890s, the process of active involvement of the region into the general Russian economic system was initiated. The tsarist government made efforts on wide-scale resettlement of sedentary peasants from Russia, Ukraine, Belorussia, and other regions to the Kazakh steppes. On August 26, 1920, the Kyrghyz Autonomous Republic was formed in RSFSR, which later, on April 19, 1925, was renamed the Kazakh Autonomous Republic. From December 5, 1936, K. acquired the status of union republic. In the 1930s, the industrial construction actively arose here, and in the 1950s virgin and fallow lands were actively developed. From 1960 to 1965, the Tselina (virgin-land) Territory on the basis of five northern regions (Kokchetav, Kustanai, Pavlodar, North-Kazakh and Tselinograd) was formed here. K. turned into one of the leading industrial and agricultural regions of the USSR.

On October 25, 1990, the Supreme Council of the Kazakh SSR passed the Declaration on State Sovereignity, and on December 16, 1991, the Supreme Council adopted the Law "On State Independence of the Republic of Kazakhstan." On May 25, 1992, the Russian Federation and R.K. signed the Treaty on Friendship, Cooperation, and Mutual Assistance.

Diplomatic relations with Russia were established on October 22, 1992, and K became a CIS member state in 1991.

R.K. became a UN member on March 2, 1992, as well as a member of ADB, EBRD, IBRD, OIC, ECO, SCO, ESCAP, and other international organizations.

R.K. belongs to the group of countries with transitional economies. After disintegration of the USSR, due to breakup of many links and economic crisis, most production ceased. At present, the privatization process in the country has been completed; progressive tax legislation has been adopted; a modern banking system has been established; and reforms in agriculture, the housing sectorm, and the social sphere have taken place. During the years of Soviet power, the republic developed a multi-sector industrial production base from its own resources. Heavy industry is dominant, prominently fuel-power, metallurgy, and food industries.

The principal industry of the Kazakh economy is the fuel-power complex, which uses its own deposits of coal, oil, and natural gas.

R.K. had been producing oil for over 100 years. Until recently, only land-based oilfields – Tengiz-Korolev, Kumkol and others – had been explored along with the gas condensate fields in Karachaganak. After becoming independent, R.K. started development of the Caspian shelf. In 2000, the major oilfield – Eastern Kashagan – was opened with reserves of approximately 7 bill tons. In 2003, the State Program on Development of the Kazakh Sector of the Caspian Sea was made public. According to an agreement signed between Russia and the Republic of Kazakhstan on the division of the bottom of the Northern Caspian for realization of sovereign rights to natural resource utilization, the Khvalynsky and Central fields will be developed jointly on the 50/50 principle. In terms of oil resources, K. is fifth in the world.

The coal industry is concentrated in the Karaganda and Ekibastuz basins; however, coal production is rather low. Of total CIS production, K extracts 40% of the uranium, 97% of the chromium, 70% of the lead, and 50% of the zinc. Considerable are also the volumes of production of other kinds of raw materials and pure precious metals, including gold. Hydropower engineering is based on large hydropower plants constructed on the Irtysh River (Bukhtarminsky, Ust-Kamenogorsky, and Shulbinsky), and high-capacity thermal power plants operated in Ekibastuz, Karaganda, Taraz, and near Almaty and Pavlodar.

Metallurgy takes a special place in the national economics. Its share of industrial production is 14% and of exports is 30%. A large, full-cycle metallurgical plant operates in Temirtau, and ferroalloy plants operate in Aktobe and Aksu (Pavlodar Region). Nonferrous metallurgy is one of the leading industries in Kazakhstan. Two large territorially united regions of nonferrous metallurgy may be distinguished: central (copper) and eastern (polymetal – lead, zinc, aluminum, copper, gold, etc.). The main centers of nonferrous metallurgy are Ust-Kamenogorsk, Leninogorsk, Balkhash, Zhezkazgan, and Shymkent; the center of alumina production is Pavlodar, and of chromium compounds is Aktobe. The Circum-Caspian mining-metallurgical plant processes uranium ore into concentrate while simultaneously producing numerous rare and rare-earth elements, alloys on their basis, as well as accompanying productions of phosphate fertilizers. One of the largest plants to produce fuel for nuclear reactors is found in Ust-Kamenogorsk.

Oil processing plants are located in Atyrau, Pavlodar, and Aktau. Their total capacity is 11 mln tons of oil. The only gas processing plant is in Novyi Uzen. The chemical industry is represented by the production of fertilizers (phosphate meal, phosphate salt, superphosphate), sulfuric acid, chemical fibers, plastics, and others. The main centers are in Taraz, Kostanai, and Aktau.

The machine-building industry was created during the World War II on the basis of the equipment evacuated here from the European part of the USSR. Prevailing were the agricultural (Astana, Aktobe), tractor (Pavlodar), electrical engineering (Almata, Petropavlovsk, Uralsk), mining and transport (Karaganda, Almata, Ust-Kamenogorsk) and other machine-building.

Light industry developed are leather, fur, cotton, and footwear productions. The food industry is represented by large enterprises producing meat (Semipalatinsk, Almata), sugar (Taraz, Taldykorgan), butter, cheese, flour milling (in regional centers of the republic).

The share of agriculture in the agro-industrial complex reaches 50%. There are two leading agricultural industries in R.K.: large-scale mechanized farming and distant-range cattle breeding. Arable lands concentrated mostly in the north of the republic occupy 18% of the agricultural lands. The virgin and fallow lands were developed mostly in the early 1950s. Smaller areas of arable lands are found in the piedmonts of the Altai and Tien Shan river valleys. Cultivated areas are over 15 mln ha. The main branch of farming is cultivation of grain crops on an area over 11 mln ha. In the Syrdarya valley, sunflower, millet, corn, and rice are also grown. As concerns technical crops, cotton is grown on small areas in the south, and sugar beet and tobacco are grown in the south-east. In piedmont areas, gardens and vines are grown. Melon crop growing is also developed.

Pasturelands account for 57% of the country's territory. The main branch of cattle husbandry is sheep raising (9.8 mln heads), primarily fine-fleece, semi-fine-fleece, and karakul sheep. Meat cattle are raised in the steppes, while milk cattle are raised on urban farms. The population of large-horned cattle is 4 mln heads. Also developed are horse, camel, and pig rearing.

The main kinds of transport are railroad (95% of the country's cargo turnover) and automobile. The operating length of railroads is approximately 15 thou km, while automobile roads extend for about 100 thou km. The largest oil pipelines are Omsk-Pavlodar-Shymkent, Aktau-Atyrau-Samara, Tengiz-Novorossiysk; gas pipelines are Zhanaozen-Makat-Saratov, Bukhara-Ural, Taskent-Almata. The large ports on the Caspian Sea are rehabilitated Aktau and Atyrau. Navigable rivers are Irtysh, Ural, and Ily.

Airports are Astana, Almata, Karaganda, Aktobe and Aktau. In the Karaganda Region near the town of Leninsk, Russia rents the Baikonur cosmodrome to launch spacecrafts. The telecommunication system of R.K. is represented by the National Joint Stock Company, "Kaztelecom."

A tendency to attract foreign, largely Western investments to the extraction and production industries is being shaped. Export revenues from the sale of mineral deposits and their processed products are considered to be the main means for overcoming the crisis in the republic. These are, first of all, projects on development of the Tengiz and other oil and gas fields, and active foreign economic relations with the USA, European, and Asian countries. K. signed agreements with foreign oil companies amounting more than 40 bill dollars that will be invested in the next four decades into the oil production industry. Major investors include "Chevron" (USA) and Elf-Akiten (France).

The main exported commodities are oil and oil products (40% of the total export), ferrous and nonferrous metals, and mineral products, including ore and fuel. A great share of imports is machinery, equipment, transport means, fuel-power resources, as well as sugar, tea, etc. The main importing countries are Germany, Austria, Morocco, Mongolia, Turkey. The deliveries from Russia account for about 82% of the import from the CIS states.

Tourism is also developing. In the territory of the republic there are such well-known recreational zones as Medeo, Chimbulak, and Kokchetav. The republic has several dozens of higher educational establishments, including the Kazakh State University in Almata, the L. Gumilev Eurasian University in Astana, the Turkestan International University, and the Karaganda University, to name but a few. The Academy of Sciences of Kazakhstan was founded in 1946. About 40 theatres are working in the republic. The Kazakh Teleradio Company broadcasts in Kazakh, Russian, German, Korean, and Uigur.

Republic of Uzbekistan – located in the middle of the Central Asian region, the territory of the republic extends between the Syrdarya and Amudarya from the northwest to the southeast. The northwestern and northern border with Kazakhstan (2203 km) is desert. In the east and southeast, in its mountainous part, Uzbekistan borders Kyrghyzstan (1099 km) and Tajikistan (1161 km). In the south, it borders Afghanistan (137 km) and in the southwest, Turkmenistan. The area of the republic is 447 thou sq. km.

Fig. 41 Republic of Uzbekistan (www.lib.utexas.edu/maps/asia.html)

Nearly four-fifths of R.U. territory is covered by plains with the mountains rising only in the easternmost part (the highest point is 4663 m). A strip of piedmont plains runs between the mountains and the lowlands. The ridges in the east refer to the Tien Shan and Gissar-Altai mountain systems separated by the Ferghana intermontane depression. The flat desert part of the country is complicated by low rising uplands (the Muruntau mountains and others).

Some regions of the country (a piedmont part of the Ferghana Depression and the southern slopes of the Gissar ridge) are affected by strong earthquakes (more than 9 Richter points).

The climate here is moderate, sharply continental, transitional to a subtropical one. Quite typical are great differences in the soil and air temperatures between the day and night; temperature differences in summer and winter are great. The relatively moist air of moderate latitudes prevail in winter, while in summer the air is replaced with warm, dry, tropical. The amount of precipitation is seldom more than 200 mm a year, and some places receive no more than 70–80 mm a year. Only in the mountains does precipitation increase appreciably to 500–600 mm, sometimes reaching 1000 mm a year. The average temperature in June even in the northern regions of the republic is higher than +26°C, while in the south it may be as high as +31 or +32°C. The average temperature in January in the north of the republic (on the Ustyurt Plateau and in the Amudarya lower reaches) varies from –7 to +12°C. Over the greater part of the flat regions winter temperature is close to 0°C, while in the south (in Termez) it is +3°C.

The surface waters are distributed over the territory very unevenly. The extensive plains have few waterways, while in the mountains a highly furcated river network forms. All rivers belong to the Aral Sea basin. The main rivers – Syrdarya and Amudarya – take their origin in the high mountains of Kyrghyzstan and Tajikistan. Due to growing irrigation needs in the recent four decades, the water flow in the lower reaches of the Amudarya has decreased nearly two-fold.

The flat plains of the republic are covered by a desert with prevailing saxaul and thistle vegetation, the low piedmonts are covered by a semi-desert with the prevailing ephemeral and ephemeroid vegetation. High piedmonts are represented by grass-wheatgrass steppes, and the medium-mountain areas are covered by tree and shrub vegetation. The high mountains are covered with sub-Alpine and Alpine meadows. Riparian (tugai) forests are found along river valleys. The plains of the country abound in rodents, reptiles, and hoofed animals. Here one can find jerboa, lizards and snakes, tortoises, koulans and saiga, while in the mountains there are goats. The carnivorous animals are represented by wolves and foxes. In order to preserve the natural environment, 15 nature preserves, both on the plains and in the mountains, were created in the republic. The largest of them is the Chatkal preserve.

More than one hundred kinds of mineral deposits are found in the republic. In gold reserves (the major gold-mining center is the Muruntau region as well as the eastern areas of the country), tungsten, silver, lead, zinc, and uranium, Uzbekistan is one of the world leaders, and in uranium reserves (explored uranium resources is approximately 55 thou tons) Uzbekistan is 7th in the world. Nonferrous metal deposits are still more valuable because they can be mined in open quarrys. At the same time, many deposits are not easily mined and processed, so their development depends, to a great extent, on world markets. All kinds of fuel resources are found in the republic: large coal mines – Angrensky, Shargunsky and Baisunsky and a major gas field – Bukhara-Khiva (the greater part of oilfields is concentrated in the Pre-Ferganie). In the recent years, 8 gas condensate fields were found in the Uzbek part of A.S. (in the Ustyurt region), and some of them are already being developed. Geological surveys conducted in this region proved a high hydrocarbon potential of A.S., and about 30 perspective oil and gas fields may be expected in the Aral area.

The territory that is Uzbekistan was home to several ancient peoples. Eencampments from the Stone Age have been found in Teshinkanda. In the

Central Asian interfluve area, ancient peoples were influenced by the peoples of neighboring regions: in the north, the Turk-speaking nomadic cattlemen and in the south, Persian-speaking, mostly sedentary farmers. The territory of Uzbekistan was a part of such slave-owning states as Sogdiana, Bactria, Khorezm, Margiana, and others. In the 2nd century A.D., this territory was crossed by the main caravan trade route from Europe to India and China, which was an important part of the well-known Great Silk Road. At that time, the cities were the main trading points. Later on, the Kushan Kingdom, the ephthalites state, and the Turk Kaganate existed here. After resettlement of large Turkish groups to the territory of Uzbekistan, a great part of the local population merged with the nomad newcomers and started speaking the Turk language. As a result of this assimilation, the Turk-speaking peoples were shaped, becoming the nucleus of modern Uzbeks.

An important event in the history of the country was intrusion in the 7th–8th centuries of the Chinese and later the Arabs. The influence of Islamic civilization has survived to the present. Periods of economic and cultural revival were in the 9th-10th centuries during the Samanide ruling, the 12th century during the rule of the Khorezm Shahs, in the late 14th–early 15th centuries during the ruling of Timur and his successors, though these were replaced with decades and even centuries of complete decline during invasions of nomads (Huns, Turks, Mongols, Djungars). In the 16th century, the mixed population of Central Asia was added with the tribes that were generally called "Uzbeks" (from the name of one of the khans) and in time assimilated with the Turk population in this territory. The Bukhara and Khiva Khanates that were formed in the early 16th century on the remnants of the great Uzbek empire were immersed in permanent conquering wars with each other. In the early 18th century, the southern regions of Bukhara separated and formed the independent Kokand Khanate. The intrusion of the Russians into Central Asia in the 19th century was connected both with their economic interests and with the attempt to oppose Britain which was waging wars to secure Afghanistan in this period. In 1865, Tashkent was conquered as a result of military actions, as were Bukhara in 1868, and Khiva in 1873. The Turkestan General-Governorship was formed over the greater part of this territory. The Bukhara and Khiva Khanates became vassals of Russia. Soviet power was established here during a long time period. Uzbekistan as a republic was formed on October 27, 1924; in 1925 it entered into the USSR. Before this, however, in 1920, the Khorezm and Bukhara People's Soviet Republics were declared, replacing the Khanates. In December 1991, after the all-people's referendum, Uzbekistan declared itself a sovereign state.

The republic pursues an independent policy while participating in many regional associations. The main principle of its economic policy is orientation to its own resources while attracting some industries through foreign capital. The head of the state and executive power is the President who appoints the Prime Minister and members of the government. Since December 1991, the President of Uzbekistan has been Islam Karimov. The legislative body is the two-house Parliament, Olyi Majlis.

The population is 25.1 mln people (2002), of which 63% live in rural areas. During the transitional period, the birthrate lowered somewhat, but it is still one of the highest in the CIS countries (over 1.8%). The traditional type of reproduction ensures a great share of young people (more than 60% of the population are younger than 25). The average population density is the highest in the Central Asian republics (54.2 people/sq. km), but the population is distributed very unevenly. Its greater part is concentrated in the regions with a farming orientation (the Ferghana Valley, Tashkent, and Samarkand Regions), where the population density increases to 450–480 people/sq. km. The desert territories do not have permanent populations. More than 120 peoples and ethnic groups live in Uzbekistan. The most numerous are the Uzbeks. After declaration of independence, an outflow of the Russian-speaking population (the second largest in the republic) was observed. The urbanization level in the republic is not high, at 37%.

The Uzbek language belongs to the Turk group of the Altai language family. At present, Uzbek is the state language. After the declaration of independence, the Latin alphabet was again in use as it was in the 1930s. The greater part of the population (88%) is Sunni Moslem. The Russians, who typically live in the cities, are Orthodox (9%). In Tashkent is the Episcopate Chair (regional department) of the Moscow Patriarchate for Uzbekistan and Central Asia.

From the time of the breakdown of the USSR, Uzbekistan formed a new economy. In the crisis period, the gross national product dropped by only 20% (compared to Kazakhstan, which dropped 45% and Russia, which dropped 40%). This may be attributed to the development of the raw materials industry, and a smaller share of the depressive early-industrial and defense industries, which ensured modernization of the economy. Low procurement prices in the early 1990s on basic products of Uzbek export, the wearing out of available equipment, and the breakdown of links with main suppliers of equipment and foodstuffs urged the republic to carry out radical economic reforms on the basis of macroeconomic and financial stability and the attraction of investments for updating the raw-material industries. The production capacity in all industries is on the rise, and the inflation level is lowering. The shares in GNP are: industry – 19.3%, agriculture and forestry – 31.3%, construction – 8%, transport – 8%, trade – 5.8%, other industries – 27.6%.

In the Soviet time, U. specialized on extraction and processing of gas, nonferrous ores, and machine-building for the cotton industry. Of regional significance was the chemical industry (production of defoliants for cotton and various kinds of fertilizers). In 1998, the structure of the industrial production was as follows: power engineering – 19.2%, fuel industry – 13.3%, machine-building and metalworking – 13.1%, food industry – 12.6%, nonferrous metallurgy – 10.9%, light industry – 8.5%, chemistry and petrochemistry – 5.2%, building material industry – 4.2%.

The gas industry (in 1998, it produced 53 bill cu.m) meets not only the domestic needs of the country, but provides a considerable quantity of natural gas for export. Russia receives 10 bill cu. m of gas annually. A gas processing plant is in Mubarek that, apart from gas, produces gaseous sulfur. By this index, Uzbekistan is among the world's top-10 gas producers. The oil industry (production – 8 mln tons) is much less significant. There are two oil processing plants in Khamza and Ferghana. A new oil processing plant is under construction in Bukhara.

Uzbekistan has 37 power plants with an installed capacity of 11.2 mln KW; 87.5% of power is generated by thermal power plants. In the Tashkent Region, power plants use the brown coals mined in the Angren basin (95% of coal production in the country) and the cascade of hydropower plants on the Chirchik River. In other regions, natural gas is used for power generation. On the basis of natural gas and oil, the chemical industry is developing. Attention is focused on meeting the needs of agriculture in fertilizers, primarily nitrogen, at Chirchik, Navoi, and Ferghana. Using Kazakh phosphates, plants in Ferghana, Kokand, Samarkand, and Almalyk produce fertilizers and means for plant protection. In Ferghana, the major center of the chemical industry, the complete cycle of hydrocarbon processing is completed.

Uzbekistan competes successfully with the USA through an assortment of equipment composing the cotton complex. Its manufacturing and repair enterprises are concentrated in Tashkent and nearby towns. Developed are also electrical engineering and instrument-making industries. The aviation association in Tashkent assembles transport aircraft. In 1996, an automobile plant was commissioned in Asaka, in the Andijan Region. In addition, automobile plants in Samarkand and Khorezm also operate. Nonferrous metallurgy is represented by production and processing of copper-molybdenum ores near Almalyk. The country is the leading world producer of gold (up to 90 t/year). Silver is produced in the Ferghana Valley, and the uranium production industry is developing. At present, with the financial support of foreign companies, spinning and textile plants as well as the cotton plant were constructed in Bukhara. The main centers for production of raw silk and silk fabric are located in the Ferghana, Namangan, and Bukhara Regions. Production of carpets and nonwoven materials has been developed in the Namangan Region and in Khiva. Leather-making enterprises are found in Tashkent, Samarkand, Kokand, and Bukhara.

Agriculture is a very important sector in the economy and provides great currency revenues. Uzbekistan is the world's second largest exporter of ginned cotton and is among the world's ten leading countries manufacturing raw silk. It is the acknowledged leader in karakul production. About 40% of the economically active population are involved in agriculture. The share of agricultural lands is 57% of the total area. In the structure of agricultural lands the dominating are pasturelands (81%). Approximately 15% of all lands (more than 4 mln ha) are under irrigation. At present, the dekhan farms provide 60% of the gross product (in the past their share was 18%). Uzbekistan is seeking food independence, so the structure of cultivated areas has changed: the share of lands under grain crops has increased to 40%. As a result, by the late 1990s, grain production reached 4.3 mln tons. In addition to wheat and barley, rice and corn are grown, but the main crop remains cotton, which is cultivated on irrigated lands. In the 1990s, the areas under cotton was shrinking. Today, the production of ginned cotton is 3 mln tons, out of which approximately a fourth is exported. Also cultivated are tobacco, ambari (its fiber is used for manufacturing of coarse fabric and ropes); production of sugar beet has been revived. One of the most ancient branches of agriculture in the country is fruit growing, the method of melon cultivation (similar to vine cultivation) having been practiced for no less than 20 centuries. The most popular fruits grown here are apricots, peaches, apples, pears, quince, and nuts as well as such subtropical varieties as figs and pomegranates. Among prevalent vines are table varieties, including sultanas and raisins. Some urban farms in Tashkent, Samarkand, and Namangan Regions grow lemons in trenches.

In sheep husbandry, the main branch is karakul sheep rearing, which is practiced mainly in the west of the country. The sheep of half-fine fleece and meat-lard breeds are grazed in the piedmont areas and mountain valleys. In oases, milk-meat cattle breeding is practiced using the legumes that are grown here as forage. The most ancient and rather efficient branch is sericulture using a rather specific forage – mulberry leaves. It combines well with the growing of cotton, vegetables, and melon crops and with horticulture.

Uzbekistan possesses a well-developed transport complex. In terms of cargo transport, the most prominent is automobile transport (91%). The country is also seeking to more actively pursue automobile transit traffic through its territory for transportation of cargo from China via Almata – Bishkek – Tashkent – Turkmenabad (Chardjou) – Ashgabad and further on by ferries to the Transcaucasus and Europe. The total length of automobile roads is 84.4 thou km.

Of great importance is also the railways, the share of which in overall cargo transport is more than 5%. The total length of railroads is 3655 km, of which 13% are electrified. The density of the railroads is the highest in this subregion – 77 km per 10,000 sq. km of the territory.

The pipeline network was created in the Soviet times. Major gas fields are linked by the main gas pipeline, Central Asia – Center of Russia. In addition, two gas pipelines are constructed from Bukhara to the Urals. One more pipeline runs from Bishkek to Almata via Tashkent. Oil pipelines are only of local significance.

In the Soviet period, Tashkent was the main airport of Central Asia. Only this airport had transit flights to foreign countries. Today, each of the Central Asian states seeks to participate in international air transport, so Tashkent has lost its former significance.

The role of river transport is not great, although the Amudarya within the country is a navigable river.

Uzbekistan has a developed telecommunication system. In 1998, the first stage of the national segment of the Trans-Asian-European fiberoptic communication line (885.6 km) long was commissioned.

Uzbekistan exports cotton, gold, natural gas, mineral fertilizers, metals, and textiles. The main imported items are machinery and equipment, chemicals, and metallurgical products. The main trade partners are Russia, European countries, Belorussia, and Kazakhstan.

The works of the pre-Islamic period have survived to modern times (Avesta and the Legends about Siyavush). Many sacred books of Islam were created on the territory of Maverannakhra (Interfluve). Treatises of medieval scientists, such as al-Khorezmi, al-Ferghani, Avicenna, Biruni, and Ulugbek, are still used by modern science.

Achievements in handicraft, architecture, and science in the Tamerlan epoch were the apex of the medieval culture of the Moslem Orient.

Resolution of the USSR Supreme Council "On Implementation of the Resolution of the USSR Supreme Council 'On Urgent Measures for Environmental Improvement of the Country' on the Aral Sea Problems" (March 4, 1991) – recognized that the "Aral problem, the most serious environmental disaster on the globe, had become most acute." In general, "the environmental situation in the region has become uncontrollable." It was also noted that the measures outlined in previous resolutions on the Aral had been, in fact, fulfilled.

Restoration of the Sudochie Lake Marshlands – activities that were carried out in the period from late 1999 to late 2002 within the framework of the WEMP Project ("Water and Environment Management Project") in the line of the Global Environment Facility (GEF). A hydrodynamic model was built that was used for design of a complex of structures, including:

- a dam on the Akkumsky ridge with a design water level of 52.2 m;
- a regulator of a water outlet structures on the Akkumsky ridge with a maximum carrying capacity of 52.8 cu. m/s and a depth 2.5 m over the foot level (49.5 m);
- a cutoff header KS-3A for collection of drainage water on the northern side of the Raushan farm with the a design water flow of 4.5 cu. m/s;
- a pumping station (Raushan), located approximately 8 km to the northwest of the Raushan farm at the end of the KS-3A header, to include 6 pumps with total carrying capacity 3.0 cu. m/s;
- a head water intake at the Ustyurt header to supply river water to the Greater Sudochie Lake with a maximum carrying capacity of 55 cu. m/s;
- an updated diversion canal from the hydroelectric complex at Altynkol to supply the required quantities of river water from the Raushan canal to the Ustyurt canal with a maximum carrying capacity of 34 cu. m/s;
- a linking canal between the Kungrad header (*KKS* – see) and *Akushpa Lake* (see).

The whole territory is affected by the flow regimes of the Amudarya River near Takhiatash, the flow regimes of the Main Left-Bank Header (GLK) or KKS, and the flow regimes of the Ustyurt header and the Raushan irrigation canal. The principles of water management facilitate development of optimal conditions in terms of the appropriate environmental functioning of the Sudochie Lake marshlands.

Although every year differs from the previous one, the first principle of water management is as much maintenance as possible of the design water level at 52.5 m over the sea level and as long as possible during a year. Only a strictly portioned quantity of water needed to raise the water over this level should be released from tributaries. The maximum water levels should be ensured in early spring (in the spawning period) and during the whole winter season (for fish survival). With the existing water supply regime, approximately 600 mln cu. m of water enter into the system via KKS and approximately 200 mln cu. m enter via the Ustyurt header.

Rudd *(Scardinius erythrophthalmus)** – fish of the carp family (*Cyprinidae*). Its length is up to 36 cm, weight – up to 1.5 kg. It lives mostly in the basins of the Baltic, Black, Azov and Caspian Seas, in the basin of A.S., the Chu River. Prefers lakes. In the rivers it keeps in oxbows, backwater with canes and water vegetation. It reaches maturity on the 3–5th year reaching a length of 12 cm. The spawning period lasts from April to June in water with temperature $+18^{\circ}C$. In the A.S. basin it spawns both in fresh and saline water. It lays eggs on aqueous plants. It feeds on thread algae, young cane sprouts, eggs of mollusks and larvae of insects. Its commercial significance is not large. Caught with seines and nets.

Rybatsky Bay* – carved for 12 km southward between the shore formed by the *Amudarya delta* (see) and *Muinak Island* (see). The shores were low, and overgrown with the cane thickets. In the northwestern part of the bay on Muinak Island, the shore was elevated. At the entrance to the bay and in its northern part, the depths were 6–7 m; to the south the bottom was smoothly rising. The Anjeyur-Uzyak arm along which ships once sailed to the Amudarya flowed into R.B. In the south of the bay was an entrance into the Parokhodnaya tropa (Ship's Path) arm that was divided into 2 channels, the eastern of which led to the Amudarya, while the western led to the *Motornaya tropa* (Motor Path) *channel* (see). In 1993 it ceased to exist.

Rybnitsa (Motorybnitsa) – deck sailing or motor boat up to 20 m long with a carrying capacity up to 60 tons. R. may be found on A.S. and on the Caspian Sea. Some R. were used as permanent bases anchored in the sea to service fishing boats. P. were also used for transportation of fresh or slightly salted fish to the processing plants.

S

Saberfish* (***P. Culturatus (L.)***) – anadromous, schooling fish of the carp family (*Cyprinidae*). Its body length is up to 50 cm, and its weight is up to 500–600 g. Lives mostly in the lower reaches of rivers and reservoirs. Propagates at the age of 3–4 years. In spring it runs for spawning to rivers from early May to mid-June. The fish eggs are pelagic. Fertility is 10–58 thou eggs. The fries feed on zooplankton, the adult fish, apart from zooplankton, feed on air insects and fish larvae. In the sea, it made regular daily vertical migrations. Most widespread in the basins of the Baltic, Black, Azov, Caspian, and Aral Seas. A commercial fish, it is caught during spawning runs.

Fig. 42 Saberfish

Sai (Turk.) – pebbles, pebble deposits; shallow, dry channel; gully, spring in a gully; river.

Salinity of the Aral Sea – an important characteristic of its regime. It depends on the nature of the water and the salt balance of the sea, the regimes of river flow, atmospheric precipitation and evaporation, water exchange between the sea and the neighboring territory, as well as hydrological and hydrochemical processes in the sea proper. The horizontal and vertical distribution of S. in the sea was influenced significantly by the morphological peculiarities of the depression, its type of circulation and mixing of waters, and the processes of ice formation and melting. The results of regular monitoring of S. by hydrometeorological stations and surveys in the open sea during expeditions, including century stations and profiles, obtained by 1985 were generalized in the

I.S. Zonn et al., *The Aral Sea Encyclopedia*, DOI 10.1007/978-3-540-85088-5_19,

monograph *"Aral Sea"* (see). The average many-year S. in the period of the sea's natural regime varied from 9.4 to 10.5%, with the minimum figures in the regions of river inflow. The annual maximum was observed in winter during intensive ice formation in shallow areas, in particular. In spring, during intensive thawing, the surface water layers became less saline. In the periods of maximum evaporation, S. became higher in shallow zones. Beginning from 1961, the general growth of S. was recorded by all stations. Especially significant growth was registered after the 1970s. In this period, the range of annual S. variations increased. The highest rate of water salinity growth was observed in the Small Sea; however, in recent years, water S. in the Small Sea was 3-fold lower than in the Large Sea due to its gradual restoration. After the sea level dropped below the threshold separating the Small and Large Seas, the most intensive increase of water salinity was witnessed in the eastern part of the Large Sea. The average annual S. of the sea increased annually in 1960–1970 by 1.8‰, in 1979–1980 by 5.5‰, and in 1981–1990 by 16‰. According to the results of single sample analysis made in 1999, the water salinity in the Large Sea nearby Aktumsuk was 62–68‰. The Small Sea, unlike the Large Sea, gradually became less saline; the average-weighted salinity in 1999 was equal to 13–15‰, while in the area of the Syrdarya inflow, it was only 1.5–8‰. In 2006, S. in the western part of the Large Aral reached 100‰, while in the eastern part it reached 150‰.

Salmon Aral (*Salmo*)* – migratory fish of the *Salmonidae* family. Large (up to 1 m long) and weighing 13–14 kg. It was found in the eastern and southern parts of A.S., in the Amudarya (as far as Turtkul) in particular. It had no commercial significance.

Salt drift, salt-dust drift – in 1960, the Aral contained 10 bill tons of salt, of which the most abundant were sodium chloride (56%), magnesium sulfide (26%), and potassium sulfate (15%). With the sea drying, these salts were deposited on the seabed. Among the causes of this were capillary rise and subsequent evaporation of highly saline ground waters along the dried coast and the seasonal variations of the water level facilitating sedimentation due to evaporation as well as winter storms blowing sulfates over the coast that were deposited due to the action of low temperatures on water. More than 28 thou sq. km of the seabed dried that in the period from 1960 to 1989 were covered with salts. Here, we speak about formation of the sandy-solonchak barren lands. Unlike earlier definitions that led to misunderstanding of the geochemical processes of drying, the salted Aral contained in the form of sediments not only calcium sulfates and carbonates, but also sodium and magnesium chlorides and sodium sulfates. Due to a high concentration of toxic salts in the upper soil layers, lack of nutrients, and fresh water, the dried out seabed could not be used for natural or artificial restoration of the vegetation cover.

The serious problem was drifting of salts and dust with wind from the dried seabed. By late 1970, the predicted formation of a crust of sodium chlorides that would prevent or mitigate the deflation process did not occur. Although storms occurred over the whole sea area, only along the northeastern and eastern coast of the sea did enormous salt strips up to 100 km wide appear; storms, however, occurred over the whole sea wherever the dried seabed was exposed. Lightweight particles rose to a height up to 4 km and were deposited to distances up to 400 km away. For the first time, the dust storms were registered by Soviet cosmonauts in 1975. In the period from 1975 to 1981, Soviet scientists on the basis of interpretation of space photos registered 29 large storms. From 1966 to 1985, the storms transferred dust and salts to the Amudarya delta, and during 60% of this period the northwestern winds prevailed, during 27% eastern winds prevailed, and during 12% the southern winds prevailed. The greatest number of days with dust and salt storms (1299) was recorded in Aralsk on the northern sea coast, followed by Muinak (965) on the southern coast. Estimates of dust and salt drift varied from 13 to 231 mln tons a year, but most probably it was from 40 to 150 mln tons. Measurements carried out in 1977–1985 have shown that about 43 mln tons of salt and dust were transferred annually from the dried seabed to neighboring territories: data of 1981–1986 provided by the meteorological stations in the Amudarya delta gave the figure of 9.5 t/ha or less than the half of the previous estimates. Dominating in the salt-dust drift were calcium sulfates, but there were also great quantities of sodium chlorides, sodium sulfates, magnesium sulfates, and calcium bicarbonates. Sodium chlorides and sodium sulfates are rather toxic for plants, especially in the blossoming period. Despite the expanding area of the dried seabed, by 2000 some decrease of the quantity of drifted salts and dust was observed at 39 mln tons a year, a result of depletion of deflation material, the leaching of salts over the soil profile, and the hardening of previously saline territories.

By 2006–2007, the water level in the Aral had dropped by more than 24 m. The size of the water surface area became twice as small. The area of the dried seabed was over 30 thou sq. km. Intensive development of the salt buildup processes and the eolian transformation of the dried seabed caused the formation of potent sources of salt and dust drifts affecting the natural environment. Their negative impact may be attributed to the following natural factors: the relief structure of the dried seabed, the lithological composition of bottom sediments, the soil salinity in the deflated layer, the level and salinity of ground waters, the temperature and wind regimes, the dynamics of eolian processes, the nature of overgrowing of the exposed seabed, and the remoteness of the sea recess.

Samarkand – ship built by the Belgian Company "Cockerill" from an order of the Shipbuilding Department of the Russian Marine Department to strengthen the Aral fleet. The body of the ship was delivered to Saint-Petersburg in 1866, and three years later it made its first voyage over the Syrdarya. In 1879, commanded by Captain-Lieutenant Bryukhov, S. went as far as Khodja-Solar, thus proving the possibility of navigation in the upper reaches of the capricious river. In 1881, near fort Perovsk, the ship was ran aground, and all attempts during nearly a year to lift it were in vain.

Sandal Bay* – located near the eastern coast of A.S. 7 km to the south-south-west of *Kashkynsu Bay* (see). On the north, the bay was covered by the low, sandy islands Ulken-Sarysh and Kishkene-Sarysh. The shores of the bay were highly incised. In the central part the depths reached 6 m.

Sarychaganak, Saryshiganak, great Sary-Chaganak Bay* – located in the north-east of A.S. It was connected with the Small Aral Sea via a strait up to 20 km long. The bay area (at a water level of 53 m) was 1135 sq. km, its the greatest length was 48 km, its width was 35 km, and its average depth was 9.2 m. An ice cover formed here in late November, was broken in mid-April, and by the end of April the whole bay was free of ice. In July-August, the water temperature in the bay reached 23–26°C. The average salinity level until 1961 varied from 9.6 to 10.3‰. By 1989 the bay area had shrunk to 50 sq. km.

Sarykamsy depression (Turk "*sary*" means yellow, "*kamysh*" means cane) – one of the largest drainless depressions in Asia. Located 200 km to the south–west of A.S. The Persians considered it "the hell of the Earth." S.D. represented a geographical paradox: a heavy, rotten bog in the middle of a desert. It was found in 1876 by geologist N.G. Petrusevich. Its length was about 150 km, and its width was about 90 km. The minimum bottom elevation was 45 m below sea level. It represented a flat oval bowl filled with the sediments of an ancient lake and was covered by solonchaks and eolian sands.

S.D. was located near the feet of chinks of the southeastern Ustyurt, while the Assake-Audan Depression and the Northern Bay of the Sarykamysh incised into the Ustyurt surface and were connected with S.D. only via narrow straits at absolute elevations of 45–50 m. Its northern and central parts were located within the Assake-Audan sag, and linked with the Upper Uzboi sag. The area of S.D. was approximately 15 thou sq. km. Its steep western slope passed into the chink of the Ustyurt, while the flatter eastern slope merged with the surface of the Circum-Sarykamysh delta. From time to time, S.D. received the waters of the Amudarya, turning into a lake. When the river turned to A.S., the lake dried out. S.D. was a lake in the Late Neogene, and Upper Quarternary Period (its level reached 58 m abs.). For the last time the waters of the Amudarya reached the depression in 1878.

S.D. was mentioned in the novel "Djan" by Andrey Platonov.

Sarykamysh Lake (Uzbek "*Sarikamish kuli,*" Antique "*Meotyi,*" sometimes "*Scythian bay*") – located in the northwest of Turkmenistan and in the south of Karakalpakstan in the Uzbek Republic in the depression of the same name, within the Amudarya basin. This was a large evaporation lake in the plains of Central Asia. In the second half of the 1st millennium B.C., the irrigated lands extended enormously with the development of the antique culture in Khorezm and other regions of Central Asia. As a result of the reduction of the flow, the Aral-Sarykamysh basin was divided into two lakes, one of which, Sarykamysh, dried out quickly. At the turn of the 4th and 5th centuries, when after the Sasanide-Hionite wars irrigated farming degraded, Sarykamysh Lake was restored for a short time. After the Mongol invasion in 1221, the lake in Sarykamysh revived again. But as a result of five marches of Timur to Khorezm (the last was in 1388) when not only cities, but irrigation systems were destroyed, Sarykamysh again ceased to existence. The anthropogenic stage in the lake regime was related to land reclamation development in the Khorezm Region (Uzbekistan) and Dashoguz velajate (Turkmenistan). The construction of main drains diverting drainage waters from irrigated lands and their connection to the ancient channel of the Amudarya, the *Daryalyk* (see), ensured a supply of 5–6 cu. km of water to the lake annually. Beginning from 1961, by a recharging regime, it was transformed into an irrigation-discharge lake. In 1967, the lake received up to 7 mln tons of salts due to the salinization of reclaimed lands; in 1981 the salt input increased to 21.2 mln tons. The area of the lake had grown significantly: in 1963 to 103 sq. km, in 1975 to 1450 sq. km, in 2000 to 2575 sq. km. Common carp and bream appeared in the lake, but they were poisoned by herbicides, defoliants, and pesticides washed off from irrigated cotton fields.

Saxaul (*Haloxylon*) – the gender of plants of the goosefoot family. It includes small trees (up to 12 m high) or shrubs with segmented shoots and leaves in the form of scales or mounds. It is found in semideserts and deserts of Central Asia. Most widespread are black and white S. The black S. may be as high as 3.5–6 m, and grow on loamy, sandy, and saline soils, forming a kind of desert "forests." The white S. grows only on sands in the form of thinned and ground thickets. The wood of S. is fragile, very compact, and used as fuel. The best shashliks are cooked on S. wood.

Seas – the fundamental reference publication in the series "Nature of the World" by Doctor of Geography B.S. Zalogin and Professor A.N. Kosarev. It was published in Moscow by the "Mysl" Publishing House in 1999. It contains descriptions of seas and among them the comprehensive geographical characteristic of A.S.

Secretariat of the Interstate Coordination Water Management Commission (ICWC) – located in Khodjent (Republic of Tajikistan). This was a permanent body. It was formed at the ICWC Meeting on October 10, 1993 in Nukus. The Secretariat organized implementation of ICWC orders and preparations, and jointly with BVO "Amudarya" and "Syrdarya," programs, events, and draft resolutions at ICWC meetings. It prepared cost estimate sheets for operating expenses and capital construction for financing BVO by ICWC member countries. Among its tasks were accounting and reporting on implementation of the plan of operating works and capital construction, control of finance input by the ICWC member countries intended to BVO for operating works and capital construction and other needs in a current year, and coordination of international ties. The Secretariat had the right to control implementation of the decisions taken by BVO "Amudarya" and "Syrdarya."

Segizsai Cape* – the southern top of the *Shubartarauz Peninsula* (see). Located 13 km to the south-west of the *Vasiliya Cape* (see). The cape is high and hillocky.

Seleuli Island* – located between the southern tip of *Kenderli Island* (see) and the eastern shore of A.S. near *Karatma Bay* (see). The low, sandy shores of the island changed their configuration depending on the sea level. In some parts, especially in places inundated with water, it was covered with reed thickets.

Shalanash Island* – located in the western part of the *Small Sea* (see), 21 km to the south-southwest of the Torangly Cape. It extends from north to south and is separated from the mainland by a narrow shallow strait.

Shalpak Island* – found 18.5 km to the southeast of the *Agurme Peninsula* (see) on the eastern coast of A.S. The island is low and sandy, having in some places separate barkhans 3–4 m high. To the north of it *Manas Island* (see) is found.

Shapankalgan Island* – located 3.5 km westward of *Ushmurza Island* (see). It was small in size with low, sandy shores. In some places, Sh. was flooded with water. Its configuration changed depending on the sea level.

Shaposhnikova Kultuk* – located to the north of A.S. Incised into the low, sandy coast for 900 m to the north-northeast, 4.5 km to the west of the Aralsk Bay (see). The depth at the entrance into kultuk was 0.5 m, while in the kultuk proper it was 1.5 m. From the kultuk, the coast ran to the southwest and became more elevated.

Sharpray (*C. Kuschakewitschi* (*Kessler*)) – fish of the carp family (*Cyprinidae*). Its body length is up to 18 cm. Lives in the plain parts of rivers and in backwater bays connected with river channels. Spawning is in late April–early May. Fertility is 3–4 thou eggs. Eggs are glued to plants. Feeds on detritus. Found in the A.S. basin. No commercial significance.

Fig. 43 Sharpray

Shevchenko Bay (former Paskevich Bay)* – named in honor of Ukrainian poet *T.G. Shevchenko* (see), one of the participants of the *A. Butakov* (see) expedition to the Aral. It incised into the coast for 13 km to the north between the *Aiderli*

Cape (see) and the Torangly Cape. It was rather large and open in its southern part. From the Aiderli Cape to the top of the bay, the coast gradually lowered and in some parts became low and flat. Further to the south-west the coast was high and steep; cliffs came up to the water edge. The eastern coast of the bay was dissected by small shallow bays, most important of which in navigation terms was Terestubek Bay in which small vessels found shelter from storms. Depths of the bay were typically 20–27 m.

Shevchenko, Taras Grigorievich (1814–1861) – the great Ukrainian poet, artist, thinker, revolutionary democrat. He was taught to write and read by a village vicar. In spring 1838, Sh. was bought out from serfdom. The first known works of Sh. are dated 1837–1838. In 1840, his collection of poems, “Kobzar,” was published. In 1846, Sh. joined the secret Kirillo-Mefodyi society where he took the most leftist position. In April 1847 by the report of an agent, he was arrested and sent as a solider to the Orsk fortress (in the Orenburg goubernia), and in 1850, to the Novopetrovsk fortress on the Mangyshlak Peninsula (presently Fort-Shevchenko) for, as it was formulated in a sentence, “making poems in the Malorossian language of revolutionary content.” Approving the exile, Nickolay I added that he was to be “under strict supervision with no writing and drawing.” Sh. was in exile from June 1847 to August 1857, freed only after the death of Nickolay I. A trooper of the Russian army, he was included into the Aral expedition of *A.I. Butakov* (see) as an artist to make sketches of all coasts, islands, straits, villages, and local people. Sh. once said, “Sea is around and grief – in the middle”.

During the exile years, he wrote the following novels in Russian: “Dutchess” (1853), “Musician” (1854–1855), “Miserable,” “Captain,” “Artist” (1856). Earlier, he wrote “Naimichka” (1844) and “Varnak” (1845); after exile, he wrote “A trip with Pleasure and without Morale” (1856–1858).

After exile, Sh. was prohibited entrance into Moscow and Saint-Petersburg; however, his friends got him permission to live in Petersburg where he arrived in spring 1858. Here, he became friends with the authors of the journal “Sovremennik” (Contemporary), in particular with Chernyshevsky, Dobrolyubov, Nekrasov, and others. The “third police division” kept the poet under strict supervision. In summer 1859 when Sh. went to Ukraine, he was arrested and forced to leave Ukraine.

In 1860, a new version of “Kobzar” (most complete of three) was published.

Sh. was also known as an artist. In 1838–1845, he studied in the Petersburg Academy of Fine Arts in Bryullov’s class. On returning from exile, Sh. worked much as an engraver. In 1860, he was awarded the degree of academician for copper etching. He created some deep psychological portraits (“Self-portrait”, 1840–1841) and emotional landscapes of Ukraine and Kazakhstan.

The name of Shevchenko was given to a bay in A.S.

Shordarya (“saline river”) – the popular name of the main drain, construction of which was initiated in 1987 on the right bank of the Amudarya beginning from the Durkhandarya Region toward A.S. This main drain was designed to collect all waters diverted from irrigated lands in the lower reaches of the Bukhara oasis, Karshi Steppe, and Dashoguz oasis as well as in Karakalpakstan and discharge them into the Aral.

Shovel-Nosed Sturgeon – see Greater Amudarya Shovel-Nosed Sturgeon.

Shukurgan Bay* – located in the eastern part of A.S. to the north of the Agurme Peninsula (see). The bay was not large. It was shallow with depths not exceeding 2 m. The coast was overgrown with reeds.

Shurbartauz (Chubar-Tauz) Peninsula* – located on the north of A.S. between *Butakov Bay* (see) and *Shevchenko Bay* (see). The coast was largely elevated and steep. On the northern side, some areas with a low, sandy coast were found; the vegetation over the whole coastal zone was scarce. Depths of about 30 m were found south of the peninsula.

Shurcha – a city-fortress that existed in the 4th–3rd centuries B.C. It was located on the northwestern margins of *Nukus* (see) on the former borders of the ancient Khorezm State. The first references to the Nukus fortess were found in the mid-12th century. It represented a fortified structure with strong defensive walls forming a square with sides 50–60 m long and an area of 0.25–0.30 ha. This fortress controlled the waterway along the Amudarya. Hundreds of years later, the city of Nukus was constructed at the place of Sh.

Siberian-Aral Canal, Sibaral – according to feasibility studies for partial transfer of Siberian rivers flow to some regions of Central Asia and Kazakhstan, it was assumed that the canal would link the Ob basin with the A.S. basin. The river flow would be withdrawn from the middle reaches of the Ob and later on from the Yenisey (along the Kas-Ket rivers). The route of the Sibaral would have started near the Belogorie site on the Ob River, went along the left bank of the Tobol River, came over the water divide along the Turgai saddle to the Syrdarya near Djusaly, crossed the interfluve of the Syrdarya and Amudarya, and after the 2550 km joined the Amudarya between Tuyamuyun and Takhiatash. The carrying capacity of the canal at its head would have been 1150 cu. m/s. The water should have been lifted to a water divide by 7 pumping stations. The width of the canal was to be 200 m, and its depth was to be 16 m. 27 cu. km of water a year would have been taken from the Ob (the Ob flow into the Kara Sea is 316 cu. km a year).

Small Amudarya Shovel-Nosed Sturgeon (***Psendoscaphirhynchus hermanni***) – fresh-water fish of the sturgeon (*Acipenseridae*) family. Its length was up to 27 cm. It was met only in channels, usually in the same places as the *great Amudarya shovel-nosed sturgeon* (see). Fed on the water larvae of insects and eggs of other fish. Had no commercial significance.

Small Aral Sea, small Aral (sometimes Kazakh sea) – the northeastern part of A.S. was formed by the large bays Perovsky, Greater and Smaller Sarychaganak (Sary-Chaganak), and Paskevich (Shevchenko). It was separated from the other part, the *Large Sea* (see), by *Kugaral Island* (see), the largest longitudinally stretching island of the Aral. Water was exchanged between the Small and Large Seas largely via the narrow 15 km *Berg Strait* (see) between Kugaral Island and the eastern mainland coast. The strait that separated Kugaral from the western coast was shallow, no less than 2 km wide. It did not in any way influence the water exchange. The depth of S.A.S. was about 28 m (1960). The greatest depths (over 20 m) were registered in three depressions in the central part of the sea that were separated by small underwater uplands. They covered 15% of the whole sea area. The prevailing depths in S.A.S. were 10 to 20 m (on 44.1% of its area).

In autumn 1989, the shipping channel that was excavated for the passage of vessels between the Small and Large seas was completely silted and represented a chain of lakes. By spring 1990, the water level in S.A.S. started rising and the shipping channel became deeper. By spring 1992, the depth of the channel was already 2 m, its length was 5 km, and its width was 100 m. The overflow discharge was equal to 100 cu. m/s. In late July–early August, a temporary dike 1 m high was constructed in the Berg Strait. When in April 1993 the water in S.A.S. rose by nearly one meter higher the dike broke. The flow returned to its old channel where the depth was about 1 m; however, the flow from S.A.S. did not exceed 100 cu. m/s. The next was the dam that existed from August 1996 to April 1999. This was a non-overflow earthfill dam 12.7 km long. Its crest length was 4 m and its crest elevation was +44 m. The water level in the sea before the breakdown of the dam was 42.8 (±0.1 m). The breaching flood (300–500 cu. m/s) flowed over the dried seabed, filling local depressions, and by several small streams reached the Large Sea. To prevent further water losses to the Large Sea, the Kazakh authorities decided to construct a solid permanent dam, which was built in 2005.

The S.A. coast compared to the Large Aral coast was more populated, including several settlements such as *Aralsk* (see), Birlestik, Tastyubek, Aksepe, Akbasty, Karateren, Karasholan, and Bugun, that played a significant role in the economy of Kazakhstan. Water rise in S.A. caused an extensive growth of canes. Many small arms of the Syrdarya delta that had been dry were again filled with water. Many more birds not only landed in the delta on their migration routes, but started nesting there. The quantity of pelicans, swans, flamingos, cormorants, and various ducks (red-crested pochard, Australian pochard, tufted duck, teal, baffle-mallard, headed duck, and others) grew. The resumed gradual growth of salinity in the direction from the delta to the sea made it possible again for fish to leave the river for the sea for maturement. Due to a rather considerable residual flow from the Syrdarya into the Small Sea, water levels drop and growth of salinity stopped, and at present the environmental situation in the Small Aral may be considered rather favorable. Moreover, it produced a positive psychological effect on those who lived in the coastal zone because they now could see a possibility for partial rehabilitation of A.S.

Smaller Barsuki – see Greater and Smaller Barsuki.

Smaller Sarychaganak Bay* – incised into the upper part of the *Greater Sarychaganak Bay* (see) for 7.5 km in the northern direction. Both bays were linked via a narrow strait that on the east was limited by the shore forming the Saryshok Cape and on the west by the low and shallow Zhainak Cape. The shores of the bay were sandy and shallow. The depths in the central part of the bay reached 5.5 m. On the western coast, fisheries were developed.

Snakehead (*Ophiocephlus argus warpachowskii)* – fish of the snakehead (*Ophiocephalidae*) family, a freshwater representative of the fish fauna of the Far East. It was accidentally brought to Central Asia, an invasive in the Amudarya and Syrdarya basin. The length of its body is up to 1 m, and its weight is 7 to 8 kg. It prefers living in calm channels and overgrown lagoons with low oxygen levels. From time to time, S. gets to the water surface to have a gulp of air. It reaches maturity at the age of 2–3 years, having a length by this time of approximately 30 cm. It spawns in June–July. A male and female make a nest in the coastal zone. To this end, they push aside the vegetation, making a free-water zone approximately 1 m in diameter. It is predatory species, feeding on fish and frogs. It was a commercial fish.

Fig. 44 Snakehead

Socioeconomic problems of the Aral and circum-Aral area – the monograph prepared by the Council on Study of Production Forces of the Uzbek Academy of Sciences. It was published in Tashkent in 1990. The monograph assessed the effect of the A.S.'s drying on the economics and social life of the Circum-Aral area and provides validation for a system of actions aimed at mitigation of the negative socioeconomic consequences of environmental changes in this region. Special attention was focused on measures for improvement of drinking water quality and achievement of better economic structures in the Lower Amudarya.

Soil reclamation, land reclamation – radical improvement of soils with a view to increase soil fertility for a long period. Depending on physiographical conditions different groups and kinds of S.R. are recommended. Hydraulic land reclamation includes irrigation and drainage of lands; chemical reclamation includes lime treatment of acidic soils, argillization of solonets soils, desalinization of saline soils, and acidification of alkaline soils; physical reclamation includes removal of stones, sanding of clay soils, argillization and colmatage of light and peaty soils, etc.; biological reclamation includes afforestation of sands and gullies, planting of forest shelterbelts and sodding of eroded and deflated soils, etc.

Solonchak – (1) Relief lowerings or a bottom of a dried temporary lake covered with a clay crust and a dust layer. It forms in areas of high occurrence of saline ground waters (about 1.5 m deep). When a lake dries out, groundwaters rise upward by capillaries and form a wet solonchak that is partially or completed covered with a salt layer. If the depth of groundwater is less than 1.5 m, the water partially wells up by capillaries and evaporates, and the salt, crystallizing in the clay crust of the drying lake bottom, loosens this crust and forms a puffy solonchak. If the groundwater level drops below 1.5 m depth, the water stops welling up by capillaries and a takyr is formed. (2) A group of saline soils covered with salt efflorescences, salt crusts, or puffy heavily saline surface horizons. It is formed in different geographical zones, but is most often found in semideserts and deserts. After land reclamation, solonchaks may be suitable for farming. With incorrect irrigation practices, secondary solonchaks are formed.

Fig. 45 Solonchak (www.dryland-biodiversity.de/aralsea/Aral7.jpg)

Sorkol (Surgul) Lake – located between Muinak city and the Ushsai settlement, this lake dried out. Its bottom was composed of black silt similar by its composition and curative features to the silt of the Karaumbet Lake (Ustyurt).

Southern circum-Aral area – covers the Amudarya delta between the Mezhdurechensk reservoir in the south to the A.S. coast in the north, and from the chink of the Ustyurt Plateau in the west to the Kyzylkums in the east. The climate here is sharply continental with a dry, hot summer (up to 44–45°C) and a cold winter (to −30°C), small amounts of precipitation (130 mm/year on the average), and high evaporation (up to 1200–1600 mm/year). The relief of this area is characterized by the absence of large uplands and depressions. The water bodies in the delta had rather large water areas and were usually shallow. The margins of the lakes were completely overgrown with semi-aquatic vegetation and had no a clear-cut coastline.

Before the 1970s, the water bodies in the Circum-Aral area existed only due to the Amudarya flow. At that time, the water salinity in the lakes did not exceed 0.4–1.25‰. Beginning in the 1960s, however, the water inflow into the river delta was gradually reduced and by the late 1980s it reached only a few percent of the average many-year figures. As a result, many freshwater lakes in the delta disappeared. At the same time, a number of main drains were constructed in Karaklpakstan which led to formation in the Circum-Aral area of a new type of water body: terminal through-flow and drainless water bodies amassing saline drainage waters.

The through-flow nature of the deltaic lakes guaranteed the maintenance and restoration of the water quality, as the original halting of the through-flow is what invariably led to a sharp rise of water salinity and water quality deterioration. In 2000-2001, the Amudarya flow was the lowest in the whole history of hydrological observations. The reduction of the river flow started from April 2000 and lasted until spring 2002. Such extreme water scarcity caused drastic reduction of the drainage flow. The drainage flow in the main drain decreased from 568 mln cu. m (average many-year value) to 46 mln cu. m; the water salinity increased to 9.7–14.4‰. The main drain and MN-1 dried out completely. Some quantity of water was still available in the downstream of MN-3, but there was no flow. The water salinity was as high as 13.2 ‰. Because of the water deficit, the water bodies in the circum-Aral area lost their through-flow nature that, in view of high natural evaporation and the lack of water inflow, led to a complete shallowing and even drying of many of them. These were the Mashakol, Khodjakol, Ilmenkol, Akushpa, Begdulla-Aidin, Greater Sudochie, Makpolkol lakes, Mezhdurechensk reservoir, and Zhyltyrbas Bay. In the remaining water bodies, such as Taily and Karateren lakes, and the Muinak and Rybachyi bays, the water area and water depth decreased significantly with a resultant water salinity growth up to 14‰ in bays and up to 50–60‰ in lakes. The most vivid example of the negative effect of the water shortage was the environmental situation in the *Sudochie wetland* (see), the largest lake system in this area. Up until 2000, the water surface area of lakes in this wetland was as large as 42 thou ha, while by the end of 2001, it had shrunk to 6.5 thou ha. In Akushpa, the biggest lake in this system (11600 ha), water salinity by 2001 had reached 90–100‰, and by the end of this year, the lake dried out completely. The growing water salinity caused degradation of the originally freshwater-brakish fauna and flora of the lakes, and their replacement with brakish-marine species with a progressive dropping of general bioproductivity. The final drying and salinization of the lakes led to the death of aquatic biota. The ichthyofauna of the lakes deteriorated everywhere, and the productive fish species, such as silver carp, grass carp, and common carp, were replaced with less productive species – crucian carp, roach, and low-value trash fish.

Southern circum-Aral area: new perspectives – monograph edited by Professor V.A. Dukhovny and engineer J.L.G. de Schutter. It was prepared in 2000–2002 within the framework of the project, "Integrated Water Resources Management in the Aral Sea Basin for Water Surface Restoration in the Southern Circum-Aral Area," using a NATO grant and receiving contributions from leading specialists of the *Scientific-Information Center of ICWC* (see), *SANIIRI* (see), "ECO of Prearalie" and "Aral Consulting." The book discusses ways to address the Aral problems on the basis of an integrated approach, taking the restoration of wetlands (near the *Sudochie Lake* – see) and some smaller water bodies as case studies. The monograph was published in 2003 in Tashkent, Uzbekistan.

Southern Karakalpak main drain (SKMD) – a project began during Soviet times, then revisited by the Uzbek Republic in 2000 due to the wide-scale development of soil salinization in three southern regions of Karakalpakia: Turtkul, Ellikalin, and Biruni. The SKMD route would run only over Karakalpakstan territory. It was designed to divert drainage waters along the old channel of the Akcha-Darya and to discharge them into the Zhana-Darya (the Syrdarya's old channel). Pursuit of financial support from the World Bank for construction of this drain has been discussed. The expected outcome was that diversion of drainage waters from the three principal agricultural regions of Karakalpakstan into the Amudarya main channel would stop and that the average water salinity would become 0.2–0.3 g/l less.

Southern seas (Aral, Caspian, Azov and Black) affected by anthropogenic stress – monograph of V.I. Kuksa that appeared in 1994 and shows that in recent decades intensive economic activity in the southern sea basins have resulted in serious disturbances of the whole complex of natural conditions, primarily including unfavorable changes in the hydrophysical, hydrochemical, and hydrobiological regimes of the seas as well as steadily growing damage to the economics of littoral countries. The author provides a detailed description of the changes observed in the recent four decades that affected both the regime, especially of the Aral, and natural conditions of the coastal zones. Stressed is that the main issue to be addressed concerning A.S. is elaboration of scientifically valid actions to ensure preservation and rational management of the natural wealth of these water bodies.

Southern white-eye – commercial fish of the carp family (*Cyprinidae*). The body is up to 40 cm long. It was found in the basins of A.S. and the Middle and Southern Caspian. Its main habitat is sea, but it runs to rivers for spawning. It reaches fertility at the age of 3–4 years when its body length is over 15 cm. The spawning grounds are usually located close to the mouth. It lays eggs in spring when the water temperature is 12–15°C and feeds on animals and plants. Its main food is filamentous algae. It is fished in autumn.

Special ICID work team on the Aral Sea basin, ST-Aral – created in 1994 within the framework of the *International Commission on Irrigation and Drainage (ICID)* (see) following the decision of a special session devoted to the problem of A.S. in Varna (Bulgaria). Among the tasks of the group were collection and exchange of information about irrigation and drainage of the Aral Basin (defining how ICID member countries faced such problems) and interest in exchange of information with the Aral basin countries; outlining the most important problems for detailed discussion by technical groups of ICID; and organization of discussions at the international and national levels in the Aral basin countries of the scientific and technical aspects of irrigation and drainage systems in the A.S. basin. A special technical session on the Aral basin problems was held in Ljubljana, Slovenia in April 1996. The Iranian National Committee of ICID convened a workshop and technical trip on the subject, "Aral and other problems of the Aral Sea," in Mashad, Iran, in 1997. The Indian National Committee of ICID organized the seminar, "Drainage and water resources management, including application of saline waters in agriculture and agroforestry on saline soils," for experts of the Aral basin in 1999. The Spanish National Committee of ICID organized a technical trip for experts of the Aral basin in 1998 to organizations involved with irrigation.

Sudochie Bay, Sudochie Lake – in the past this was a vast intra-deltaic shallow basin occupying the Aibugir Depression (the *map of A. Butakov* (see) showed the waterlogged Aibugir or Laudan Lake) extending for 24 km to the southeast from the coast to the southwest of the *Urga Cape* (see). Dense reed thickets separated S.B. from *Adjibai Bay* (see). This was the largest inland water body in the Amudarya delta. Its water surface area reached 350 sq. km, its average width was up to 15 km, and its length was 250 km. The average depth was 2 m. Water salinity ranged from 0.6 to 1.7‰. In the 1960s, due to the drying of A.S. and the actual stopping of the water inflow after damming the Amudarya arm, Raushan, the lake became shallow, and by 1968 was broken into small water bodies. In this period, the waters from the main drains were diverted into the

Fig. 46 Sudochie Lake in November 2007. Photo by Pavel Kosenko, http://pavel-kosenko.livejournal.com

lake, and from this time on, the fate of the lake was fully dependent on these waterways. At present, the wetlands of S.L. are represented by four water bodies – *Akushpa* (see), *Karatepen* (see), *Begdulla-Aidin* (see), and *Greater Sudochie* (see), with an area of 43.8 sq. km.

Sultanuizdag – the ridge in the Amudarya-Sarykamysh Region of Karakalpakstan with a maximum absolute altitude of 473 m. Its length is 50 km, and its width is 10–25 km. Its southern slope is steep, in some places cliffy, and heavily broken by sai, gullies, and gorges. The northern slope is more flat and less broken. It is composed of ancient Paleozoic dislocated rocks like quartzites, gneiss, marble, and granitoid intrusions. S. is surrounded by plains with dry riverbeds. To the south of the ridge, the plains have absolute elevations of 100–150 m, while to the north they go under eolian sands of the nearby upland and reach elevations of 200–250 m. In this region, erosion and denudation processes in mountains and erosion-accumulative processes on piedmont plains dominate. S. divides the Karakalpakstan territory into the northern and southern zones, which differ by climatic conditions.

Sulu Bay* – located east of Uyaky Island in the northeastern part of A.S. Two passages led from the sea to the bay – the northern and southern. The northern passage was found between *Altai Island* (see) and *Uyaly Island* (see). It was shallow, with prevailing depths of 1.4–1.6 m. The southern passage is wide and straight, and up to 6 m deep.

Syrdarya (Antique – "*Yaksart, Silis,*" Ancient Greek "*laxártēs,*" in the Middle Ages Arabian "*Seikhun, Djeikhun*") – the largest river in Central Asia, and second after the Amudarya in terms of water availability. It was formed by the confluence of the Naryn and Karadarya Rivers in the eastern part of the Ferghana Valley and flowed into A.S. The Naryn was formed from the confluence of the Greater and Smaller Naryn, its headwaters being located in the Northern Tien Shan. The Karadarya was formed from the confluence of the Tara and Karaduldji Rivers, collecting water from the slopes of the Ferghana and Altai ridges. The Syrdarya is 2137 km long (3019 km from the Naryn origin, Arabelsu). The S. basin forms a part of the A.S. basin. The basin area is 443 thou sq. km, of which 219 thou sq. km are covered by watershed (mountains), amounting to 32% of the whole territory of Central Asia. The S. flow is formed largely in the mountains (upper reaches), while the mid-channel runs over steppe regions that are replaced in the lower reaches by the sands of the Kyzylkums. In its upper reaches, the S. waters are used for irrigation of the Ferghana Valley. Most of its tributaries did not reach the river, forming a debris cone with a fan-shaped irrigation network. The largest of the S. tributaries are: left – Isfairam, Shakhimardan, Sokh, Isfara, Khodjabakirgan, and Aksu; and right – Padshata, Kassansai, Gavasai, Chaadaksai. None of them reach the river because their waters are diverted for irrigation. The main water resources of the S. basin were formed in the Ferghana Valley. In 1948, the Farkhad hydropower plant was constructed at an outlet from the Ferghana Valley where S. cut through the Farkhad Mountains. This hydraulic unit included the head structures of the Dalaverzin (on the right bank) and Golodnaya Steppe (on the left bank) irrigation systems, after which the river flowed over a wide valley which borders downstream Chirchik and are not clearcut. Running out of the Ferghana Valley, S. received right tributaries Angren, Chirchik, and Keles, which supplied water to the complicated irrigation network, and still lower, from the last tributary, Arys. The most water abundant rivers in the S. basin are Naryn and Chirchik. Downstream, at Chardara, where the Chardara reservoir was constructed, S. passed over the eastern margin of the Kyzylkum sands where the riverbed and low, flooded banks are composed of easily eroded loess-like loams and sands. Downstream, at Kyzyl-Orda (near Karaozek station), nearly half of the river flow passes into the Karaozek arm branching off from S.

S. has mostly snow and, to a less extent, glacier and rainfall recharge. The water salinity in S. before the Ferghana Valley is about 1‰ and in downstream stretches it reaches 3.5‰. The S. regime after the confluence of the Naryn and Karadarya is close to the regime of other rivers with glacier-snow feeding: a low-water period in October–March and the highest flow in June–July. The graph of annual flow variations reveals its cyclical nature, alternating between low- and high-water years. Low-water periods occur every 3–4 years and last for 5–6 years in succession. The high-water periods are often only a single occurrence.

The total flow from the river watershed is equal to 38 cu. km, out of which 33.2 cu. km comes upstream of the Chardarya reservoir. Its greater part (over 60%) is formed by the runoff from the mountains. The flow of the Naryn, Karadarya, Chirchik, and S. in the stretch from the Toktogul to the Chardarya reservoir, a total length of 1000 km, is regulated by the Naryn-Syrdarya cascade of reservoirs. Five of them are most important: three upstream – Toktogul (design full capacity – 19.5 cu. km), Charvak (2 cu. km), and Andijan (1.75 cu. km); and two are in-channel of seasonal regulation – Kairakkum (4.2 cu. km) and Chardarya (5.7 cu. km).

In the S. basin are found 196 hydraulic structures, 225 km of canals of the interstate significance, and 190 water gauging stations. In addition, 9 hydropower plants with a total installed capacity of 3.72 mln KW are also found there. Water resources are divided among the main water users in the basin as follows: about 92% is used for irrigation needs, 3.4–4% for drinking and municipal water supply, 2% for industrial-technical needs, and 1.5% for the agricultural water supply.

By 2000, approximately 3 mln ha were under irrigation in the S. basin. The most extensive irrigated lands were found in the Ferghana Valley and in the middle reaches of the river (Golodnaya Steppe). In 1939, the Greater Ferghana Canal was constructed in the Ferghana Valley, receiving water from the Naryn and Karadarya. Till 1960 about 13.2 cu. km of runoff was disposed into A.S. every year. After the division of the A.S. into the Large and Small Aral Seas, the

S. flow went to the Small Sea, which is, at present, separated from the Large Sea by a dam.

After 1991, the S. basin was shared by four sovereign states – Kyrghyzstan, Uzbekistan, Tajikistan, and Kazakhstan. On the territory of the S. basin are found 6 regions of the Uzbek Republic: Andijan, Namangan, Ferghana, Tashkent, Djizak, and Syrdarya; two regions of Kazakhstan: Kyzyl-Orda and South-Kazakhstan; and one region of Tajikistan: Leninabad.

Large cities located on the river are Khodzhent (former Leninabad), Begovat, Kyzyl-Orda, and Kazalinsk.

Syrdarya Shovel-Nozed Sturgeon (***Pseudoscaphirhynchus fedtschenkoi***) – freshwater fish of the sturgeon (*Acipenseridae*) family. Its length (without caudal filament) was up to 27 cm. Its way of life was not studied. Lived in the plain parts of the river. Spawning was in spring (April) on stony ground. Fertility of approximately 1.5 thou eggs. Fed on larvae of aquatic insects. Had no commercial significance. Lived in the Syrdarya, but was found quite seldom.

T

Taijegen Island* – located in the eastern part of A.S., 15 km to the south–east of Kuzjatepes Island. The island is low, with a flat coast surrounded by the reed thickets. Fishery activities were practiced in its northern part.

Tailakjegen Islands* – belonged to the *Akpetkinsky Archipelago* (see). They were found in its western margins and represented a group of low sandy islands with broken shores overgrown with reeds. In some places, dunes and barkhans up to 10 m high were found. The islands were separated from each other by narrow straits with depths varying from 0.6 to 5 m. A large fishing base was found here. At present, they are linked with the mainland.

Takhiatash (means "stone tyubeteyka" *) – the youngest city in Karakalpakstan. It appeared in 1951 with construction of the *Takhiatash waterworks* (see). It received its name from the cape located nearby on the left bank of the Amudarya where a hill shaped as a "tyubeteyka" rises. The city has machine repair and concrete plants, food, and light industry factories.

Takhiatash waterworks – constructed in 1974 near the Takhiatash Cape on the Amudarya to ensure a reliable water supply of 930 cu. m/s, of which 457 cu. m/s flows to the right-bank canal Kyzketken in Karakalpakstan, 426 cu. m/s flows to the left-bank canals Lenin and Sovetyab in Turkmenistan, and 47 cu. m/s flows to other needs. At the same time, the irrigated area was increased to 650 thou ha. No pumps were needed to supply water for irrigation, and the need to clean irrigation canals of sediments became much less. A bridge crossing was constructed over the Amudarya. The average many-year river flow at the site of T.W. was 46 cu. km, with average monthly discharges ranging from 5760 cu. m/s (maximum) to 186 cu. m/s (minimum). The annual flow of suspended sediments was 120 mln tons, and including bottom sediments was 130 mln tons. T.W. is comprised of a sluice concrete dam, a structure to allow fish to pass through, an earth dam, water intake structures, settling basins, navigation locks, and a levee. Later, the construction of T.W. was discovered to have led to the drying of water bodies, the drying of pastures, and the degradation of *tugai* (see).

* Tyubeteyka – a skullcap.

I.S. Zonn et al., *The Aral Sea Encyclopedia*, DOI 10.1007/978-3-540-85088-5_20,

Fig. 47 Takhiatash waterworks

Takyr (Turk – "smooth, flat"). (1) Flat clay depressions in deserts of Central Asia periodically (seldom) filled with flood or storm waters. After water evaporation, T. turns into a hard crust, broken at drying by cracks a là parquet blocks and nearly devoid of vegetation. Its area varies from 1–2 sq. m to 10–12 sq. km. (2) Type of soils formed in the described conditions. These soils have two clear levels: the upper (up to 8–10 cm thick) composed of compact clay and laminated crust that do not contain salts and an underlying level composed of slightly-changed soil-forming ground. (3) The first Kazakh rock-opera devoted to A.S.

Tastaubek Peninsula* – located to the north of A.S., in the northern part of the Small Aral. It stretches for 7 km to the south-southwest of the western coast of *Kokturnak Island* (see) and restricted entrance into *Butakov Bay* (see) on the east. The peninsula is covered by low, sandy hills that in some places slant steeply towards the sea. A gradual lowering of cliffs was observed near the southeastern end of the peninsula, which was rimmed by shallows.

Tasty Island* – located in the eastern part of A.S. It is covered in the south by *Bozkol Bay* (see). The island was low and surrounded by shallow waters. On the eastern coast, a fishery was developed.

Terraces of the Aral Sea – traces of seven sea levels were found in the Aral Depression. They indicated that transgressions and regressions of A.S. have been observed through its history. L.S. Berg wrote for the first time about Aral terraces (the Aral coastline). In 1908, he found only one of them at an elevation of 54 m abs. Later, T. were discovered at different heights from the sea surface (up to 75–80 m). In addition, there were discovered some more underwater T. below the sea surface at depths down to 16 m (+35 m abs.). The most recent Late Aral T. (elevations +53–55 m) were found at many points along the Aral coast in the west, north, and east.

Numerous ancient Aral T. are found at different places and different elevations at A.S.: elevations 57–58 m found in the southeastern coast of the sea, 58–60 m – along the southeastern sea, 65 m – on the shores of the Small Sea and in the south-west of the Aral, 72 m – in the south-west of the Aral and westward of the Small Sea and nearby Aralsk, 72 and 73 m – on the Muinak Peninsula, near Aralsk and in the west of the Small Sea, 75 and 80 m – in the south-west of the Aral, etc.

Different authors suggest different dates of formation of the Aral T.

Territorial redistribution of the river flow – see Partial Transfer of the Siberian River Flow to the Aral Sea Basin.

The Aral Sea basin management model, ASB-MM – made on a compact disc (CD). Includes hydrological and socioeconomic models developed by regional specialists. It demonstrates the complexity of decision-making in Central Asia. The CD contains two versions of the management model for two different purposes: The model "ASB-MM" facilitates better understanding by the wide public of the A.S. problems and their solutions, while the model "ASB-MM Expert" is intended to show politicians what strategies should be applied and what the consequences might be. The management model prototype was considered by the UN Development Program. The IFAS Agency and GEF through the WEMP Project provided the financial support to develop the model "ASB-MM Expert." This product was produced in 2002 by Resource Analysis (the Netherlands) and SIC ICWC of Uzbekistan.

The Aral Sea crisis – generalized proposals on assistance in addressing the crisis prepared by the World Bank (WB) in March 1993 on the request of the Republics of the Aral basin – Kazakhstan, Kyrgyzstan, Turkmenistan, Tajikistan, and Uzbekistan – after the WB mission visited these republics in September 1992. The mission offered a program of likely assistance from the Bank, including the following interrelated components:

A. Restoration of the water level in A.S. within reasonable limits.
B. Restoration and development of the A.S. disaster zone.
C. Strategic planning and efficient management of water resources of the Amudarya and Syrdarya.
D. Establishment of organizations on planning and implementation of these programs.

The Aral Sea: diagnostic study for development of a plan of action for conservation of the Aral Sea – a report prepared in 1992 by an international working group of experts from USA, France, Germany, Poland, Australia, Czechoslovakia and the representatives of WMO, UNEP, and ILEC in response to the dramatic situation unfolding in the Aral Sea region. The purpose of this report is to provide UNEP with a comprehensive diagnostic document on the environmental, demographic, and economic problems underlying the A.S. crisis. To discuss this report, the Working Group met four times: in Nukus (Uzbekistan) in 1990, Moscow (Russia) in 1991, Almaty (Kazakhstan) in 1991, and Geneva (Switzerland) in 1992. The report was prepared using numerous publications from Soviet times and on the basis of consultations with academic, scientific, and investigation-design organizations of the Central Asian republics. The studies were supported by administration leaders of Karakalpakstan, the Khorezm Region in Uzbekistan, the Tashauz Region in Turkmenistan, and the Kyzylorda Region in Kazakhstan. The report comprised 10 chapters covering the following issues of the Aral basin: nature and history of development; manpower, natural resources, and economic potential; basic changes in land ecology; changes in the water-salt balance; changes in water ecology; environmental pollution; state of public healthcare; basic reasons of the Aral crisis; and conclusions and resume.

The Aral Sea: navigation, geographic, and hydrometeorologic essays – a small book written by sea captain P.P. Tum and published in Moscow in 1960. It provides a navigation and geographic description of A.S. This hydrometeorological essay presents meteorological characteristics, specific meteorological events, hydrological characteristics of the sea, and its hydrobiological phenomena. Later, this essay was included into the *"Pilot Directions of the Aral Sea"* (see).

The Aral Sea: Selected bibliography – a book prepared and published in English in Moscow in 2002 and edited by Professors Jacques Nihoul (Belgium), A.N. Kosarev, A.G. Kostianoy, and I.S. Zonn (Russia) within the framework of the *Project "Aral-Kum"* (see) under the Program "INCO-Copernicus-2." It contains 1540 publications beginning from the 19th century discussing a wide range of problems of A.S. the and Aral region, including historical and geographical descriptions; water balance and hydrology; thermal, salt, and ice regimes; water chemistry, geology, and biology; variations in the Amudarya and Syrdarya flows; salinity buildup processes; dust storms; flora and fauna degradation; the effect of sea drying on the climatic changes in the region; and others. The name of the publication is given in the original language and in English.

The Aral Sea: water, climate, and environmental change in central Asia – booklet No. 982 prepared by M.H. Glantz (USA) and I.S. Zonn (Russia) and published by the World Meteorological Organization (UN WMO) in Geneva in 2005 in both Russian and English. It consists of two parts. The first part describes the history and geography of A.S. and its surrounding territories as well as the consequences of the Aral crisis. The second part includes 40 color maps prepared on the basis of digitized bathymetric maps of the sea, indicating changes in the Aral Sea area and level for the period from 1960 to 2004.

Tiger Turanian or Caspian (*Panthera tigris virgata Illiger, 1815*) – one of 9 subspecies of a tiger (Amur, Chinese, Indochina, Sumatra, Bengal, Malay,

Fig. 48 Nihoul J.C.J., Kosarev A.N., Kostianoy A.G., and Zonn I.S. "The Aral Sea: Selected Bibliography", Noosphere, Moscow, 2002

Fig. 49 Glantz M.H., Zonn I.S. The Aral Sea: Water, Climate and Environmental Change in Central Asia, WMO Publ. N 982, 2005

Bali, Java and Turanian – the last 3 are extinct). The tiger is the honorary king of large cats, bypassing even the lion. The largest are Amur tigers (the so-called Ussurian): their males weigh more than 300 kg. The length of a tiger body without a tail varies from 1.3 m (in females) to 2.3 m (in males); the tail is approximately half as long. It's height at the shoulder is from 0.6 m to 1.1 m. The front paws have 5 toes each. The body is elongated and well-muscled, and the tail is long and evenly downy. The head is round with short ears and whiskers. The background color of the back and sides is yellow or orange, and the neck and belly are white or yellowish with brown or black cross strips. The typical coloring of a tiger helps him to remain invisible in thickets. It varies in different geographical regions, which is used to identify tiger subspecies. Tigers are an exclusively Asian species. Historically, the tiger habitat area extends from the Tien Shan piedmonts to the west along river valleys in Central Asia (Turkmenia, Uzbekistan, Afghanistan, Kazakhstan) to the Caucasus. In 1848, A.I. Butakov, in a letter to his parents, described a tiger hunting in the Syrdarya mouth. The Kirghiz people called T. "Djulbars" (Turkestan tiger). In his time, well-known Russian traveler P.A. Chikhachev wrote, "Tiger is a real curse of Perovsk and its surroundings. The devastation it causes is so great that the people often have to change their camping place to get rid of this terrible beast of prey which attacks their cattle practically near the gates to Perovsk." It is interesting to note that in his memoirs, *I.F. Blaramberg* (see) wrote how he met in the reed thickets of Yaskart (Syrdarya) "the Bengal, king of tigers" who often attacked people. By some data, the last Tiger was killed in 1949 in the Takhtakupyrsky Region of Karaklpakstan, while other sources claim the last tiger was killed in 1968.

Fig. 50 Turanian tiger (www.smoking-room.ru/data/pnp/anim_mort6.jpg)

Tigrovy Khvost Cape – the western tip of the Muinak Peninsula. Its shape is reminiscent of a tiger's tail. Tigers lived in these places in the early 20th century. In 1942, a hydrometeorological station was opened here that monitored the sea water level. It existed until 1979.

Tokmak-Aty Peninsula* – see Muinak Peninsula.

Tokuzarkan Strait* – located on the northwest of the Akpetkinsky Archipelago. It originated east of the *Tailakdjegen Islands* (see) and stretched for about 10 km, first to the south, then turning to the southeast within the *Akpetkinsky Archipelago* (see). The entrance into the strait was located between the *Tailakdjegen Islands* (see) and Mayajun Island. The depth of the strait was up to 8 m, but a 2–3 m depth prevailed.

Tolmachev Island* – located near the east coast of A.S., 24 km to the southwest of *Kendyrli Island* (see). This is a low, sandy island stretching as a narrow strip. It was often eroded by water and consisted of several parts. At present, it is linked with the mainland.

Transborder waters – in accordance with the Convention on Protection and Utilization of Transborder Waterways and International Lakes, TW are understood as "any surface or ground waters that define cross-borders between two or more states or are located on such borders; if the transborder waters flow directly into the sea, the limits of such transborder waters are delineated by a straight line crossing their mouth between the points located on a low-water line on their coast." After breakup of the Soviet Union, the Syrdarya and Amudarya as well as other rivers became transborder.

Tropa – navigable canals, such as the Kamyshovaya and the Motornaya, found in the Amudarya delta and on the A.S. coast.

Tugai, Tugai forest (Turk) – forests and riparian forests growing in river valleys in Central Asia. T. are often called Central Asian jungles. They distinguish trees and shrub Tugais. This is specific type of plant community, and its geographical occurrence is rather narrow. Apart from Central Asia, tugai is found in Middle Asia and China. In floral terms, the Amudarya T. is considered unique. It is very rich and contains over 550 varieties, of which 29 are Central Asian endemics. Only in the Amudarya delta do 54 relic and endemic varieties grow. T. flora abounds in medicine, honey, and essential oil plants (more than 200 varieties). Thanks to copious moisture, a specific biocenosis developed in T. that differs drastically from the nearby steppes, deserts, or dry mountains. In such biocenosis, poplar, *Asiatic poplar* (see), oleaster, Dzungarian willow, tamarisk, Russian olive, buckthorn, and cane prevail. All these plants intermingle, as if by liana, with the long and flexible stems of dogbane. In the Turanian Lowland, the giant cereal plume grass is found in some places among trees, while in open areas the camel's thorn grows. In recent years due to the aridization of the Amudarya floodplain and delta, the territories under tugai has shrunk significantly: by 2001 from 600 to 30 thou ha. As a result, the population of birds, especially birds of prey, has also decreased. The hangul (Bukhara deer) is no longer found in the wild in T.

Turan, Turan Depression – *see* or Turanian Lowland

Turanga – the local name of the Asian poplar (*Populus ariana*). Together with the bloomy poplar (*Populus pruinose*), it is an edificatory of the Turang formation (*Populeta pruinosae et arianae*). In *tugai* (see) forests of the Southern Circum-Aral area, Turanga growths dominate. The Asian poplar grows also in the northern part of Balkhash Lake. As a rare and endangered species, it is currently preserved within small protected areas, groves and wooded areas. Their groves are subjected to destruction primarily by unattended cabins, cattle grazing, wood wreckers, and population growth. At present, the poplar groves are not specifically under the control of any nature protection bodies, so they have no official protective status. As a result, the poplar groves are being destructed and the Asian poplar species could disappear completely.

Turanian Lowland, plain – the flat and low part in the northwest of Central Asia represented largely by sandy and clay deserts. In the south it is confined by the Kopetdag and piedmonts of the Paropamiz, in the west by the eastern coast of the Caspian Sea and the southern and eastern piedmonts of the Mugodjar and Southern Urals, on the north by the southern margins of the Kazakh hummocky area, and on the east by the piedmonts of the Tien Shan and Pamir-Altai. In the north of T.L., a great area is taken by the Circum-Aral area with the center being the *A.S. depression* (see). In the northwest, it is linked via the Turgai hollow with the vast West-Siberian Lowland separated from it with a water divide poorly distinguished in relief. The Caspian Sea lies at 27 m below sea (ocean) level; some drainless depressions on T.L. occur much lower than the Karagie Depression on the Southern Mangyshlak (minus 138 m) and the Akchakai in the northwestern Karakums (minus 80 m). Only some areas of T.L. rise higher than 300 m; these are some massifs of the Kyzylkums (up to 1000 m), the southwest of the Ustyurt, the mountains in the inner Mangyshlak, the south of the Krasnovodsky Plateau, Tuarkyr, and the massifs of the Greater and Smaller Balkhans. T.L. is the northernmost area in Asia with a continental, subtropical climate where wormwood-thistle and ephemeral vegetation prevails on the carbonate gray-brown and gray soils. Among agricultural crops, only cotton is cultivated here under irrigation.

Turkestan (Turk – "country of the Turks") – the ethno-historical term in the historical and geographical description of the vast territory extending from the Caspian Sea in the west to the border with China in the east and from the Aral-Irtysh water divide in the north to the border with Iran and Afghanistan in the south (Western or Russian Turkestan). It also included some provinces of Western China (Chinese or Oriental Turkestan) and the northern part of Afghanistan (Afghan or Southern Turkestan). Its northern border went along the line of Caspian Sea – Aral Sea – Chu River – Balkhash Lake. The total area of T. was over 3 mln sq. km. On the territory of the Western or Russian T. were found the Bukhara and Khiva Khanates and the Turkestan Governor-General Territory (after 1867), which was transformed in 1886 into the Turkestan Territory. After 1917 and the formation in 1924–1925 of the Soviet Socialist

Republics, and Autonomous Areas, the notion "Russian Turkestan" was renamed "Central Asia."

Turkmen circum-Aral Area – included the territory of the Dashoguz velajat and the Darganatinsky etrap (region) of the Lebap velajat located in the north of Turkmenistan. The area of T.C.A.A. was more than 90 thou sq. km. Its territory was directly affected by the consequences of the A.S. ecological crisis. Up to 90–95% of irrigated lands were affected by salinity to different degrees. By estimates, every year the territory of T.C.A.A. received 600 thou tons of solid aerosol particles, of which 430 thou tons fell on irrigated lands and about 170 thou tons fell on deserts.

Turkmen Lake "Golden Century" – Lake that accumulates the collector-drainage waters (CDW) of Turkmenistan and partially Uzbekistan. It was formed in the natural depression Karashor (–35 m) located in the north-west of the Karakum Desert in Turkmenistan. The depression area ranges from 3500 to 4000 sq. km, and its maximum depth is 70–100 m. In 2000, the President of Turkmenistan approved the state project on creation of a man-made sea with a storage capacity of 132 cu. km to take CDW supplied by two feeding canals, the northern and the southern, called the Dashoguz input canal and the Main Drainage Header "Golden Century," respectively. Their total length is 720 km. The CDW flow will exceed 10 cu. km, about half of which is from Uzbekistan.

The northern system will divert drainage waters from irrigated lands in the Dashoguz velajat and a part of drainage waters coming from irrigated lands in Uzbekistan that are at present discharged into Sarykamysh Lake over the Ozerny (150 cu.m/s) and Daryalyksky headers (60 cu.m/s). The maximum flow of the Dashoguz input canal is 210 cu.m/s.

The southern system will divert all drainage waters from the irrigated lands in the Akhalsky, Maryisky, and Lebapsky velajates in Turkmenistan via the Main Header, uniting the drainage systems of the mentioned velajates. This system will receive drainage waters from the both the right- and the left-bank areas of the Amudarya middle reaches. The water area of the lake at the maximum level should be 3460 sq. km. The design performance of the collection systems will ensure diversion into T.L. of salt flow amounting to 23–28 mln tons a year from irrigated lands in all velajates of the country.

Turtkul – the "mountains-witness," a flat table residual mountain composed of sedimentary rocks and rising high as an isolated hill or hummock separated from the main plateau and other uplands as, for instance, Djiltau residual massif from the *Ustyurt* (see); an earth bank, either from remnants of ancient fortresses or cities on the *Turan lowlands* (see). T. formed whole agglomerations and strips of the erosional relief. T. presents a marvelous view on the bank of the Amudarya upstream to Khorezm. Also, Turtkul City (former Petro-Aleksandrovsk), the former capital of Karakalpakia.

Tuschibas Lake – located on the right bank of the Syrdarya.

Tuyamuyun reservoir – built in 1978, 450 km from the mouth of the Amudarya River, near the entrance to the Southern Khorezm oasis. The waterworks and reservoir for seasonal regulation were constructed to provide water in the spring low-water periods to more than 500 thou ha of irrigated lands in Karakalpakstan, the Khorezm Region of Uzbekistan, and some regions of the Dashoguz velajat of Turkmenistan. Apart from guaranteed water intake into the irrigation system, T.R. made it possible to significantly reduce the sediment input during water intake into the left- and right-bank main canals to ensure required water releases for the *Takhiatash headworks* (see), to create conditions for control of such phenomenon as *"deigish"* (see), and to construct a hydropower plant with capacity 150 thou KW, etc. As a result of the backup created by the waterworks, the in-stream reservoir was constructed. It is the largest by area and one of the largest by the volume in Central Asia. Its storage capacity is 7.8 cu. km (full) and 5.3 cu. km (effective). The water surface area is 650 sq. km (65 thou ha). Via canals and water intake structures, it was connected with the off-channel reservoirs Sultansandjar, Kaparas (for drinking water supply), and Koshbulak located in depressions on the Amudarya left bank that made it possible to regulate the seasonal flow: Kaparas – full storage capacity 1 cu. km, Sultansandjar – 2.7 cu. km, Koshbulak – 1.5 cu. km as well as a canal for filling and drawdown of these reservoirs 18 km long with carrying capacity of 210 cu. m/s, ensuring better water supply of approximately 200 thou ha of irrigated lands.

Tyup (Turk.) – a peninsula, cape, bay. On A.S., the Karatyup Peninsula.

Tzar Islands* – a chain of islands opened in 1848 by *A.I. Butakov* (see). It was comprised of Nickolay I Island (about 160 km from the Syrdarya mouth), *Konstantin Island* (see) (named in honor of Great Duke Konstantin Romanov, the President of the Russian Geographical Society (RGS)) and Naslednik Island. In the Soviet period, Nickolay I Island was renamed *Vozrozhdenia Island* (see) and Naslednik was renamed *Komsomolsky* (see).

U

Ulken-Karatyup Bay* – secondary bay in the eastern part of *Sarychaganak Bay* (see). It projected insignificantly eastward from the shore. The shores of the cape were high and abrupt.

Ulkentumsyk Cape* – projected insignificantly eastward from the shore. The shores of the cape were high and abrupt.

Uprak-Kum – local name of sand mounds on the ancient alluvial plains in the Amudarya and Syrdarya deltas.

Urga Cape* – projected significantly eastward from the southwestern coast of *Adjibai Bay* (see). An steep coast near the cape was over 100 m high. The cape was surrounded by shallow water completely overgrown with reeds. There were several passages in the reed thickets; the largest of them was the Urginsky path. The fishworks Urga and a settlement was located nearby on the eastern part of the cape under a cliff. An airport for small aircraft was located on the upper plateau 2 km from the settlement. Here, planes from Muinak Island landed several times a day. Small jetties found here were capable of receiving boats and barges. There was a passenger jetty from which regular transport of people to Muinak Island was organized.

Ushkol Bay* – cut into the northeastern part of the *Shubartarauz Peninsula* (see), 7 km to the east-southeast of Akbidaik Bay. *Aitek-Aral Island* (see) was located at the entrance into the bay.

Ushmurza Island* – located in the north of A.S., 18 km southward of Kendyli Island. The island is low and sandy, and its elevation is 9 m. The fishery base was located in south.

Ushsai, Uch-Sai – city and port once located in the southwestern part of the Muinak Peninsula. The second largest port on A.S. after Aralsk. Between these cities there was large cargo traffic. The port was the northernmost point in Karakalpakstan. The total length of the wharf walls was about 800 m. The principal cargo handling operations in the port included reloading of cargo from river bases to ships and back. Port U. had regular traffic with Aralsk; cargo and passenger steamers of the *Central Asian State Shipping Company*

I.S. Zonn et al., *The Aral Sea Encyclopedia*, DOI 10.1007/978-3-540-85088-5_21,

(see) plied between them. Timber, grain, fertilizers, food, and manufactured products were exported from Aralsk, while cotton, fish, and canned goods were imported. Regular passenger traffic was organized by the routes: Ushsai-Fishworks-Muinak and Ushsai-Urga. In 1932, a stationary sea monitoring post was opened that functioned unitl 1964.

After the drying of the Aral, the port population decreased drastically from 10 thousand to 1 thousand people. Many people became environmental refugees.

Ustyurt, Ustyurt region (Kazakh) – a plateau confining A.S. on the west. Its area is about 200 thou sq. km, and its maximum altitude is 370 m in the southwest. U. breaks down by steep benches (*Chink*, see) 60–150 m high. It is composed largely of Neogene limestones and other sedimentary rocks that form gentle folds mostly of the latitudinal strike. Oil and gas-oil deposits are found here. U. has a slightly rolling relief. The most prominent negative forms are represented by the North-Ustyurt Depression, which is covered by solonchaks and sands, the Assake-Audan Depression going into the Sarykamysh Depression, and the bay-like depression in the southeastern margins of the plateau with the Kara-Shor solonchak and Kum-Sebshen sands. Less significant flat lowlands with takyrs are found as well as forms with karst-piping origin. The climate is sharply continental here. The atmospheric precipitation is slightly more than 100 mm. The soils are brown, semidesert and desert, and gypsum-containing gray soils. Wormwood-thistle prevails here.

In the plateau relief are distinguished the Northern and Central Ustyurt. In the northern part of the Northern Ustyurt is the North-Ustyurt table, a gently sloping plateau (120–170 to 220 m abs.) that confines the Samsko-Kosbulaksky Lowland (100-70 m abs.) of the sublatitudinal east-northeast strike on the north and east. The southern part of the Northern Ustyurt is covered by the Aktumsyk table plateau (up to 200–220 m abs.) of the latitudinal extension, and still more southward is the Barsakelmes Depression (100–120 m abs.). The Central Ustyurt is represented by the Central Ustyurt table plateau (up to 250 m abs.) with a system of ouvals (up to 280–340 m abs.). Here the strike of large relief elements changes from east-northeast to east-southeast. This reflects the changes in the strike of tectonic structures in the region: for the Northern Ustyurt – the North-Ustyurt sag, Aktumsyk projection and Barsakelmes sag; for the Central Ustyurt – the Central Ustyurt zone of uplifts.

Many large and some small structures of the Ustyurt are reflected in the relief. The exception is the North-Ustyurt sag within which in recent times the North-Ustyurt Plateau was formed, while the Samsko-Kosbulaksky Lowland is confined to the axial and is the lowest part of the sag.

The Ustyurt territory was uplifted and its continental stage of development started in the Late Miocene, while the greater part began in the post-Pontian. As a result, over the former seabed a system of rather high structural-denudation plateaus with depressions was formed repeating, in general, the correlation of ancient structures in the modern relief. An example is the depressions and

Fig. 51 Ustyurt plateau: a view on the roads from the tower at Aktumsyk meteo station. Photo by Dmitry Soloviev, June 2008

ouvals of the Ustyurt that are confined to the ancient negative and positive structures.

In the modern Ustyurt Plateau, armored, arid-denudation plateaus dominate. On the north, from the Circum-Caspian Lowland and on the east from the Aral Depression, the plateau is confined by steep (in the east up to 250 m high) cliffs-chinks with hollows, which are furrows of deflation in the piedmont areas. A flat relief of the plateau is slightly diversified by flat ouvals and solonchak lowlands with rare massifs of eolian sands at the bottom. The giant gas pipeline Bukhara-Ural runs via U.

Considerable natural gas and oil deposits are found in U. In the field Shakhnahty, gas production is about 3 bill cu. m a year. In mid-1980, an oilfield in Barsakelmes was explored. The predicted reserves of natural gas are 250 bill cu. m and of oil are 50 mln tons.

Uyaly Island* – located approximately 2 km to the west-southwest of *Altai Island* (see) and once separated from it by a shallow strait. The northern part of the island was low, while the southern part was covered by dunes 5–6 m high. The depths near the western shore of the island gradually increased seaward. In 1940–1941, a sea level monitoring station was opened and in 1983 observations ceased.

Fig. 52 Derrick exploring for gas and oil at former bottom of the Aral Sea. Photo by Dmitry Soloviev, June 2008

UZBOI (Turk "uz" + "boi" – "along water") – (1) name for the dead valleys and dry channels in desert regions of Central Asia; channel-like elongated depression having flow only during short periods or having no flow at all. (2) The name of the ancient river connecting Sarykamysh Lake with the Caspian Sea; presently – a dry valley. Its ancient channel was 550 km long, up to 4 km wide, and 55 m deep and ran along the northwestern margins of the Karakum from the *Sarykamysh Depression* (see) to the Caspian Sea, and from the southeastern end of the *Ustyurt Plateau* (see) to the Kelkor solonchak, which was still a bay in the Caspian Sea at the end of the last century. In the Neogene Time (approximately 10 mln years ago), a tectonic fault was formed along the present U. Later it turned into a river channel. About 9 thousand years ago, the *Amudarya* (see) flowed into the tectonic Sarykamysh Depression. Its waters, having filled the depression, flowed into the tectonic depression of U. and ran along it to the Caspian Sea. At present, U. is dry. In some places, saline lakes and thick salt formations are found. Many ancient authors from Herodotus to Ammian Marcellinus and other scholars of the 10th, 11th, and 15th centuries asserted that the Oxus (the name of the Amudarya in ancient times) flowed into the Caspian Sea. Studying the Caspian Sea, Patroclus (285–282 B.C.) came to a conclusion that the Oxus (Amudarya) and Yaksart (Syrdarya) flowed into the Caspian. Proceeding from Patroclus' conclusions, Eratosphen (275–194 B.C.) and Strabon spoke about the Oxus and Yaksart flowing into the Caspian from the east. According to Strabon, the Oxus was the largest river to his knowledge, apart from the Indian rivers. Eudox (3rd century B.C.) mentioned an enormous waterfall at the Oxus inflow into the sea. Polybius (208–127 B.C.) knew about this, too. Historian Arrian (2nd century B.C.) wrote about inflow of navigable rivers into the Caspian. The Greeks considered the Oxus and Yaksart to be among these. A. Biruni (971–1048) described the Zheikhuna (Amudarya) flowing through the deserts to the Khazar Sea. Idrisi (12th century) called U. the "greatest world river both by the volume and depth of waters and by the riverbed width." It should be remembered that by the 8th century, the Arabs went around the Caspian from the north, moving from Turkestan to the Caucasus and, consequently, they could compare their personal impressions about the Volga and U.

In the work "Amu and Uzboi" (Samara, 1879), it was mentioned that before Peter I the Amudarya was presented on all maps as a tributary of the Caspian Sea and only Peter I informed the French Academy of Sciences about diversion by the Khiva Government of the Amudarya waters into A.S. Kaulbars stated that the Amudarya diversion may be dated between 1470 and 1575. In some periods it was proposed to connect once again the Amudarya with the Caspian Sea via U. A hypothesis that in the ancient times the Oxus flowed into the Caspian Sea was refuted by many. Regarding the present flow of the Amudarya (42 cu. km/year), it can be assumed that it could play a significant role in variations of the water level in the Caspian Sea. Disappearance of the flow along U. about 3 thousand years ago was most likely connected with development of irrigated farming in Central Asia.

Uzynkair Island (meaning "long brade")* – located near the eastern coast of A.S., 16 km south of the Karashokat Cape and 1.5 km from *Kaskakulan Island* (see). It was a low, sandy island with dissected shores, partially overgrown with reeds and inundated with surge waves in some places. In its northern part, a fishery base existed. In 1940–1941, a station was organized here for monitoring the sea level; it existed until 1961.

Uzyak – a narrow strait; an arm in the Amudarya and Syrdarya delta (see *Ozek*).

V

Vasily Cape* – a cape near the western entrance into *Butakov Bay* (see) that represents a vast projection of the *Shubartarauz Peninsula* (see). The cape is high and steep, rimmed with a sandy bar extending for 1 km eastwards of it.

Vetlandy – a loan translation (cognate) of the English "wetlands." These are marshlands and over-saturated areas. Perhaps due to a considerable number of international projects in the Circum-Aral area, Uzbek specialists decided to use this Western sounding name. The V. ecosystem is widely used by the local population for a source of plant raw material (cane), for a fishery, and for hunting muskrat, which is why creation of a wide wetland management system is the most effective way for elimination of the negative environmental and social consequences of the Aral crisis. By the water supply regime permanent and non-permanent lakes, short-living water bodies, and floodplains may be distinguished.

Voeikov, Alexander Ivanovich (1842–1916) – outstanding Russian climatologist and founder of Russian climate science. Also a geographer. Corresponding Member of the Petersburg Academy of Sciences (from 1910). In 1865, the Gottingen University awarded him the degree of Doctor of Philosophy, and in 1880, he became the Doctor in Physical Geography at Moscow University. In 1886, he was elected a member of the Russian Geographical Society, and his membership lasted for thirty years. In 1870, on the initiative of B., the Russian Geographical Society organized the Meteorological Commission where he took the position of Secretary, and from 1883, he headed this commission. In 1891, he organized publication of the first meteorological journal "Meteorological Bulletin." In 1872–1876 he traveled over Western Europe, North, Central and South America, India, Central Asia, China, Ceylon, Java, and Japan. In 1884, he published his spacious work, "Climates of the World, and Russia in Particular," for which he was awarded the RGS Great Gold Medal.

At times when there were disputes about the feasibility of marshland drainage, he went to Polesiye and proved convincingly the high economic and natural conservation significance of drainage works.

I.S. Zonn et al., *The Aral Sea Encyclopedia*, DOI 10.1007/978-3-540-85088-5_22,

Fig. 53 A.I. Voeikov (1842–1916)

For the first time in the geographical science, he applied a balances method in studying geographical events (water balance of glaciers, moisture balance in air, etc.); he laid the foundations of paleoclimatology, agricultural meteorology, and phenology. He also proposed the idea of the world river classification on the basis of water regimes.

In his work, "Artificial Irrigation and its Application in the Caucasus and Central Asia" (1884), he proved the need for extensive irrigation in these regions. He considered irrigation as one of the most potent means of climate improvement. He advocated the construction of the canal to link the Amudarya and Caspian, thinking that as a result a whole chain of oases would appear in an area from Khiva to the Caspian Sea. "Russia faces a multitude of problems," he wrote, "but some time these works will be carried out and, perhaps, in the not so distant future... The objection against such project is that Russia should not indulge itself too much in Central Asia..."

B. asserted that the Aral would never die or dry out because it was a part of a water cycle: the water evaporated from its surface fell back on the glaciers and snowfields of Tien-Shan and Pamir from which rivers flowing back into A.S originated.

B. supported the idea of wide irrigation development. Thus, in 1908, he wrote: "In the far perspective with the awaited success of hydraulic construction and agriculture development we must use the whole stock of water of the Aral basin in dry years for artificial irrigation. In wet years, the lake must take in the excessive waters." Unfortunately, 50 years later specialists in irrigation and drainage did not follow this advice and regulated the whole flow of water-abundant years, thus dooming the Aral.

In 1903–1904, B. published a vast and original course in meteorology. In 1910, he was elected corresponding member of the Petersburg Academy of Sciences.

In 1912, he made a scientific trip to Central Asia, and in 1915, he traveled to the South Urals and Crimea. In the same year, he took the post of the director of the Higher Geographical Courses, Russia's first geographical higher education establishment.

B. was a member of many Russian and foreign scientific societies. In 1949, the Main Geophysical Observatory in Saint-Petersburg was given the name of Voeikov.

Among his works there are: “Climates of the World, and Russia in Particular” (1884), “Artificial Irrigation and its Application in the Caucasus and Central Asia” (1884), “Irrigation of the Trans-Caspian Area in Terms of Geography and Climatology” (1908), “Climatic Conditions and their Relationships with Irrigation and Cotton Growing in Turkestan” (1913), Selected Works, vols. 1–4 (1948–1957), and “Man’s Impact on Nature” (1963).

Voropaev, Grigory Vasilievich (1932–1999) – Russia’s most prominent scientist and science promoter for land hydrology and water management problems. Doctor of Technical Sciences (1972), Professor (1973), Corresponding Member of the USSR Academy of Sciences (1976). Graduated from the Moscow Institute of Water Management Engineers. In the period from 1954 to 1971, he carried out extensive research and training work in Tajikistan and Kazakhstan, studied the problems of water resources, and explored options for their utilization and land reclamation. After 1971, he worked at the Water Problems Institute (WPI) of the USSR Academy of Sciences (USSR AS), and from 1976 to 1988, he was the Director of this Institute, editor-in-chief of the journal “Water Problems,” editor-in-chief of the academic series “Caspian Sea,” permanent member of the State Expert Committee (SEC) of the USSR State Committee for Planning (Gosplan) (1975–1994), Chairman of SEC of USSR Gosplan (1983–1987), Director of the Research-Coordination Center “Caspian” (1989–1998), Chairman of Scientific Councils of the State Committee for Science and Technology and the USSR Academy of Sciences on integrated studies of the Caspian Sea problems (1992–1998) as well as the “Scientific Foundations of Management of the Regime and Resources of Continental Waters,” Co-Chairman of the Russian-Iranian Working Group on the Caspian Sea within the framework of the scientific-technical cooperation between the Islamic Republic of Iran and the Russian Federation, and Vice President of the International Commission on Hydrology.

His main works written in his first period of activities were devoted to the problems of irrigation, water application regimes and practices, and the economics of land reclamation and water management. During his work at WPI of USSR AS, he was engaged in studies of territorial re-distribution of the river water flow (partial transfer of the Siberian river flow to the USSR southern slope) and led the scientific development of the problem on territorial re-distribution of water resources in the USSR. He was the editor, together with D.Ya. Ratkovich, of a monograph published in 1985 that synthesized the investigative results on the scientific justification of transfer of the northern river flows to the Caspian and Aral Sea basins.

V. developed methods for modeling natural water management systems that are helpful for solving the problems of rational management of water and land resources and environmental sanitation of water objects and territories. He was the first to propose a method to estimate the water, land, and labor resource needs to validate planning-design solutions in land reclamation and water management. V.'s scientific investigations included such fields as the formation of continental waters and the development of systems to control their regimes as the most important complex of the biosphere and an indispensable element of production forces. He was engaged in studies of a water balance of irrigated areas and irrigation systems, primarily concerning reserves of irrigation, and developed the physical and geographical foundations of water management balances. Headed the All-Union Scientific School of "Theory and Practice of Land Water Resource Management". V. actively supported the creation of a single water management system in the country.

He was the author of over 200 scientific publications, including 7 monographs. He wrote the following books: "Irrigation in Some World Countries" (in co-authorship with B.S. Niyazov, 1970), "Physical and Geographical Foundations of a Water Management Balance" (in co-authorship with V.B. Mestechkin, 1981), "Modeling of Water Management Systems in the USSR Arid Zone" (in co-authorship with G.H. Ismaiylov and V.M. Fedorov, 1984), "Problem of Partial Withdrawal, Transfer, and Distribution of the Siberian River Flow in Western Siberia, Urals, Central Asia, and Kazakhstan" (in co-authorship with G.V. Bastandjoglo, 1984), "Economic-Geographic Aspects of Territorial Unit Formation in the Country's Water Economy" (in co-authorship with B.G. Blagoverov and G.H. Ismaiylov, 1986), and "Water Management System Development" (in co-authorship with G.H. Ismaiylov and V.M. Fedorov, 1988). After his death, the book, "Problems of Water Resource Management in the Aral-Caspian Region" (in co-authorship with G.H. Ismaiylov and V.M. Fedorov, 2003), was published.

Vozrozhdenia Bay* – located in the eastern part of Vozrozhdenia Island (1960). At the entrance of the bay were the Taz-tumsuk Cape on the west and the Tastyubek Cape on the east. On the east, the bay was confined by the steep Fonarnaya bar. At the entrance into the bay was found the Chagala bank which formed as a result of island erosion. The depth of the bay was over 10 m.

Vozrozhdenia Island* – one of the largest islands in A.S., located centrally, 61 km eastward of the *Keinchiyak Cape* (see) (1960). Its area was 216 sq. km. Its greater part belongs to Uzbekistan, while the northern part belongs to Kazakhstan. The Kazakh sector is shared by the Aktyubinsk (western part) and the Kyzylorda (eastern part) Regions. In Soviet times, the jurisdiction of Uzbekistan and Kazakhstan to this part of the mainland was purely formal. It was discovered in 1848 by Russian hydrographer *A.I. Butakov* (see) and named in honor of the Russian Tsar Nickolay. It was a part of the *Tsar Islands* (see). In the Soviet times, it was renamed to V. Its northwestern shore is low and sandy, gradually rising up to the south-east, and in some places the shores go down steeply into the sea. The shores are broken; here capes and overwater bars going far into the sea are found, with bays and gulfs cutting in between them. The elevated and abrupt bar Fonarnaya extends from the eastern shore of V.I. for 12 km northwards. It confines on the east the large *Vozrozhdenia Bay* (see). In the

Fig. 54 Shoreline at the former Aral Sea bottom eastward of the former Vozrozhdenia Island. Photo by Dmitry Soloviev, June 2008

northwest of the bay is the Udobnaya Gulf. Near the southwestern end of V.I., 500 m to the north, the small inlet Alga cuts into the shore. The shores of V.I., in particular the eastern and southern, are steep, and in some places areas of hummocky sands are found. When it was discovered, the island was overgrown with saxaul where many steppe antelopes (saiga) were found. By the early 20th century, the saiga were driven out completely and all saxaul growths were cut.

In 1936, a hydrological station was opened here to observe the sea regime but in 1949 it was closed.

Near V.I. two small islands were found: to the north, Komsomolsky Island (see), and to the south, Konstantin Island (1960).

W

Water balance of the Aral Sea – the water balance of the sea depends on the correlation between the inflow of river and groundwaters, atmospheric precipitation, and water losses to evaporation. The structure of the A.S. water balance is a function of various levels of economic development. From 1930 through 1990, the available water resources in the basin were estimated at 115 cu. km annually. Natural water losses in 1930 were 36 cu. km, in 1960 – 13 cu. km, and in 1990 – 7 cu. km. Water consumption in 1930 was 25 cu. km, in 1960 – 48 cu. km, and in 1990 – 103 cu. km. Water inflow into A.S. in the period from 1930 to 1960 was 54 cu. km, while in 1990 it was 5 cu. km. The reduced water inflow was caused by flow regulation with reservoirs and diversion of excess waters into deserts for irrigation; drainage waters did not return into rivers. Since 1961, the water level has dropped by 23.6 m, which has led to disastrous socioeconomic consequences for the population, numbering an estimated 3 mln people, in the Circum-Aral area. The WB that was earlier nearly at equilibrium is now characterized by deficit.

Water division – defining water consumption quotas on the basis of schemes for integrated utilization and conservation of water resources and water management balances. WD is carried out on various levels depending on the objectives of water management activities: on the basin level, among the states of a water basin, and on the subject level, between municipal authorities and major water users. It takes into consideration the water use regime by months (for irrigation and drainage systems, by decades) for the years of different water availability for all kinds of water use. When changing water use conditions gives rise to additional needs in water resources (appearance of new water users), the basin or federation subject authorities on water resource management and conservation revise the existing water use quotas for issuance of a license to a new kind of water use. Estimation of the future water management balances (WMB) enables planning of WD. In conditions of WMB deficit, the water use conditions specified in a license for water use is subject to more careful control, and violation of the water use quotas involves punishment in accordance with the water legislation. WD is used as a basis in preparation of various official documents on joint water resource management by the Central Asian countries.

I.S. Zonn et al., *The Aral Sea Encyclopedia*, DOI 10.1007/978-3-540-85088-5_23,

Water-related vision for the Aral Sea basin for the year 2025 – this document is the first phase of the "Aral Initiative" declared by UNESCO. This document states: "Vision is a practical picture of the future that we are eager to create. Vision shapes our world." At the UNESCO General Conference in October 1997, the governments of the Central Asian countries applied to the UNESCO Secretariat for advice on how to act on the conditions of the Aral crisis. In January 1998, UNESCO established within its framework the *Scientific Advisory Board on the Aral Sea (SABAS)* (see). In September 1998, SABAS elaborated a long-term vision plan. In November 1998, in Tashkent, the UNESCO Director General at the Executive Council meeting initiated the UNESCO Project on the Aral Sea Basin. In early 1999, national groups for the preparation of regional visions were established in the Central Asian countries. After discussion of the draft document, the new visions were elaborated. The figures on water resources included into the document were verified in terms of their attainability. In late 1999, at the workshop "Water and Peace in Central Asia" held by UNESCO in cooperation with International Fund for saving the Aral Sea, the document was distributed among the water management ministers of five countries, and in March 2000, the "Vision of the Aral Sea Basin" was presented at the Second World Water Forum in the Hague.

This document is a contribution into the world advisory project's "Long-Term Vision for Water, Life, and Environment" of the World Council on Water Resources.

Water Resources – a scientific journal established in 1972. It addresses issues of natural water quality, conditions of aqueous ecosystems, integrated management of continental waters, and interactions of these with the natural environment. Chief editors of this journal were at different times A.N. Voznesensky (1972–1974), V.N. Kunin (1975–1976), G.V. Voropaev (1976–1995), M.G. Khublaryan (1995–2004), and V.I. Danilov-Daniljan (from 2004). It is published 6 times a year by the Water Problems Institute of the Russian Academy of Sciences. Some issues of the journal have published articles on the problems of A.S. and its basin, in particular No. 1-1972, No. 2-1992.

Water resources management and agricultural production in the central Asian Republics project (WARMAP) – the basic program of EU assistance to the states in the A.S. basin, launched in 1995. This program was targeted to rendering assistance to 5 former Soviet republics in development of policies, strategies, and formations of programs for utilization, distribution, and management of water resources in the basin and facilitation on the regional level of the establishment of institutional structures for interstate water utilization and management. The first two phases of the project were completed by mid-1997. The principal achievement of the program was development of geographical information systems (GIS) based on the database of land and water resources in the basin (WARMIS).

"WARMAP-1," financed by the European Union, includes 3 subprograms initiated by the *SIC ICWC* (see).

The first phase of the project was implemented in 1995. It ensured cooperation of many local and foreign organizations. In cooperation with more than 160 local experts, the WARMAP Project prepared a detailed report on some problems, including legal and organizational, strategy of land and water use, and agricultural production and consolidation of some institutes.

At present, the following results of the adopted program are available: creation of a regional information system of water and land resources in the A.S. basin, including a database, interface, modules, and models; GIS with the central base in ICWC Scientific-Information Center, five national bases, and two BWMA; establishment of a system of observation and monitoring of water, land productivity, and economic indices in typical areas in all countries of Central Asia; analysis of dynamics of fertility and water use in irrigated farming WUFMAS (Water Use and Farm Management Monitoring Survey).

Water resources, problems of the Aral and environment – a collection of works showing the results of discussions of the mentioned subject in which more than 30 leading specialists from various water management departments of the Uzbek Republic, the Uzbek Ministry of Higher and Vocational Secondary Education, the Tashkent University named for Mirzo Ulugbek, the Uzbekistan State Committee for Science and Technology, the Republican Ministry for Emergency Situations, and the Civil Defense Institute took part. It was published in Tashkent in 2000. It contains an accurate analysis of river flow formation and of water resources of Central Asia, primarily providing their quantitative and qualitative characteristics, problems of strategic and operative management, the condition of the drinking water supply, the environmental situation in the water basin, the Aral problem, and rational inter-basin water redistribution. This collection is very interesting because it provides a wide range of opinions of different scientists on addressing the burning issues of water economy development and on saving the Aral Sea.

Water-salt balance of the Aral Sea basin – by estimates of some specialists, the input and output balance is composed of the following:

- input: inflow of salts to the irrigated lands in the A.S. basin with irrigation water – 117 mln tons; buildup in the zone of aeration – 10 mln tons; inflow of salts with the collector-drainage waters (CDW) directed to recycling – 7 mln tons. Total: 134 mln tons;
- output: salt removal with CDW – 64 mln tons and beyond the irrigated zone – 72 mln tons. Total: 136 mln tons.

Of the above salt input, the share of the Amudarya basin is over 88 mln tons and of the Syrdarya basin is approximately 46 mln tons.

Water world day – in 1992 during the UN Conference on Environment and Development (Rio de Janeiro, Brazil), the international community initiated World Water Day. The UN General Assembly declared observance of WWD for March 22 of each year. This Day was observed for the first time in 1993.

Watershed basin of the Aral Sea, water basin of the Aral Sea – comprises a considerable part of the territories of the Central Asian countries (Kazakhstan, Uzbekistan, Tajikistan, Turkmenistan) and also the northern provinces of Afghanistan (257 thou sq. km) and Iran (65 thou sq. km). The total area of the basin within the Central Asian borders is 2.4 mln sq. km. This is a drainless region comprising the large, independent basins of the Amudarya, Syrdarya, Murghab, Tedjen, and other less significant rivers. WBAS is one of the most ancient centers of civilization in the world, originating with irrigated farming and the development of water resources thousands of years ago. In antiquity (4th century B.C. to 2nd century A.D.), when irrigation development was increasing, the irrigated areas in the lower reaches of the Amudarya, Syrdarya, and Zarafshan exceeded the present-day area of irrigated lands. The total irrigated area in the lower reaches of the Amudarya and Syrdarya was 3.5– 3.8 mln ha, including the Amudarya lower reaches (1.3 mln ha) and the Syrdarya lower reaches (2.2 to 2.5 mln ha). Thus, in antiquity, the irrigated area in the lower reaches of the Amudarya and Syrdarya was four-fold greater than at present; it should be remembered, however, that, in fact, smaller land areas were irrigated than now. Although the main canals that time were very long, the network of smaller irrigation canals was rather sparse, and with great water consumption, no more than 10–15% of lands being in command zones of irrigation systems were sown. In feudal times, irrigation in Central Asia was developed further, in particular in the 9th – early 13th centuries when such large feudal states as Samanides (9–10th centuries), Karakhanides (11–12) and Khorezmshakhs (11 – early 13) were formed. The 12–14th centuries were the period of the highest irrigation development in the lower reaches of the Amudarya and Syrdarya. The total area of cultivated lands reached 2.4 mln ha, out of which 1.4 mln ha were irrigated on a regular basis. In the Middle Ages, especially in the 10–12th centuries, irrigation was practiced not only on flatlands, but also in piedmont areas of Central Asia. For irrigation of waterless steppes and semideserts in the piedmont areas, the engineering-type reservoirs with a capacity up to 1.5 mln cu. m were built on large *sais* and rivers.

The joining of Central Asia to Russia in the second half of the 19th century facilitated further development of irrigated farming and related sectors of the economy. The tsar's civil servants, seeking to yield as much cotton as possible, actively pushed the extension of areas under cotton. For this purpose, they needed information about water resources, likely regions of irrigation development, soil conditions, etc. that might be a basis for design of large irrigation systems. In this period, the first surveys were carried with a view to define areas suitable for irrigation on the Golodnaya Steppe and in the Ferghana, and Zarafshan valleys in the Amudarya lower reaches; however, despite all efforts during the whole colonial period, only 80 thou ha of land was developed. True large-scale development of land irrigation and reclamation in Central Asia started only after the October revolution of 1917. Despite economic collapse and Civil War, the Soviet government focused great attention on irrigation development in Turkestan. On May 17, 1918, V.I. Lenin signed the Decree "On the Allotment of 50 mln rbls. to Irrigation Development in Turkestan and on Organization of these Works." Based on the economic development mandates of the 5-year plans, large irrigation canals and reservoirs and main headers with high efficiency were built in the Central Asian republics. As a result, irrigated areas increased nearly two-fold, from 1339 thou ha in 1913 to 2474 thou ha in 1960.

The resolution of the May Plenum of CPSU CC (1966) defined a new stage of land irrigation development that was principally different from all past ones. For the conditions of Central Asia and Southern Kazakhstan, it meant, first of all, that better, more efficient irrigation was the most important precondition for further increase of the yields of technical, fodder, and forage crops as well as the extension of areas under irrigation for greater yields of raw cotton. In 1966–1986, the construction of water management projects and commissioning of Central Asia's largest irrigation canals (such as Karakum, Karshi, Amu-Bukhara, and others) was pursued aggressively. Their head-water intake varied from 200 to 500 cu. m/s and more. Such large reservoirs as Andijan, Charvak, Chardarya, Tuyamuyun, and Nurek were constructed and put into operation (all in all there were 80 reservoirs with a total capacity over 100 mln cu. m). This enabled regulation of river flows in this region. In 1965–1988, as a result of wide-scale irrigation works in the Central Asian republics and Kazakhstan, over 3 mln ha of new lands were developed and water supply to a greater part of previously irrigated areas was improved. For transportation of drainage and diversion waters from oases, main headers were constructed. A part of this flow is discharged into the Syrdarya and Amudarya for replenishment of their flows, and the other part is discharged to drainless depressions in the Kyzylkums and Karakums. In 1988, a total of 9354 thou ha were developed in Central Asia and Kazakhstan. The total capacity of reservoirs exceeded 60 cu. km. The total length of the irrigation inter-farm network was about 50 thou km, and the on-farm network was about 270 thou km. After the breakup of the Soviet Union, the newly independent Central Asian countries, on their transition to market economies, decided to revise their plans of agriculture development to reduce water consumption, so they significantly cut irrigated areas.

Western (Aral) Sea – the deep-water section that remains after the drying of the Large Aral Sea. It is located between the Ustyurt chink and the former *Vozrozhdenia* (see), *Lazareva* (see), *Komsomolsky* (see) *Islands*.

World Day to Combat Desertification and Drought – following the adoption of the Convention to Combat Desertification of June 17, 1994, the UN General Assembly declared this day the World Day to Combat Desertification and Drought. This Day was observed for the first time in 1995.

World Environment Day – Annually celebrated on June 5. On this day in 1972, the UN Conference on the Human Environment was opened in Stockholm. In order to draw attention of the world public to the problem of environmental protection it recommended observance of this day as World Environment Day.

Y

Yuzhsibreka – conventional name of a project proposed in the 1950s by A.A. Shulga, a specialist from SOPS of the USSR Academy of Sciences, to design of a large waterway running from the east to the west and involving the Caspian Sea and its basin. This canal would get water from the Ob upper reaches, the Irtysh middle reaches, and other rivers that would be met on its way. In general, water withdrawal from these rivers would be 100 cu. km a year (nearly 5 times more than the entire combined flow of the Kura and Terek rivers) into the Caspian Sea.

Already in the first stage of construction, Yu. would be turned into a network of deep waterways quite suitable for navigation and linking Kuzbass with the Urals, the North Caspian (via Emba), and the A.S. The second stage assumes water intake from the Tom River (with a retaining dam near Kemerovo City) and its transfer via the Severn, Unga, and Innyuz rivers to the Novosibirsk reservoir and further on along the main Yu. route to the Circum-Caspian Lowland. Downstream of the North-Caspian HPS, the Tom' River waters would be taken for irrigation, watering, landscaping, and creation of fishery lakes in the trans-Ural part of the Northern circum-Caspian Lowland. In this way, about 5 mln ha of desert and semi-desert territories would be improved.

I.S. Zonn et al., *The Aral Sea Encyclopedia*, DOI 10.1007/978-3-540-85088-5_24,

Z

Zhalanash Bay* – cuts into the Kokturnak Peninsula 7.5 km west of the Baishura Cape. This bay is 3.6 km long. It has the steeply dropping shores (500–750 m from the shore the depth is 5–6 m). In the past, fishery activities were practiced here.

Zhidely Bay (former Djidely)* – projects into the southern shore of the *Shubartar-auz Peninsula* (see) between the Segizsai and Aiderly Capes. Moving to the north-west of the Segizsai Cape, the shores of the bay become more and more flat, and cliff heights gradually diminishe until a strip of sandy beaches appears.

Zhingyldastubek Island* – located to the east of A.S. 15 km to the southeast of Kuzzhenes Island. The island is low and sandy with flat shores and is surrounded by reed thickets. The island is covered with sandy mounds on which rare shrubs grow.

Zhiltyrbas Bay (former Iskekuk Yzyak)* – incises into the southern sea shore for 24 km to the south between *Erzhan Island* (see) and the Uzynkair Cape. The shores of the bay are low and shallow. The bay itself is shallow; its prevailing depths do not exceed 3 m. The entrance into the bay is obstructed by two bars. In 1967–1968, the bay dried out completely. In 1980, the coastline receded for 30 km, though in relation to the former A.S. water level of 53 m. In subsequent years, the recession rates of the coastline were as follows: 1990 – 53.3 km, 1995 – 65 km, 2000 – 75 km. Later, the waters that usually went to the Amudarya delta were used to fill Z.B., the water surface of which stretched from north to south for 20–22 km and its lower part approached close to the KKS header. The water salinity in the lake was 3.4–3.5 g/l (2000).

The territory of Zh.B. abounds in cane thickets and floodplains, thus attracting many water fowl and near-water birds. In the dried territory of Zh.B., several artesian hot springs were constructed that give slightly saline (1.5–2.0 g/l) hot water (38–40°C). Camps of shepherds grazing cattle and stocking cane for hay concentrate around these wells. After drying, an irrigation-disposal lake was formed in the bay.

I.S. Zonn et al., *The Aral Sea Encyclopedia*, DOI 10.1007/978-3-540-85088-5_25,

Chronology of the Aral Sea Events from the 16th to the 21st Century

Years	General Events
	16th century
1558	An English merchant and diplomat, Anthony Jenkinson, travels through Central Asia and observes the medieval desiccation of the Aral Sea. He writes that "the water that serveth all to country is drawn by ditches out of the river Oxus [old name for Amudarya] into the great destruction of the said river, for which it falleth not into the Caspian Sea as it gath done in times past, and in short time all land is like to be destroyed, and to become a wilderness foe want of water when the river Oxus shall fail."
	A. Jenkinson crosses the Ustyurt and visits Khiva and Bukhara, preparing a map of Central Asia.
1573	"Turn" of the Amudarya from the Sarykamysh to the Aral; in other words, the rather regular flow of part of its waters into the Sarykamysh ceases, the waters from this time running only to the Aral.
	17th century
1627	In the book, "Knigi, glagolemoy Bolshoy Chertezh" ("the big sketch"), the Aral Sea is named "The dark blue sea."
1670	German geographer Johann Goman publishes the map "Imperium pereicum," on which the Aral is represented as a small lake located 10 German miles from the northeastern margin of the Caspian Sea.
1697	On Remezov's map of the Aral Sea (*more Aral'sko*), it is for the first time represented as an internal lake completely separated from the Caspian Sea and into which the Amundarya (Amu Darya, Oxus), the Syrt (Syr Darya, Yaksart), and many small rivers flow.
	18th century
1714	A unit commanded by Colonel I.D. Bukhgolts tries to enter into the Aral Sea from the Irtysh River.
1715	The first expedition by A. Bekovich-Cherkasskiy to the Caspian Sea. The most correct map of the Caspian Sea and a description of all east coasts almost right up to Astrabad Gulf is drawn up. It is discovered that the Amudarya does not flow into the Caspian Sea.
1717	Peter the Great's trip to France where, in particular, he met G. De L'Isle and assured him that the Oxus (Amudarya) does not flow into the Caspian Sea, but flows into the completely unknown Aral Sea.
1723	On the map of French court geographer G. De L'Isle, the Sea for the first time is named "Aral."

Years	General Events
1729	Marine officer Mark Dubrovin, commanded to Central Asia, visits the Aral Sea, Uzboi, and the Khiva Khanate and prepared a map.
	Ivan Kirillov publishes a map of the Aral Sea prepared by M. Dubrovin.
1730	On the request of Khan Abushair the Kirghiz (Kazakh), the people between the Yaik (Ural) and Syrdarya Rivers are given Russian citizenship.
1731	Interpreter of the Russian Foreign Affairs Board Megmet Tevkvel and two officer-geodesists make the first topographical survey of the Aral Sea coast.
1732	Greek Vasilyi Vatatsi brings the first data on the Aral Sea to Europe.
1740	Russian Lieutenant Dmitry Gladyshev and geodesist Ivan Muravin perform a topographical survey from Kuvan Darya (the left tributary of Syrdarya) up to Khiva.
1741	Ivan Muravin creates a map, "A new map of the way from Orenburg through Kyrghyz, Karakalpak, and the Aral territories up to the city of Khiva and of the Aral Sea part and the rivers flowing into it, a part of Syrdarya, Kuvan Darya, and Ulu Darya" (Uludarya is a name for the Amudarya).
1753	English merchant Hanvey is the first European to map the dried channel of the Amudarya.
	19th century
1819	N. Muraviev travels from the Balkhansky Bay to Khiva but does not see the Aral, thus, on the "Traveling map of Captain Muraviev from Tiflis to Khiva, 1819," the Aral Sea is shown as rectangular "with quite a fantastic coastline."
1820	Russian scientist Meindorf travels through areas of the Aral Sea region that at the end of 18th century were flooded by sea water.
1823	Expedition of Russian Colonel Fedor Fedorovich Berg (Friedrich Wilhelm Rembert von Berg) to the western coast of Aral. For the first time, he makes a number of astronomical observations and a survey of the western coast for military-topographical descriptions.
1825	Aral-Caspian expedition of F.F. Berg.
1831	The map of the Aral Sea, drawn up on the basis of new data, is published in A. Levshin's book, "Description of Kyrgyz-Kaisaks hordes and steppes."
1839	G. von Helmersen publishes "Nachrichten über Chiwa, Buchara, Chokand . . ."
1839–1840	Expedition of Lieutenant-General V.A. Perovsky to Khiva.
1840	Topographical surveys and astronomical observations on the northern and northeastern coasts of the Aral Sea are carried out.
	Well-known zoologist and Professor of Kazan University E.A. Eversman publishes his book, "Natural History of the Orenburg Territory," in which he describes the Aral Sea and the physiographical and geological characteristics of its coast, and predicts its possible future drying.
1844	A. von Humboldt publishes "Central-Asien. Untersuchungen über die Gebirgsketten und die vergleichende Klimatologie."
	In a report presented at the meeting of railroad engineers in Saint-Petersburg, Sveridov dwells on the fact that the Uzboi was the ancient channel of the Amudarya and proposes the possibility of returning the river to it, in this way connecting it again with the Caspian Sea.
1847	Near the mouth of Syrdarya a fortification Raim is established.
	Near the Syrdarya mouth (64 km), Infantry General V.A. Obruchev discovers the Raim fort (later on renamed into Aralsk) and then the Kos-Aral fort in the river delta.

Years	General Events
	Under command of retired sea officer Mertvago, surveys of the Syrdarya mouth from Raim to its inflow into the Aral Sea are carried out.
	Following the order of Chief of the Orenburg Territory V.A. Obruchev, 2 ships are constructed: the military ship "Nickolay I" and the private ship "Michael." The first one is designed for investigation and survey of the Aral Sea, while the second one is for fishery development. Both ships are dismantled in 1854.
1848	The map, "Plan of the Aral Sea and Amudarya Delta" (Skizze des Aral-Sees und Amu Delta), is prepared by natural scientist F. Baziner and published in German.
	Russian officer Aleksey Ivanovich Butakov is commissioned to carry out a topographical survey and physico-geographical description of the Aral Sea. He is recommended by well-known round-the-world seafarer admiral F.F. Bellinsgausen. In Orenburg city, under supervision of Butakov, two-gun schooners are built. The "Konstantin" is transported to Syrdarya to fortify the Raim, which is located 64 km from the mouth. The crew consists of 27 persons, including the exiled poet T.G. Shevchenko, who served as a sketch artist. Under A. Butakov's leadership, during 2 months of general reconnoitring of the Aral Sea, measurements of depths and determinations of latitudes are made. The sea expedition also collects some geological and biological samples. An island named in honour of the Russian Emperor Nickolay I is discovered (in Soviet time it is renamed Vozrozhdenia Island). It was a part of the Imperial islands, together with the island "Konstantin," which is named in honour of Grand Duke Konstantin Romanov, the president of Russian Geographical Society. Also discovered is the island, Successor (during the Soviet period – Komsomolsky).
1850	In "Geographical Proceedings," issued by the Russian Geographical Society, Ya. Khanykov publishes for the first time, "The map of the road from fortress Orsk through Kirgiz, Karakalpak, and Aral territories up to the city of Khiva, drawn and composed by Geodesist-Ensign Muravin in 1743."
	The Hydrographic Department of the Russian Navy Ministry prints an Aral Sea map based on the surveys of Butakov and Pospelov from 1848 to 1849.
	Hydrographic map of the Aral Sea is published in St.-Petersburg.
	A.I. Butakov is sent to Sweden to order ships for the Aral fleet.
	Topographer Yakovlev surveys the territory from the Aralsk fort via the Barsuki sands to Emba.
1851	Fort Raim is moved to Kazalinsk city (Fort N1).
	A work by A.I. Maksheev, "Description of the Aral Sea" is published in the "Proceedings of the Russian Geographical Society."
	"The explanatory note for the map of the Aral Sea and Khiva Khanate with their vicinities" compiled by Ya.V. Khanykov is published in "Proceedings of Russian Geographical Society."
	Proceedings of the Russian Geographical Society" publishes the work of A.I. Maksheev, "Aral Sea Description."
1852	A.I. Butakov transports to Raim the disassembled parts of two iron steamships, the "Perovsky" and the "Obruchev," which were built in Sweden.
1853	A.I. Butakov, commanding steamship "Perovsky," excels during the seizure of the Kokand fortress Ak-Mechet (white mosque).
	By the nomination of A. Humboldt, A.I. Butakov, for his research of the Aral Sea, is elected as an honorary member of Berlin Geographical Society.

Years	General Events
	A.I. Butakov makes the first voyage over the Syrdarya on the ship "Perovsky," during which topographers Rybin and Yakovlev make a detailed description of the meanders of the river channel.
	Ak-Mechet is restored as the Perovsky Fort.
	The Aral military fleet is established.
	Captain-Lieutenant N.L. Ivashintsov makes the "Map of the Syrdarya lower reaches from Perovsky Fort to the Aral Sea prepared on the basis of recent surveys."
1854	A.I. Butakov transfers the Aral shipyard to fort No. 1 (Kazalinsk city).
1855	A.I. Butakov describes the Syrdarya from Kumsuat for 85 km above Perovsk city (Kyzylorda). He is promoted to captain of the 2nd rank.
1858	A.I. Butakov navigates on the ships of the Aral flotilla along Amudarya up to Kungrad city to assist the embassy in Khiva.
	A.I. Butakov prepares a detailed inventory of the Amudarya delta up to Nukus.
	Secret mission of Count N. Ignatiev to Khiva, one of the purposes of which is to negotiate a possibility for the Russian ships to use the Aral Sea and Oxus.
1859	A.I. Butakov with a troop of 140 men engages in military actions at Kungrad; then, having delivered troops to Kazalinsk on the steamship "Obruchev," he returns to the delta of Amudarya and surveys down to Nukus city.
	A.I. Butakov prepares a handwritten map, "Flat map of the southern coast of the Aral Sea and arms of the Amudarya River."
1860	A.I. Butakov is sent to Britain and the USA to order two iron ships.
1861	A.I. Butakov conveys two steamships, which he ordered in England and the USA, the "Aral" and the "Syrdarya" to Kazalinsk.
1863	A.I. Butakov describes the Syrdarya from Perovsk up to the tract of Baildyr-Tugaj for 807 versts, an obsolete Russian unit of length equal to 1.0668 km.
1864	Russian armies under command of generals Tchernyaev, Skobelev, and Golovachyov cross the Russian border and act in the campaign against Kokand and Khivan khanates.
	A.I. Butakov is awarded the rank of Rear Admiral.
1865	A.I. Butakov is awarded by the King of Italy the St. Maurice and Lazarus Order for hydrographic works in the Aral basin. Russian armies occupy the city of Tashkent.
1866	A.I. Butakov prepares a short essay, "Delta and mouth of the Amudarya River," which is published in "Otechestvennye Zapiski" No. 1.
1867	A.I. Butakov publishes "The Delta and Mouths of the Amudarya, or Oxus" (1866–1867).
	At the Annual Assembly of the Royal Geographical Society of London, the chairman, Sir Roderick I. Murchison, awards the medal of the founder of the Society to Admiral of the Russian fleet, A.I. Butakov, "the first to launch ships on the waters of and navigate the Aral Sea; and also for successfully researching the main mouth of the Oksus (Amudarya) in Khivan Khanate, and proving that along the Syrdarya that flows into the northern Aral Sea, steamships can travel 500 miles upstream, locating for the first time a safe route to China through western Turkestan."
1868	Russian armies occupy the city of Samarkand.
1869	Russian engineer-technologist N.F. Uljanov designs the project to construct a channel from Syrdarya to irrigate the Hungry Steppe.

Years	General Events
	A.I. Butakov dies. In the obituary, he is called "a real sea officer who would honor any country and any people."
1870	R. Lenz publishes "Unsere Kenntnisse über den früheren Lauf des Amu-Daria."
1871	Engineer Ya.G. Demchenko presentes the essay, "On the flooding of the Aral-Caspian lowland to improve the surrounding climate."
1872	A first attempt is made to build a main irrigational channel (named "Kaufman") from Syrdarya (the work ceased in 1881 because of the vast expenses).
	H. Yule publishes, "Papers connected with the Upper Oxus Region."
	H.C. Rawlinson publishes, "Monograph on the Oxus."
1873	The contract is signed according to which the Bukhara Emirate recognizes Russia as its protectorate.
	P. Lerch publishes, "Khiva oder Khârezm. Seine historischen und geographischen Verhältnisse."
	R. Roesler publishes, "Die Aralseefrage. Noch einmal geprüft."
	A.V. von Kaulbars and A.I. Glukhovsky lead the Urun-Darya expedition and carry out serious investigations of the Amudarya lower reaches, its old channels, and the water system of the Khiva Khanate.
	Khiva is conquered.
1874	N.A. Severtzev (Russia) passes along the southern coast of the Aral Sea.
	A.A. Tillo, on behalf of the Russian Geographical Society and its Orenburg Department, surveys in the Aral-Caspian region.
	I.A. Strelbitskiy makes the first calculations of the water surface area of the Aral Sea (65780.1 km^2, without including the area of the islands).
	The Russian Geographical Society sends an expedition headed by the famous scientist N.G. Stoletov to Khivan Khanate to survey the Kunya Darya and Daudan from Amudarya up to the Sarykamysh depression.
	A.A. Tillo, by request of the Russian Geographical Society and its Orenburg Branch, carries out a survey of the Aral-Caspian region and installs a benchmark on the northwestern coast of the Aral Sea (in Kara-Tamak). The Aral Sea is determined to be 74 m higher than the Caspian Sea.
	The President of the Russian Geographical Society (RGS) sends the "Amudarya Expedition" to the Amudarya delta and the Aral Sea. The expedition is comprised of 3 units: The first is organized by the Natural Scientist Society and includes geologist N.P. Barbot-de-Marni and zoologists V.D. Alenitsyn and Bogdanov. They study the geology of the Aral coasts and the Aral and Caspian fauna. The is second is a surveying unit led by General A.A. Tillo, and the third is a natural science and historical unit led by General N.G. Stoletov. The latter unit also includes N.A. Severtsov, meteorologist F. Dorandt, hydrographer Zubov, and Major of the British Engineering Troops Herbert Wood.
1875	In Paris, the monograph "Das alte Bett des Oxus - Amû-Darja" by M.J. de Goeje is published about the Amudarya crease.
	In St.-Petersburg, the work of N.P. Barbot-de-Marni, "On Geological Investigations in the Amudarya Area," is published.
1876	Russian geologist N.G. Petrusevich discovers the Sarykamysh depression. Its survey mark is determined on the basis of surveying from Novy Urgench.
	In London, H. Wood publishes his large monograph, "The Shores of Lake Aral."

Years	General Events
1877	The second expedition of the Russian Geographical Society to Khivan Khanate departs under the leadership of Filippov to survey in the Amudarya delta, for surveys of horns and floods, and to draw up hydrographical maps. Their work lasts 3 years.
	The office of the Turkestan General-Governor issues the "Temporary rules concerning irrigation in Turkestan Territory."
	In St.-Petersburg, Professor K.F. Kessler publishes his monograph, "Fish of the Aral-Caspian-Pontian Ichthyological Area," from the results of studies of Aral fish collected by zoologist V.D. Alenitsyn during the 1874 expedition.
1878	During floods, the water of Amudarya reaches the Sarykamysh depression and raises the level of lakes located at its bottom by more than 5 m.
1879	A member of the Russian imperial family, Grand Duke Nickolay Romanov, orders the destruction of the dam separating the leftbank channel of Kuna Darya from the main channel of Amudarya.
	N. Tcharykov publishes « Un voyage dans l'Ouzbekistan en 1671» (1879–1880).
	The "Proceedings of the Russian Technical Society" publishes the article of N.G. Petrusevich, "On Hydraulic Conditions of the Amudarya River and Possible Direction of Its Flow to the Caspian Sea."
1880	Expedition headed by A.I. Glukhovskoy (Russia) investigates Khoresm, the Amudarya crease, the Sarykamysh depression, and Uzboy.
1881–1883	A.V. von Kaulbars publishes "Description of territory in the lower Amudarya" in "Proceedings of the Imperial Russian Geographical Society."
1882	Known Russian climatologist and geographer A.I. Voeikov, in his speech "Rivers of Russia" made at the meeting of the Department of Physical Sciences of the Society of amateurs of natural history, anthropology, and ethnography, writes: "...lower reaches in the low and even in the middle stream of the rivers flowing into Aral are so dry that the existence of the Aral Sea in its present boundaries is evidence of our backwardness and inability to take sufficient advantage of the current water volume and fertile silt that the Amu and Syr bear. In a state capable of using gifts of nature, the Aral would serve as a drain of water in winter (when it is not necessary for irrigation) and also of high-level waters in summer."
1883	Russian geologist A.M. Konshin, studying areas of Sarykamysh and Uzboy, comes to the conclusion that Sarykamysh was filled with water and connected to the Aral in prehistoric times.
	On the request of the Ministry of Agriculture, zoologist A.M. Nikolsky studies the situation for a fishery in the Aral Sea and also in the Amudarya.
1884	A hydrometeorological station opens nearby Aralsk.
1885	At the court of the Bukhara Emir, the Russian Imperial political agency is established to help Russian citizens take concessions for irrigation of fallow grounds.
	Russia imports 67 tons of cotton from Turkestan.
1886	"The statute about governing Turkestan territory" is issued.
	Russian Grand Duke N.K. Romanov organizes construction of the Bukhara canal in the Hungry Steppe to irrigate new grounds in the steppe and supply water to the possession of the Bukhara Emir (construction lasts 6 years). In Syrdarya, a stone dam named by N.K. Romanov "Tsar-dam" is built.
	A.M. Nikolsky travels to the Aral Sea, Syrdarya, and Amudarya.
	"Newsletters of the Geographical Society" publishes the article by N.M. Yadrintsev, "Depletion of Waters in the Aral-Caspian Lowland."

Years	General Events
1886–1887	William Bateson from Cambridge, Britain, visits the northern and northwestern coast of the Aral Sea between the Chumyshkul and Djaksy-Klych Lakes among other places to collect data about the fauna of Salt Lakes.
1886–1888	Travel of V.A. Obruchev over Central Asia; study of the Karakum Desert and the Amudarya ancient channel (Uzboi).
1887	At the Turkestan General-Governor, the post of regional irrigator "for management of large irrigation canals in the region" is established and temporary rules about the irrigation of Turkestan Territory are issued.
	Newsletters of the Emperor's Russian Geographical Society" publishes the work of A.M. Nikolsky on fisheries in the Aral basin.
1888	The office of the Turkestan General-Governor von Rosenbach issues "Instruction on the rights and duties of irrigational officials, district chiefs, canal-aksakals, and mirabs for the management of irrigation in Turkestan Territory."
	Geologist G.V. Abikh describes specimens of fauna collected by A.I. Butakov, stressing for the first time the availability of nummulite limestones of the Lower Eocene on the Aral eastern coast.
1889	I.A. Strelbitsky (Russia) once again carries out calculations of the Aral Sea water surface area on the basis of the map of Asian Russia drawn by him to the scale of 1:4200000 and of maps of Asian Russia of the same scale that were issued in 1883 by the Military-Topographical Department of the General Staff of Russia. The Aral Sea's area is 65252.4 km^2; the area of its islands is 2517.0 km^2.
1890	"Revue Britannique" publishes the work of renowned geographer P.A. Chikhachev, "Aral-Caspian Depression" (published in Russian in 1982 in the book "Page about the East").
1891	The priest from Vladimir city, A. Tchaykovskiy, publishes a historical hypothesis based on the Bible and Herodotos about the formation of the Amudarya River, "Turkestan and its river."
	Water flows through the canal Buhararyk; however, after the dam at its head breaks, it is abandoned.
	Great Duke N.K. Romanov, with his own funds, starts the construction of a new canal named by him, the canal of Nickolay I.
1892	The post of Official for Special Missions in irrigation is established by the General-Governor of Turkestan.
1893	V.I. Glukhovskoy (Russia) proposes a project to connect the waters of the Amudarya to its old channel to the Caspian Sea to establish a continuous Amudarya-Caspian Sea waterway from the frontiers of Afghanistan via the Amudarya, Caspian Sea, Volga River, and Mariinskaya hydrotechnical system up to St. Petersburg and the Baltic Sea.
1896	Ya.A. Rekhtazamer, through the Russian Imperial Political Agency, applies for the construction of a channel of 300 versts (320 km) in length from the Amudarya to the lands of Bukhara Emirat.
	The Nickolay I canal (throughput – 11 m^3/s, length – 70 versts (75 km), area of irrigation – 7 thousand dessiatinas (one dessiatina is a Russian measure of land, roughly 1.1 hectares)) begins to function.
1897	Administration of agriculture and state property of Turkestan Territory is established.
	The Turkestan Branch of the Russian Geographical Society is established in Tashkent.

Years	General Events
1899	L.S. Berg, the prominent Russian researcher of the Aral Sea and future Academician of the USSR Academy of Sciences, president of the USSR Geographical Society, travels along the coast of the sea and does scientific research in the sea.
1900	L.S. Berg publishes the book, "Fishes and fishery in the mouth of Syrdarya on the Aral Sea."
	L.S. Berg together with topographer K.A. Molchanov organizes the first instrumental observations over the Aral Sea level with the help of sounding rods in the Greater Sarychaganak Bay and instals 4 such rods near the mouth of the Syrdarya.
	20th century
1901	L.S. Berg publishes "Studies of the Aral Sea" in the journal "Zemlevedenie" ("Physical Geography").
	The construction of the main canal in the project of F.A. Elistratov (Russia) to irrigate of 45 thousand dessiatinas in the northeastern part of the Hungry Steppe begins. In 1913, the construction finishes.
1902	V.V. Bartold, based on historical archaeological data he collected, finds that waters of the Amudarya flowed into the Uzboy from the second half of 13th century until 1573; moreover, he discovers facts testifying to the use of the Uzboy for navigation in 1392.
	V.V. Bartold publishes the monograph, "Data on the Aral Sea and the lower reaches of the Amudarya from the most ancient times up to the 17th century."
	In 1899, Captain K. Girshfeld publishes the results of his investigations of the Amudarya delta, providing data on the most recent changes in the delta arms and also about the sea water rise (revised by General Galkin).
1903	In Russian Turkestan, the American "archaeological" expedition works. Among the members is E. Huntington, who later became a noted geographer.
1904	On the initiative of L.S. Berg, a depth gauge is installed near the railway station "Aral Sea."
1905	Russian scientists A.A. Tillo and Yu. M. Shokalskiy publish their study in which the area of the Aral Sea is included.
1905–1906	The Tashkent Military-Topographical Division produces a ten-versta map of the Lower Amudarya on three sheets according to the 1890 survey results.
1906	L.S. Berg surveys the coasts, depths, structure of water, and the fauna of the Aral Sea in detail.
	L.S. Berg visits the northern coast of the Aral Sea by the route: railroad station "Aral Sea" – Kulandy Peninsula –the Greater Barsuki sands – railroad station "Chelkar."
1907	American geographer E. Huntington issues the book, "The Pulse of Asia," in which the hopelessness of development of irrigated agriculture in Central Asia is demonstrated.
	L.S. Berg publishes the "Journal of Hydrological and Meteorological Observations in the Aral Sea in 19900–1902: Scientific Results of the Aral Expedition."
1908	L.S. Berg publishes the book, "The Aral Sea: Experience of the physico-geographical monograph."
	A.I. Voeikov publishes the paper, "Irrigation in the Trans-Caspian region from the point of view of geography and climatology," in the News of the Imperial

Years	General Events
	Russian Geographical Society. L.S. Berg names this paper "the best enrichment of world geographical literature".
	A.I. Voeikov writes an article, "Cotton-growing in Turkestan Territory and conditions of its development," emphasizing in it that "In Amu there is so much water that it is possible to irrigate more than 4 millions hectares."
1910	The regulation "About access to private businessmen to make surveys for irrigation of the grounds in Turkestan" is issued.
	An expedition to the Amudarya is organized to study the natural conditions in its basin until 1917. The materials obtained are generalized by V.V. Tsinzerling.
1911	At the Turkestan Board of agriculture and state property, the Hydromodular Department is organized, headed by land-reclamation engineer A.N. Kostyakov (later Academician of VASKhNIL (the All-Union Academy of Agricultural Sciences of the Soviet Union in 1929–1992), Corresponding Member of the USSR Academy of Sciences and the founder of Soviet ameliorative science).
	M.N. Ermolaev (Russia) works out the schematic project of irrigation of the lower reaches of Kashkadarya in two turns: (1) to irrigate the ground in the east area of the Kerkinsk District with waters from the rivers Kashkadarya and Guza-Darya; (2) to irrigate the southern area of the Kelifo-Kerkinsk District with waters of Amudarya.
	Department of ground improvements under the direction of E.N. Blumberg (Russia) organizes a field expedition to investigate irrigation in Bukhara.
	K.K. Gilzen, having analyzed samples taken by L.S. Berg, publishes the first data about the ground composition of the Aral Sea.
1912	A.G. Ananyev receives the first irrigational concession in Bukhara Emirate (72.5 thousand dessiatinas of grounds).
	A.I. Voeikov writes the book, "Le Turkestan Russe" (Russian Turkestan), especially for foreign readers.
1913	Opening of the Romanov's canal – the first successful irrigational project in all Turkestan. The area of irrigation is 32 thousand dessiatinas with waters from the Syrdarya.
1914	V.V. Bartold publishes the book "History of irrigation in Turkestan."
1915	F.P. Morgunenkov (Russia) proposes a project of irrigation on the empty lands of the southeastern coast of the Caspian Sea using winter and superfluous waters of the Amudarya. Water supply to the canal from Amudarya would be organized above Nukus city at the mouth of the Takhiatash. After realization of this project, F.P. Morgunenkov asserts, the Trans-Caspian Territory will be turned into "Russian California and Russian Egypt."
1918	Resolution on the organization of the Turkestan Autonomous Republic is accepted in Tashkent.
	V.I. Lenin, Russian leader, signs the decree of Sovnarkom (government) "About the assignment of 50 million roubles for irrigational works in Turkestan and about the organization of these works."
	Turkestan Board of Water Management "Turkvodkhoz" and Technical Irrigational Committee are organized.
	All main canals and irrigation construction in the Turkestan Republic goes under jurisdiction of the National Commissariat of Agriculture of Russia.
	Resolution on establishment of the Turkestan Autonomous Republic is adopted in Tashkent.

Years	General Events
1919	Formation of the Belo-Cossack front in Chimbae (Khorezm).
	Commander of the Kustanai Peasantry Unit Zhilyaev organizes a counter-revolutionary revolt on the railroad "Aral Sea."
	Fierce battles with Cossack troops of ataman Dutov on the railroad "Aral Sea."
	Nukus fortification is conquered by the Red Army in Khorezm.
1920	The proclamation of the former Khivan Khanate as Khorezm People's Soviet Republic.
	Bukhara People's Soviet Republic is established.
	Decree of Council of People's Commissars of RSFSR (Russian Federation) "About restoration of cotton culture in the Turkestan and Azerbaijan Soviet Socialist Republics."
	The Council of People's Commissars of RSFSR approves the Head Committee for Water Management and Land Improvement of VSNKh (Supreme Council of the Russian (1917–1932) and the Soviet Union (1963–1965) National Economy), which is responsible for the management of irrigation works in Turkestan.
	Hydrographic works are initiated in the Aral Sea to ensure safe navigation of the Aral military fleet. The Chief Hydrographic Department at the Fleet Headquarters organizes the Flying Hydrographic Squad to conduct reconnaissance in the northern part of the sea.
1920–1921	The fishery expedition of F.A. Spichakov works on the Aral Sea.
1921	G.K. Rizenkampf (Russia) proposes a schematic civil-engineering project of the Trans-Caspian canal with water draw-off from Amudarya near the confluence of Vakhsh and Pianj rivers at a length of 1500 versts (1600 km) for irrigation of 300 thousand dessiatinas in Afghanistan and 2200 thousand dessiatinas in the Trans-Caspian territory.
	To continue hydrographic works on the Aral Sea, the Aral-Balkhash Hydrographic Expedition (led by F.F. Kotelnikov) is formed to replace the Flying Hydrographic Squad. After implementation of certain works in autumn of the same year, the expedition is dissolved.
	In view of hunger in Russia, V.I. Lenin appeals with a letter "To Workers and Fishermen of the Aral Sea," in which he asks to render assistance to 8 mln deprived people and 7 mln children. 14 railroad cars of fish are transported to the hungry people.
	"Short Sailing Directions over the Aral Sea and the Amudarya Delta" by A. Malinin is published.
	Expedition of S.D. Muraveisky to Kamyshlybash Lake.
1924	Session of the Central Executive Committee of Soviets of the USSR accepts the decision to form new Soviet Socialist Republics in Central Asia.
	Perovsk City is renamed Kyzyl-Orda and is the capital of Kazakhstan until 1929.
1925	At the First Turkmen Congress of Soviets, a proposal is made to transfer the Amudarya waters to the southern oases of the Republic and to the western part of Turkmenistan.
	Observations over the Aral Sea level in its southern part are initiated at the Muinak station; they are carried out with some interruptions until 1964.
	First soil studies in the Karakalpak part of the Amudarya delta under the supervision of S.S. Neustruev are conducted.
1927	The water from Amudarya are directed to Kelif Uzboi. A trial flow passes through the chain of Kelif shallow gullies for 100 km.

Years	General Events
	Department of Amudarya delta irrigational systems is established in Novo-Urgench.
	V.V. Tsinzerling (USSR) publishes the book, "Irrigation on Amudarya."
	F. Kolácek publishes «Était l'Ousboï pendant les temps historiques un ancien lit de l'Amou-Daria?»
1928	"Preliminary Sailing Directions over the Aral Sea" by A. Malinin is published.
1929	The opening of the Aral fish industrial station.
1930	Technical and economic report, "The Problem of supplying the western areas of Uzbekistan with waters of Amudarya," is issued.
	The work of M.T. Khailov, "Aral Basin," is published in the Proceedings of the Institute of Fishery Problems.
1932	New phase in the studies of Aral Sea and its fauna begins in the 1930s under leadership of A.L. Bening (Russia). Hydrological and hydrobiological surveys in the all water areas of the Aral Sea are performed, enabling a complete listing of species of aquatic invertebrates in the Aral Sea. According to these data in zoobenthos there are about 50 species of invertebrates.
	In the "GGI Proceedings" L.S. Berg publishes his work "On the Aral Sea Levels."
1933	Construction of meat-packing factory in Muinak begins.
1934–1935	The Proceedings of the Aral Branch of the Fishery Institute publishes the work of A.L. Bening, "Hydrological and Hydrobiological Materials for Preparation of the Aral Sea Fishery Map," which is prepared from the results of the 1932–1933 expedition, Vol. III (1934), Vol. IV (1935).
1935	In the Karakalpak Republic, the main canal, Kyzketken, with a length of 25 km with a charge of 210 m^3/s at the head, is constructed. The water draw-off is from the Amudarya.
1936	The Hydrographic Division of the RKKA (Russian Red Army) Navy Department reprints the map of the Aral Sea of A.I. Butakov and K.Ye. Pospelov at a scale of 1:606546 along parallel 45°.
	A station to observe the sea hydrological regime is opened on Vozrozhdenia Island (existed until 1949).
1937–1939	Canals Su-Eli and Leninyab are united into one large canal with engineered head constructions.
	The new canal is named for V.I. Lenin.
	Decision of the Central Committee of VKP (b) (All-Union Communist Party of bolsheviks, 1925–1952) and SNK (Council of People's Commissars, 1923–1946) of the USSR "About measures on the further increase of the cotton industry in Uzbekistan."
1939	A nature preserve is organized on Barsakelmes Island to protect the desert complex and restore the population of typical animals: saiga, gazelle, yellow gopher.
1940	Decision of the USSR Government and Central Committee of VKP (b) "About measures on the further rise of agriculture and, in particular, growing cotton of the soviet long-staplesort in Turkmen SSR." By the decision it has been told: ". . . to begin after 1940 to pass the Amudarya waters through the canal of Kelif Uzboi and to supply Amudarya water to the basins of the Murghab and Tedzhen rivers to further develop of irrigation there."

Years	General Events
	Canal "Lenin" (110 km long with a head charge of 240 m^3/s) is constructed for irrigation of the grounds on the left coast of Amudarya in the Hodzhely and Kungrad regions.
	The monograph of G.V. Nikolsky, "Fish of the Aral Sea," is published.
1941	The meat plant in Muinak is constructed. At first it manufactured beef and tortoise canned meat, later on it changed over to fish products.
1942	The station "Tigrovy" is opened on the southern coast of the Aral Sea to monitor the sea level; it exists until 1979.
1943	On Syrdarya the Kafakum water reservoir is built.
1944	Central Committee of VKP (b) accepts "Decision about measures for reconstruction and development of the cotton industry in Uzbekistan."
1945	Decision by Sovnarkom (SNK) of the USSR "About measures on restoration and the further development of the cotton industry in Uzbekistan."
1946	B.D. Zaykov (USSR) publishes the book, "Modern and future water balance of the Aral Sea."
	Decision of 1945 is promulgated by special decision of Sovnarkom of the USSR "About the plans and actions for restoration and further rise of the cotton industry in Uzbekistan for the period of 1946–1953."
1947	GOIN Proceedings publishes the work of V.S. Samoilenko, "The Near Future of the Aral Sea."
	The work of G.V. Nikolsky, "Irrigation Construction in the Aral Basin and Fishery Problems," is published.
	V.P. Zenkovich publishes a detailed investigation of "Bottom Sediments of the Aral Sea."
	The work of E.M. Murzaev, "Central Asia," is published.
1948	Farhad hydrosystem, a hydroelectric power station and water reservoir of daily regulation with useful volume of 0.15 km^3, is constructed, providing a main water draw-off from Syrdarya for irrigation of all the Hungry and Dalverzinsk steppes.
1949	L.S. Berg in the academic series, "Results and problems of a modern science," releases the book, "Studies of the history of Russian geographical discoveries," in which he includes the chapter, "A.I. Butakov – researcher of the Aral Sea."
	Decision of Council of Ministers of the USSR from March 19, 1949, No. 1140 "About actions for further development of cotton production in Tadjik SSR for 1949–1952."
	In the Takhtakupyr Region of Karaklpakstan, fishermen kill the last Turanian tiger.
	The hydrometeorological station is opened on Barsakelmes Island.
1950	Under I.V. Stalin's initiative, the decision of the Council of Ministers of the USSR from September 11, 1950 No. 3906 "About construction of the Main Turkmen canal Amudarya - Krasnovodsk, about irrigation and watering of the grounds of southern areas of the Caspian plains in the western Turkmenistan, and about the lower reaches of Amudarya and the western part of Karakum desert" is accepted.
	Construction of the Main Turkmen canal begins.
	Fries of the Aral pike perch are transferred to Balkhash Lake.
	The work of G.V. Nikolsky and M.A. Fortunatov, "Irrigation Construction and Fishery Problems of the Aral," is published.
	The article "Aral Sea" is included into Vol. 2 of the Great Soviet Encyclopedia.

Years	General Events
1951	Decision of Council of Ministers of the USSR from April 30, 1951 No. 1426 "About measures on support of implementation by the Ministry of Cotton Industry of the USSR, the Ministry of Agriculture of the USSR and the Ministry of State Farms of the USSR in design and survey and building of works on irrigation and watering the grounds in the connection with construction of Main Turkmen canal Amudarya – Krasnovodsk".
	Beginning of construction of Amu-Bukhara canal.
	Beginning of historical-morphological works in the Circum-Aral Area under supervision of S.P. Tolstoy.
	In the Amudarya delta, wide-scale soil studies are carried out related to the design and construction of the Main Turkmen Canal.
	The USSR Navy Hydrographic Department publishes a map of the Aral Sea with the mark, "Prepared on the basis of the 1850 Butakov's map as amended in 1921, 1929, 1931, 1933 and 1934."
1952	Decision of the Council of Ministers of the USSR from September 2, 1952 No. 3975 "About irrigation and land-reclamation for the further development of cotton production in the Andizhan, Namangan, Fergana, and Surkhandarya regions of Uzbek SSR."
1953	Decision of Council of Ministers of the USSR from October 22, 1953 No. 2673 "About reorganization of department of construction of "Sredazgidrostroy" (construction company) of the Ministry of Agriculture and Purchases of the USSR into the Trust for building of water-economic constructions in cotton-growing areas of Central Asia."
	In connection with the discovery of the large fresh water lens of "Yaskhan" (sufficient to satisfy requirements of the Western Turkmenistan), the construction of the Main Turkmen canal ceases.
	The Publishing House of the Uzbek Academy of Sciences in Tashkent publishes the results of the Butakov expedition "Daily Records of A.I. Butakov Sailing on the Schooner "Konstantin" for Study of the Aral Sea in 1848–1849" 100 years after written.
	The Moscow Society of Natural Scientists publishes the work of A.L. Yanshin, "Geology of the Northern Circum-Aral Area."
1954	Decision of Central Committee of the CPSU (Communist Party of the Soviet Union) and Council of Ministers of the USSR "About the further development of cotton production in Uzbek SSR in 1954–1958."
	Decision of Central Committee of CPSU and Council of Ministers of the USSR from April 21, 1954 No.747 "About the further development of cotton production in Turkmen SSR for 1954–1958."
	Decision of Central Committee of the CPSU and Council Ministers of the USSR from June 5, 1954 No.1114 "About the further development on cotton production in Tadjik SSR for 1954–1960."
	Beginning of Kara Kum canal construction.
	The Proceedings of the Lake Study Laboratory of the USSR AS publishes the work of N.Z. Khusainova, "Zoobenthos of the Aral Sea."
	The work of S.P. Tolstoy and A.S. Kes, "Problems of Ancient Amudarya Flow in Light of the Most Recent Geomorphological And Archeological Data," is published.
1955	The work of V.A. Kovda and V.V. Egorov, "Regularity in Salt Accumulation in the Aral-Caspian Depression" is published.

Years	General Events
1956	L.K. Blinov (USSR) publishes the book, "Hydrochemistry of the Aral Sea."
	Decision of the Central Committee of CPSU and the Council of Ministers of the USSR from August 6, 1956 No.1059 "About irrigation of virgin lands of the Hungry Steppe in Uzbek and Kazakh SSR for increasing cotton production."
1957	M.M. Rogov (USSR) published the book "Hydrology of Amudarya delta".
1958	G.V. Lopatin, R.S. Dengin, and V.V. Egorov (USSR) publish the book, "Delta of Amudarya".
	Decision of a Central Committee of the CPSU and Council of Ministers of the USSR from June 14 1958 No.645 "About the further expansion and acceleration of works on irrigation and land-reclamation in Uzbek SSR, Kazakh SSR and Tadjik SSR."
	Resolution of the CPSU CC and the USSR Council of Ministers No.841 of July 25, 1958 "On actions for development of irrigated lands in the zone of the first stage of the Karakum canal and construction of the second stage of this canal in the Turkmen SSR."
	The Caspian Expeditionary Hydrographic Unit begins functioning on the Aral Sea. The works are completed in 1960. The maps of the Aral Sea and the Sailing Directions are published.
	The monograph of P.A. Letunov, "Soil and Land Reclamation Conditions in the Lower Amudarya" is published.
1959	V.I. Lymarev (USSR) publishes the book, "The Aral Sea."
	The Karakalpak Branch of the Uzbek Academy of Sciences is established.
1960	The average absolute level of the Aral Sea is +53.40 m and the inflow of water to the sea via Syrdarya and Amudarya is 56.0 km^3, with mineralization of water at 7.2 g/l.
	The small work of Sea Captain P.P. Tum, "Aral Sea. Navigation: Geographical and Hydrometeorological Notes," is published.
	In Moscow, the work of S.P. Tolstoy and A.S. Kes, "The Lower Amudarya, Sarykamysh, Uzboi: History of Formation and Settlement" is published.
1961	Decrease of the Aral Sea level begins.
	Self-flowing South Hungy Steppe canal (length 127 km, charge in the head 360 m^3/sec) named for A.A. Sarkisov is constructed.
	Karshinskiy canal is constructed.
	The Aral Sea level is +53.3 m.
1962	Decision of Central Committee of the CPSU and the Council of Ministers of the USSR from June 29, 1962 No.747 "About measures for liquidation of backlog in development of cotton production in Turkmen SSR."
	The Aral Sea level is +53.0 m.
1963	Hydrographical Service of the USSR Navy publishes the "Sailing directions in the Aral Sea."
	Order of the Council of Ministers of the USSR from December 24, 1963 No.2540 "About spade-work for irrigation and land-improvement of steppe Karshinsk in Uzbek SSR and steppes of Kyzyl-Kum in Kazakh SSR."
	On the basis of Glavgolodnostepstroy (construction company) of the Ministry of Agriculture of the USSR, the Main Central Asian Board for irrigation and construction of state farms is established.

Years	General Events
	On Amudarya, the construction of the Takhiatash hydrounit begins which guarantees water-security for up to 900 thousand hectares of grounds in the lower reaches of Amudarya.
	In Leningrad, the "Integrated Hydrometeorological Atlases of the Caspian and Aral seas" are published in one volume.
	The Aral Sea level is +52.6 m.
1964	Uzbekistan produces 4 million tons of raw-cotton.
	In Tashkent, the work of V.L. Shulz and L.I. Shalatova, "Water Balance of the Aral Sea," is published.
	The work of O.M. Zhitomirskaya, "Climatic Description of the Aral Sea Region," is published.
	The "Oceanographic Tables for the Caspian, Aral and Azov Seas" is published.
	The Aral Sea level is +52.5 m.
1966	Plenum of the Central Committee of CPSU accepts the Decision "About the wide development of land reclamation for reception of high and steady crop grains and other agricultural crops." In its frameworks, the plan of irrigation is developed and drainages of the grounds are designed for 10 years (1966–1976).
	Karakalpakrissovhozstroy (construction company) is organised. All rice state farms from various ministries and departments are included in it.
	The Aral Sea level is +51.9 m.
1967	V.I. Lymarev (USSR) publishes the book, "Coasts of the Aral Sea: an internal water reservoir of the arid zone."
	The Dzhizhak water reservoir of seasonal regulation with a full (and useful) volume of 0.09 km^3 is constructed on the Sanzar River.
	The Central Committee of Uzbekistan's Communist Party and the Council of Ministers of the Uzbek SSR accept the Decision "About the urgent measures on increases in production of rice, increases of water-security, and ameliorative improvements of the grounds of collective farms and state farms in the Karakalpak SSR."
	The Aral Sea level is +51.6 m.
1968	The work of M.M. Rogov, S.S. Khodkin, and S.K. Revina "Hydrology of the Amudarya Mouth Area" is published.
1969	The Institute of Geography of the Academy of Sciences of the USSR publishes collected articles, "Problems of the Aral Sea," edited by S.Yu. Geller.
	B.V. Andrianov publishes the book, "Ancient irrigating systems of Priaralye (in accordance with the creation and development of irrigation)."
	GOIN Proceedings publish the work of V.P. Lvov, "Aral Sea Level Fluctuations for the Past 100 Years."
	The Aral Sea level is +51.3 m.
1970	Decision of the Central Committee of the CPSU and the Council of Ministers of the USSR from June 25, 1970 No. 482 "About acceleration of works on irrigation and land development of the Karshinskaya steppe in Uzbek SSR."
	For the first time, the future of Aral Sea is considered in the report "Perspectives of development of land reclamation for 1971–1985, regulation and redistribution of the rivers run-off," prepared by the State Planning Committee, the Ministry of Agriculture, Minvodkhoz of the USSR, and VASKhNIL and approved by the Central Committee of CPSU and the Council of Ministers of the USSR on 24.07.1970 (No. 612).

Years	General Events
	Beginning of the construction of the Tuyamuyunskiy hydrosystem, 452 km from the mouth of the Amudarya, which will enable irrigation of 500 hectares of land. With the hydrosystem, three coastal water reservoirs are constructed: Kaparas - with full capacity of 1 km^3, Sultansanjar at 2.7 km^3,and Kosibulak at 1.5 km^3.
	Navigation across the Aral Sea ceases.
	"The Atlas of Ice of the Aral Sea" is published.
	The Institute "Sredazgiprovodkhlopok" prepares the "Scheme of Integrated Water Resources Management of the Aral Sea Basin."
	The "Master Scheme of Water Resources Management of the Amudarya" is prepared
	The Aral Sea level is +51.4 m, surface – 60 300 km^2, volume – 964 km^3.
1971	State Planning Committee of the USSR on the basis of Decision of Central Committee of CPSU and Council of Ministers of the USSR from 16.04.1971 "About measures on the further development of land reclamation and their agricultural development on 1971–1975" instructs the Ministry of Water Management of the USSR to develop actions to organize research and design work to transfer part of the run-off of northern rivers into the riverine basin of the Volga River and of Siberian rivers into the basins of the Syrdarya and Amudarya Rivers.
	Decision of the Council of Ministers of the USSR on May 13, 1971 No. 284 "About measures for further development of national economics of Turkmen SSR" (in 1971–1975 to have 105 thousand hectares of irrigated grounds, 200 thousand hectares in improvements of ameliorative conditions, 80 thousand hectares of reconstruction of irrigation systems, and 2 thousand hectares of capital planning).
	Beginning of construction of Karshinskiy canal with water draw-off 5 km^3 annually. Construction of Large Karshinskiy canal began. Six pump stations lift water for 150 m above the Amudarya with the charge of 240 m^3/s for about 5 km^3/yr.
	Beginning of adverse impact of increased salinity in the Aral Sea on zooplankton and zoobenthos.
	The Aral Sea level is +51.1 m.
1972	In the Russian scientific journal, "Water Resources" No. 1, a scientific report on the problem of the Aral Sea is published by the group of leading scientists of the Institute for Water Problems of the USSR Academy of Sciences.
	"Scheme of complex use of water resources of the Aral Sea Basin" is developed.
	Decision of Central Committee of Uzbekistan Communist Party and Council of Ministers of Uzbek SSR "About measures on the further rise of agriculture in the Karakalpak SSR."
	The Aral Sea level is +50.5 m.
1973	M.N. Novozhilova (USSR) publishes the book, "Microbiology of the Aral Sea," in Alma-Ata (Kazakhstan). Since 1973 amphipods have not been found in the Aral Sea.
	Amudarya's water reaches the steppe of Karshi.
	The work "Modern and Future Water and Salt Balance of the USSR Southern Seas" (Monograph) is published.
	The Aral Sea level is +50.2 m.
1974	The "Atlas of invertebrates of the Aral Sea" is published. According to this book, in 1971 in the Aral Sea there were 186 species of aquatic invertebrates.

Years	General Events
	Opening of the Takhiatash hydrosystem, the last one constructed on the lower Amudarya before the Aral Sea.
	The Aral Sea level is +49.9 m.
1974–1976	Disturbance of natural reproduction processes due to the Aral Sea salinity growth result in large-scale development of euryhaline species of zooplankton, loss of fresh-water species of zoobenthos, threefold reduction of phytoplankton numbers and biomass, loss of most of fresh-water planktonic algae.
1975	Under the leadership of academician I.P. Gerasimov (USSR), the temporary scientific and technical commission of GKNT (the USSR State Committee for Science and Techniques) for estimation of the influences of changes of the Aral Sea level on environment and economy of the surrounding region works.
	In Tashkent, the First Coordination Meeting on studying the influence on the environment and on estimating the socio-economic consequences of the Aral Sea level decrease takes place.
	A.N. Kosarev publishes the book, "Hydrology of the Caspian and Aral Seas," through the publishing house of Moscow State University.
	The SARNIGMI Proceedings publishes the work of V.L. Shultz, "Water Balance of the Aral Sea."
	The Aral Sea level is +49.0 m.
1976	The plans of the USSR State Committee for Science and Technology (SCST) includes unit 05.14 of scientific-technical problem 0.85.01 "Study of the environmental impact and assessment of the socioeconomic consequences of the Aral Sea level drop, elaboration of the scientific basis of actions on prevention of negative consequences of such drop, and their submission to the USSR State Committee for Planning (SCP), SCST,USSR Ministry of Water Management, USSR Academy of Sciences, Council of Ministers of Uzbekistan, and Council of Ministers of Kazakhstan." The head organization for unit 05.14 is appointed as the Institute of Geography of USSR AS.
1977	In Alma-Ata (Kazakhstan) the Second Coordination Meeting on studying the influence on the environment and on estimation and socio-economic consequences of the Aral Sea level decrease is held.
	The All-Union meeting, "Scientific bases of actions for prevention of negative consequences of the Aral Sea level decrease" is held in Moscow.
	The visiting meeting of the Scientific Council on the "Comprehensive study of desert development of Central Asia and Kazakhstan" is held in Nukus and considers the Aral issues.
	Moscow holds the All-Union Meeting on "Variations in wetting in the Aral-Caspian region in the Holocene" and is organized by the SCST Scientific Council and USSR AS on integrated study of the Caspian Sea problems, the Institute of Water Problems of USSR AS, and the Commission of the USSR AS on application of historical data in the practical economic activities of the USSR.
	The Construction Trust "Priaralvodstroy" is established within the framework of the Ministry of Water Management of Uzbekistan.
	The Aral Sea level is +47.6 m.
1978	Decision of the Central Committee of the CPSU and Council of Ministers of the USSR to "carry out research and design work on the problem of

Years	General Events
	transferring northern and Siberian rivers to the southern areas of the country" is accepted.
	The collection, "Aral Sea Problems and Nature Conservation Activity," is published in Tashkent.
	SOPS of the Uzbek Academy of Sciences in cooperation with the Karakalpak Branch of the Uzbek AS develop the "Scientific basis for development of production forces of the Lower Amudarya territorial-production complex (KKA SSR and Khorezm Region)."
	The Aral Sea level is +47.1 m.
1979	Filling of the Tuyamuyun reservoir with water from the Amudarya begins.
	A national park is established at Barsakelmes Island.
	The Aral Sea level is +46.5 m.
1980	In Nukus, a session of the Presidium of the USSR Academy of Sciences convenes in the form of scientific-practical conference, "Problems of the Aral Sea and Amudarya delta."
	At the GKNT meeting, the decision on the preparation of a special technical and economic report on the Aral Sea problem, entrusted to Soyuzgiprovodhoz, is accepted. In the framework of this task, research activities are planned and formulated "To investigate the influence of water-economic actions on the state of the Aral Sea and the social and economic developments of Priaral'e connected with the decreasing level. To develop scientific bases and actions for rational use and protection of natural resources under conditions of anthropogenous desertification of Priaral'e." A leading scientific organization, Institute of Geography of the USSR Academy of Sciences, is appointed.
	In the Northern Aral the last fish is caught.
	Academician I.P. Gerasimov together with N.T. Kuznetsov and M.T. Gorodetskaya write the paper, "Modern Tasks of Investigations on the Aral Sea Problem" (Newsletters of USSR AS, Geographical Series).
	In Moscow, the Institute of Water Problems of the USSR AS publishes the collection, "Variations in wetting in the Aral-Caspian region in the Holocene."
	The Aral Sea level is +45.8 m, surface – 51 700 km^2, volume – 644 km^3.
1981	In Moscow there convenes the All-Union Coordinating Working Meeting under the task of GKNT.
	A.A. Rafikov and G.F. Tetyukhin publish the book, "The decrease in the Aral Sea level and change of environment of lower reaches of Amudarya" in Tashkent (Uzbekistan).
	In Almata, the collection, "Natural Resources of the Modern Circum-Aral Area," prepared by Kazakh AS and SOPS of Kazakhstan (only for internal use), is published.
	Resolution of USSR SCST and USSR SCP No. 76/61 of 25.03.1981 "On prediction of socioeconomic development of the Circum-Aral Area for 2000–2020 with regard to prevention of undesirable consequences of man-made desertification" is adopted.
	In Tashkent, the book by R.T. Tleulov, "New Regime of the Aral and Its Effect on the Ichthyofauna," is published.
	The Aral Sea level is +45.2 m.
1982	In Tashkent the monograph of A.A. Rafikov, "Natural Conditions of the Drying Aral Seabed," is published.

Years	General Events
	Amudarya waters did not reach the Aral Sea for the first time.
	The Aral Sea level is +44.4 m.
1983	Institute of Geography of the USSR Academy of Sciences with participation of SOPS (Council for investigation of productive forces) of the USSR State Planning Committee and Soyuzgiprovodhoz prepare and transfer to the planning departments and Central Committee of the CPSU a special "Report concerning the degradation of the Aral Sea ecosystem and deltas of Amudarya and Syrdarya and the anthropogenous desertification of Priaral'e caused by irrevocable withdrawal of run-off of the Central Asian rivers with the purpose of intensification of irrigated agriculture."
	The All-Union coordination working meeting on the Aral Sea problem under the task of GKNT is held in Moscow.
	In Moscow, the collection "Paleography of the Caspian and Aral Seas" is published.
	In Almata, the book of U.M. Akhmedsafin et al., "Underground Water and Salt Runoff in the Aral Sea Basin" is published.
	The Aral Sea level is +43.6 m.
1984	The book "Problems of the Aral Sea and Amudarya delta" is published in Tashkent.
	The Aral Sea level is +42.8 m.
1985	The regional session of the Central Asian branch of VASKhNIL devoted to questions of the Aral and the Amudarya delta convenes in Nukus (Uzbekistan).
	The final all-Union coordination-working meeting under the auspices of GKNT is held in Moscow.
	Technical and economic report by Soyuzgidrovodkhoz on regulation of a water mode of the Aral Sea is prepared with the help of the Institute of Geography of the USSR Academy of Sciences and other organizations of the Academy of Sciences of Uzbekistan, Kazakhstan, and Turkmenistan.
	The Aral Sea level is +41.9 m.
1986	Decision of the Central Committee of the CPSU and Council of Ministers of the USSR "About stopping the work to transfer parts of the northern and Siberian rivers run-off."
	Session of the USSR Academy of Sciences on the problem of the Aral, at which 78 organizations from Moscow, Leningrad, Uzbekistan, Turkmenistan, and Kazakhstan participate, convenes in Nukus (Uzbekistan).
	Decision of the Central Committee of the CPSU and the Council of Ministers of the USSR from March 17, 1986 No. 340 "About measures on acceleration of economic and social development of Karakalpak SSR."
	The work of F.I. Khakimov and N.F. Deeva, "Formation of Soil Cover in the Amudarya Delta," is published in Moscow.
	The Aral Sea level is +41.1 m.
1987	Writer and journalist Yu.D. Chernichenko, in his TV programs entitled "Selsky chas" and "Prozhektor perestroyki," for the first time highlights the problems of the Aral Sea desiccation.
	Within the system of the Ministry of Land Improvement and Water Conservation of the USSR, the basin water-economic companies (BVO) "Amudarya" and "Syrdarya" are created.
	Session of specially created Governmental Commission on the ecological situation in the Aral Sea basin under academician Yu.A. Izrael's presidency.

Years	General Events
	The Government Commission for preparation of offers on improvement of water supply for national economics is created under the head of academician V.A. Koptyug, Vice-president of the USSR Academy of Sciences.
	State Commission headed by Yu.A. Izrael, Head of Goskomgidromet (Meteorological Service) of the USSR, works in Nukus to study the ecological and sanitary state of the Aral Sea coast.
	Group of scientists of the Institute of Water Problems of the USSR Academy of Sciences sends a letter to the Central Committee of the CPSU expressing their anxiety about the destiny of water supply in the republics of Central Asia.
	NGO "International Association to save Aral and Balkhash lakes" under the leadership of Kazakh writer Mukhtar Shakhanov is established.
	G.I. Molosnova, O.I. Subbotina, and S.G. Chanisheva publish the book, "Climatic consequences of economic activities in the zone of the Aral Sea."
	Rubanov I.V., Ishniyazov D.P., Baskakov M.A., and Chistyakov P.A. publish the book "Geology of the Aral Sea."
	NGO "Soviet cosmonauts to save the Aral Sea" under the leadership of cosmonaut G. Beregovoy is established.
	The Small Sea separates from the Large Aral Sea and a dam appears.
	The Aral Sea level is +40.3 m, surface – 41 100 km^2, volume – 401 km^3, mineralization – 20 g/l.
1988	Decision of Central Committee of the CPSU and the Council of Ministers of the USSR "About measures on radical improvement of ecological and sanitary conditions in the Aral Sea region, increasing the efficiency of use and strengthening protection of water and ground resources in the basin."
	Report of a Governmental Commission, "Modern condition and offers on cardinal improvement of ecological and sanitary and epidemiologic conditions in the Aral Sea region and lower reaches of Amudarya and Syrdarya rivers," is prepared.
	Complex scientific-journalistic expedition "Aral-88," headed by G.I. Reznichenko, is organized under the initiative of the editorial boards of the journals "Pamir" and "Novy Mir" (USSR).
	First All-Union meeting on the problems of the Aral basin, "Ecology and the literature," is organized by the Public Committee on rescue of the Aral at the Union of Writers of Uzbekistan. The meeting is mobile, being held in Tashkent, Nukus, Muinak, Uchsai, Tashauz canal, and Horezm.
	The "Days of Aral" are observed in Moscow in the Central House of writers and in editorial offices of some newspapers and magazines under the initiative of the Uzbek Public Committee on the Rescue of the Aral, the Kazakh Public Committee on problematic of Aral and Balkhash, and also the international movement of poets "20th Century. The world and Ecology".
	A company, "Aralvodstroy," directed towards rescuing the Aral, is created in Nukus, Uzbekistan.
	The crosspiece between the Small and Large Aral disappears because of the abundance of water.
	Moscow journal "Ogonek" publishes selected materials on the Aral Sea issues.
	In Nukus, the meeting of Party and economic officials of Uzbekistan is held to discuss the implementation of the Resolution of CPSU CC and USSR Council of Ministers of September 19, 1988.
	The Institute "Soyuzgiproris" prepares a feasibility report, "On priority actions in the Syrdarya delta and coastal zone of the Aral Sea nearby Aralsk."
	The Aral Sea level is +39.7 m.

Years	General Events
1989	N. Aladin and V. Khlebovich edit the book, "Hydrobiological problems of the Aral Sea," which is published in the Proceedings of Zoological Institute of the USSR Academy of Sciences.
	Decision of the Supreme Soviet of the USSR "About urgent measures on ecological improvement of the country." which ascertains that "... the situation in the region of the Aral Sea is practically out of the control. Priaral'e has become a zone of ecological catastrophe."
	Decision of the Council of Ministers of the USSR creates Research Coordination Centre "Aral," headed by V.M. Kotlyakov, Director of the Institute of Geography of the USSR Academy of Sciences.
	Decision of the Council of Ministers of the USSR "About the organization and work on artificial increases in deposits in mountain areas of Central Asia with a view of updating the water-stocks of the Amudarya and Syrdarya rivers and the Aral Sea."
	The Aral problem is discussed at a specially called Academic Council of the Institute of Geography of the USSR Academy of Sciences.
	A group from the public Aral movement addresses the Second Congress of People's Deputies of the USSR, and also the Political Bureau of the CPSU, Supreme Soviet of the USSR, and the Council of Ministers of the USSR with an appeal to reconsider proposed solutions to the Aral ecological crisis and to move the focus to the solution of urgent social problems.
	The symposium, "Development of the concept of socio-ecological development of the lower reaches of the Amudarya (Priaral'e) as a special economic zone" is convened in Shavat city (Uzbekistan).
	Selected papers are published in the magazines "Novy Mir" ("New World") and "The Aral Accident."
	Well-known geographer N.T. Kuznetzov publishes in the News of the USSR Academy of Sciences (in the Geographical Series) an "Open letter to scientists, writers, water managers, and everyone who is worried about the ecological situation in Priaral'e."
	The meeting, "Problems of the Aral and Priaral'e," convenes in Nukus (Uzbekistan).
	Scientists, writers, poets, and people in charge of cultural and educational activities of Uzbekistan create a special fund for the rescue of the Aral.
	The Aral problem is discussed at the specially called Academic Council of Institute of Water Problems of the USSR Academy of Sciences.
	The Large Aral Sea separates from the Small Aral Sea.
	Journal "Science" (USA) in September publishes an article by Ph. Micklin, "Desiccation of the Aral Sea: A Water Management Disaster in the Soviet Union."
	Journal "Pamir" publishes selected articles of the participants of the "Aral-88" expedition, in particular of D.B. Oreshkin ("Aral in the Flow of Time"), V. Selyunin ("Burden of Actions"), G. Reznichenko ("We know/knew What Is on Balance. . ."), as well as the shorthand report of the "round-the-table" meeting of journals "Novy Mir" and "Pamir" on the results of the "Aral-88" expedition.
	The Aral Sea level is +39.1 m.
1990	N. Aladin and L. Kuznetzov edit the book, "Modern state of the Aral Sea under conditions of progressing salinization," which is published in the Proceedings of Zoological Institute of the USSR Academy of Sciences.

Years	General Events
	Supreme Soviet of the USSR recognizes the Priaral'e as a zone of ecological catastrophe.
	The Supreme Soviet of Turkmen SSR accepts the decision "About the ecological state of territories of Priaral'e of Turkmen SSR and measures for its improvement." It recognizes that territories of the Tashauz region and the Dargan-Atinsk district of the Chardzhou region are zones of ecological catastrophe.
	Union-republican consortium "Aral," whose founders will become the governments of the Uzbekistan, Kazakhstan, Kirgizstan, Tadjikistan, Turkmenistan, and Karakalpak republics, and state concern "Vodstroi," Khorezm, Kyzyl-Orda, and Tashauz Executive Committees is created.
	Government Commission for development of measures to restore the ecological equilibrium in Priaral'e declared competition on the development the "Concepts of preservation and restoration of the Aral Sea, and normalization of the ecological, sanitary-and-hygienic, medical and biological, and socio-economic situation in Priaral'e."
	Under the initiative of the Kazakh Association for the United Nations "The International round table" "How to save Aral" convenes in Alma-Ata (Kazakhstan).
	Z.M. Akramov and A.A. Rafikov publish the book, "Past, present, and future of the Aral Sea," in Tashkent (Uzbekistan).
	Under the invitation of the executive director of UNEP M. Tolba, a delegation from the Committee on questions of ecology and rational use of natural resources of the Supreme Soviet of the USSR visit Nairobi (Kenya). In the signed Report, a number of arrangements are fixed, in particular, the management of UNEP agrees to speed up the realization of the Aral project and to increase its status. Within the framework of UNEP, the Center on the Aral Sea is to be founded.
	In Nairobi, Kenya, the consultative meeting on nature conservation and diplomatic departments of the UNEP member-countries convenes and signs the draft "Assistance in Preparation of the Plan Actions on Preservation of the Aral Sea" – on the Soviet side, the USSR's Deputy Foreign Minister, V.F. Petrovsky, and on the UNEP side by, the Executive Director, M. Tolba.
	The book by N.F.Glazovsky "Aral Crisis" is published.
	In Bloomington, USA, the International Conference, "The Aral Crisis: Environmental Issues in Central Asia," including independent public organizations and experts is held at Indiana University.
	The Soviet-Japanese expedition of the newspapers "Izvestia" and "Iomiuri" to the Aral basin. The route of the expedition: Alamata, Kyzyl-Orda, Aralsk, Nukus, Muinak; and the Kazakh and Karakalpak settlements: Akespe, Bugun, Uchsai, and Khodjeily.
	The book by V.A. Popov, "Aral Problems and Landscapes in the Amudarya Delta" is published in Tashkent.
	The collection "Aral Today and Tomorrow" edited by S. Nikanov is published in Almata.
	The book by D.B. Oreshkin, "Aral Disaster," is published in Moscow by "Znanie" Publishers.
	The Institute of Geography of USSR AS holds in Nukus the International Scientific Conference on the Aral Problem.

Years	General Events
	VINITI publishes the special collection, "Environmental Disaster in the Aral Basin," in the series, "Problems of the Environment and Natural Resources."
	Al Gore (US Vice-President, 1992–2000) visits the Aral Sea and described his trip in the book "Earth in the Balance," 1992.
	The American journal, "National Geographic," publishes an illustrated article by W.S. Ellis and D.C. Turnley, "The Aral: A Soviet Sea Lies Dying."
	Uzbek AS and the Council on the Study of Production Forces issues the book, "Socioeconomic Issues of the Aral and Circum-Aral Area."
	The book "Perestroika: Glasnost, Democracy, and Socialism. Environmental Alternative" provides a detailed analysis of the critical environmental situation existing in the USSR at the time. Among other articles is the work by N.G. Minashina, "Environmental Consequences of Land Reclamation and Water Management development in the Aral Sea Basin."
	In Leningrad, the collection, "Hydrometeorological Problems of the Circum-Aral area" is published.
	The series, "Hydrometeorology and Hydrochemistry of the USSR Seas," publishes the Project, "Seas of the USSR," vol. 7 "Aral Sea," prepared by the USSR State Committee on Hydrometeorology and GOIN.
	The judges of the competition of the governmental commission for elaboration of actions on restoration of the environmental equilibrium in the Circum-Aral Area and control of their implementation form a working group the formulate a concept on addressing the Aral crisis. The group is headed by Corresponding Member of USSR AS, V.N. Kotlyakov (Chairman), Doctor of Geography N.F. Glazovsky, and Doctor of Philosophy N.K. Mukitanov (Deputy Chairman).
	The Aral Sea level is +38.2 m, surface – 32 000 km^2, volume – 350 km^3, mineralization exceeds 28–30 g/l.
1991	N. Aladin and N. Glazovskiy edit the book, "Modern state of separating bays of the Aral Sea," which is published in the Proceedings of Zoological Institute of the USSR Academy of Sciences.
	Decision of Council of Ministers of the USSR "About a course of action of the Council of Ministers of the USSR Decision "About urgent measures of ecological improvement of the country on the Aral Sea problems." In the document, the Aral problem is named "the largest ecological catastrophe on our planet."
	Protocol on preparation of intergovernmental agreement on the problems of Aral is signed at the first meeting of leaders of the Commonwealth of Independent States (CIS) in Minsk (Belorussia).
	The second meeting of the international working group of experts of the project of the USSR/UNEP convenes in Moscow. The purpose of the meeting is to discuss substantive provisions and "Concepts of preservation and restoration of the Aral Sea and normalization of an ecological, sanitary-and-hygienic, medical and biological, and social and economic situation in Priaralye."
	The third session of the international working group of experts under the project of the USSR/UNEP "Assistance in the preparation of the Plan of action on preservation of the Aral Sea."
	At the Meeting of ministers of land improvement and water management of the Central Asian republics in Tashkent, the "Application on sharing water resources of the Aral Sea basin" is accepted.

Years	General Events
	Company "Sovintervod" develops "Basic Regulations of the Scheme of complex use and protection of water and ground resources of the Aral Sea basin until 2010."
	Ministry of Nature of the USSR, the USSR Academy of Sciences, Ministry of Health of the USSR, and Ministry of Agriculture and Food Production of the USSR develops the "Concept of preservation and stage-by-stage restoration of the Aral Sea in its coordination with conditions of social and economic development of the republics of Central Asia and Kazakhstan."
	The USSR Academy of Sciences and the Goskompriroda (State Committee for Nature Preservation) of the USSR with participation of representatives of the Republics of Central Asia and Kazakhstan develop "Substantive Regulations of the concept of preservation and restoration of the Aral Sea, normalization of ecological, sanitary-and-hygienic, medical and biological, and social and economic situation in Priaral'e."
	The Institute of Geography of the Kazakhstan Academy of Sciences develops a "Conceptual bases for the interstate program of liquidation of consequences of the Aral Sea crisis."
	The Central Asian Research Institute for Irrigation (SANIIRI, Tashkent, Uzbekistan) develops he "Concept of the solution of the problem of the Aral region in view of social and economic development of the republics of Central Asia."
	The construction of new, large irrigating systems and the launch of new areas of irrigated grounds in the Aral Sea region is suspended.
	N.N. Miklukho-Maklay Institute of Ethnology and Antropology and Scientific Research Coordination Center "Aral" publishes the collection, "The Aral crisis (historical-geographical retrospective)."
	In the US journal, "Environment," a paper is published by the director of the Institute of Geography of the USSR Academy of Sciences V.M. Kotlyakov, "The Aral Sea basin – a critical ecological zone."
	The president of Worldwatch Institute, Prof. Lester Brown, published in the US a paper, "Aral Sea: Going, Going…".
	In the US, the monograph, "The Water Management Crisis in Soviet Central Asia," by Ph. Micklin, is published.
	The conference, "Crisis of the Aral Sea," is held in New York, USA.
	In the US magazine, "World and I," M. Glantz and I. Zonn publish an article about the situation in the Aral Sea Region - "A Quiet Chernobyl."
	V.P. Zuev (USSR) publishes the book, "Aral deadlock."
	In the series of monographs, "History of the lakes of the USSR," is published the book, "The History of lakes Sevan, Issyk-Kul, Balkhash, Zaisan, and Aral."
	Independence of the republics of Uzbekistan, Turkmenistan, and Kazakhstan is proclaimed.
	In the US, the monograph by Philip P. Micklin, "The Water Management Crisis in Soviet Central Asia," is published.
	An international study of salt drift, "The Aral Sea – Owen Lake, USA" is launched.
	"Gidrometeoizdat" Publishers release the collection, "Monitoring of the Natural Environment in the Aral Sea Basin."
	In Nukus, the book by B. Zholdybekov, "Variations in the Soil Cover in the Amudarya Marine Delta due to Aridization," is published.
	The Aral Sea level is +37.7 m.

Years	General Events
1992	Meeting of the heads of water-economic bodies of the new Central-Asian republics, where the agreement between Kazakhstan, Kirgizstan, Uzbekistan, Tajikistan, and Turkmenistan about cooperation "in the sphere of joint management of use and protection of water resources of interstate sources," is signed, convenes in Almaty (Kazakhstan).
	In Tashkent, the heads of the water-economic organizations of five countries of the Central Asia sign "Regulations about the Interstate Coordination Water-Economic Commission" (ICWC).
	Extraordinary 8th session of the Supreme Soviet of Republic Karakalpakstan of the twelfth convocation accepts the decision "About the recognition of the territory of the Republic Karakalpakstan as a zone of ecological catastrophe."
	The University of the United Nations and the Global Infrastructure Fund Research Foundation (Japan) organizes in Tokyo the International Symposium "Management of the environment in the Aral Sea region."
	G. Reznichenko publishes the book, "The Aral Catastrophe" (a diary of the expedition "Aral-88").
	UNEP publishes a "Diagnostic Study for the Development of an Action Plan for the Aral Sea Basin."
	UNESCO, together with the Federal Ministry of Education and Research of Germany, begins the first phase of the project on ecological studies and monitoring of the deltas of Aral Sea as a bases for restoration (the project proceeded until 1996).
	International scientific-practical conference convenes in Nukus (Uzbekistan) on development of the basic directions for the solution of the problems connected with the ecological catastrophe on the Aral Sea, at which the presidents of the Academies of Sciences of the Central Asian states and over 100 prominent scientists and experts from Russia, USA, Japan, Germany, Netherlands and other countries gather. The resolution and the reference of the conference to the United Nations, presidents, governments, and parliaments of the Central Asian republics declare the Priaral'e zone an ecological catastrophe and create an international organization to coordinate actions of the world community for the Aral rescue.
	R.M. Razakov publishes the book, "The Aral Sea and Priaral'e: Problems and Solutions," in Tashkent.
	In the USSR journal, "Water resources", No. 2, a series of articles on the problems of the Aral Sea basin by G.V. Voropaev, D.Yu. Ratkovich, A.I. Budagovskiy, and L.V. Ivanova are published.
	In the USA, at the University Villanova in Pennsylvania, the seminar, "Central Asia: its strategic value and perspectives," convenes. At the seminar, the report of M. Glantz, A. Rubinstein (USA) and I. Zonn (Russia), "Tragedy in the Aral Sea Basin. Looking back to plan ahead," is submitted.
	Stockholm Environmental Institute-Boston (USA) publishes the results of modelling of the existing water balance and an estimation of strategies for water resource management in the region of the Aral Sea.
	At the bottom of the former Berg Strait, the first dam between the Large and the Small Aral is built.
	Mission of the World Bank in five republics of Central Asia.

Years	General Events
	Creation of the scientific body of the ICWC on the water-economic problems of the Aral Sea.
	The "Map of Anthropogenic Land Degradation in the Aral Sea Basin" at a scale of 1:2500000 is published. It is prepared by the Institute of Deserts of the Turkmen Academy of Sciences.
	The book by M. Feshbach and A. Friendly Jr. (USA), "Ecocide in the USSR: Health and Nature under Siege," is published in Moscow in the Russian language. It contains a chapter about the Aral named "The Sea of Alarms."
	The Aral Sea level is +37.3 m.
1992–1993	"The Program of the Aral Sea Basin" is presented to the World Bank.
1993	Protocol signed between the ICID (President, John Hennesey) and the authority of the Uzbek Republic (Minister of Land Reclamation and Water Management, R.A. Giniyatulin) on ICID support and assistance in addressing the Aral Sea problems in cooperation with SANIIRI.
	Preparation of the report on the Project "Colorado–Amudarya" by M. Glantz (USA) and I.S. Zonn (Russia).
	In Almata, the collection, "The Aral Sea Problems: The Condition of a Water Area and Dried Bottom of the Aral Sea," is published.
	David R. Smith from the National Center of Atmospheric Research (NCAR) prepares the work, "Climate Change, Climate Variability and Inland Deltas: Case Studies from the Aral Sea Basin."
	O. Skarlato and N. Aladin edit the book, "Ecological crisis in the Aral Sea," which is published in the Proceedings of the Zoological Institute of the Russian Academy of Sciences.
	In Kyzyl-Orda (Kazakhstan), five Presidents of the Central Asian republic sign the "Agreement on joint actions under the decision of the problem of the Aral Sea and Priaral'e for ecological improvement and maintenance of social and economic development of the Aral region."
	The heads of the states of Central Asia initiate creation of the International Fund for Saving the Aral Sea. N.A. Nazarbaev, President of the Republic of Kazakhstan, is elected president of the Fund.
	Statement of the president of Uzbekistan I. Karimov on the 48th session of the General Assembly of the United Nations in New York, USA. In particular, he says: "Taking into account the really global scales of the catastrophe, Uzbekistan would welcome the creation of the special Commission of the United Nations on the Aral, which, as agreed with the governments of region and through opportunities of the United Nations, would involve international forces and means in the decisions on this ecological tragedy. Among the first steps this Commission could take on this problem is an International conference under aegis of the United Nations in Nukus."
	By the decision of the ICWC, Scientific-Information Centre (SIC) of ICWC is created. The main office is in Tashkent.
	By the decision of the ICWC, the permanent body Secretary of the ICWC is formed with an office in Khodzhent (Tajikistan).
	Interstate Council for the problems of the Aral Sea Basin is organized. The ICWC is included into the structure of the Council.
	R. Lettolle and M. Mainguet (France) publish the book, "Aral" (Springer Verlag, Paris).
	During the 43rd session of the International Executive Committee (IEC) and the 15th Congress of the International Commission on irrigation and drainage (ICID), held in Hague (Netherlands), the Protocol between the

Years	General Events
	Republic of Uzbekistan and the ICID about the support and help of the ICID in the solution of the Aral Sea problems is signed.
	In Almaty, the government of Kazakhstan organizes the seminar, "Cooperation between the countries of the Aral Sea Basin – a necessary condition of efficient control and steady development of water resources in the basin."
	Joint mission of the World Bank, UNEP, and UNDP in the Central Asian republics for assistance in the establishment of priority projects of five states of the Aral Sea basin on studying urgent problems of the Aral Sea.
	The World Bank organizes the International Seminar, "Crisis on the Aral Sea," in Washington, USA.
	The dam of the Takhiatash hydrosystem, on the right bank of the river, and the museum of the Amudarya basin open.
	The second seminar of the United Nations University and Global Infrastructure Fund Research Foundation (Japan), "Management of the environment of Aral sea region," convenes in Tokyo, Japan.
	V.P. Sergiev, S.A. Beer, L.I. Elpiner, and V.G. Vinogradov (Russia) publish the book, "Medical-ecological problems of the Aral Sea."
	The Aral Sea level is +36.9 m.
1994	Decision of the President of Turkmenistan "About an increase in the capacity of water basins of Turkmenistan."
	In Nukus, the heads of five states of Central Asia, with participation of the government of the Russian Federation, sign "The Program of concrete actions in the Aral Sea basin" consisting of 8 basic directions.
	Mission of the World Bank to the countries of Central Asia with the purpose of defining and preparing special projects under the program of the Aral Sea phase.
	The meeting of countries-donors takes place in Paris under the Program of the Aral Sea Basin, organized by the World Bank, UNDP and UNEP. Eight projects of the Program of the Aral Sea basin are conceived and approved in the first phase in the sum of 40 million dollars at the meeting of donors in Paris.
	The Interstate Commission on Social and Economic Development, Scientific, Technical, and Ecological Cooperation of Interstate Council on Problems of the Aral Sea basin is organized.
	In Varna, Bulgaria, the International Commission on Irrigation and Drainage (ICID) organizes a Special session devoted to the problem of the Aral Basin. Republic of Uzbekistan becomes a member of the ICID.
	In the structure of the ICID, a special group on the Aral Basin is created. Its tasks are determined and a Plan of action is accepted.
	V.I. Kuks publishes the book, "The Southern Seas (Aral, Caspian, Azov, Black) under conditions of anthropogenous stress" in "Gidrometeoizdat" (Russia).
	In Brussels, Belgium, the department of E-3 of the European Union (EU) organizes a meeting on the realization of the project, "Water resources management in Central Asia."
	T. Sayko and I. Zonn (Russia) publish an article about the Aral Sea, "Deserting a dying sea," in "Geographical" (UK).
	The Uzbek Republic becomes a member of ICID.
	The Resolution of the leaders of the Central Asian republics, "Concept on Improvement of Socioeconomic and Environmental Conditions in the Circum-Aral Area," is approved.

Years	General Events
	The Aral Sea level is +36.6 m.
1995	A. Alimov and N. Aladin edit the monograph in 2 books, "Biological and environmental problems of the Aral Sea and Aral region," published in the Proceedings of Zoological Institute of the Russian Academy of Sciences.
	In Nukus, under the aegis of the United Nations, the International conference on steady development of the states of the Central Asia convenes. The culmination of the conference is a meeting of the heads of the states of Central Asia and the signing of the Nukus declaration.
	Consortium "Aral" is created.
	In Tashkent, a seminar on the project of Aral Sea, organized by UNESCO and the Federal Ministry of Research and Technology of Germany, convenes.
	A meeting of the presidents of the 5 states of Central Asia convenes in Dashoguz (Tashauz) on the problem of Aral.
	In Tokyo, the International symposium, "Management of ground and water resources in the Aral Sea basin," organized by the Japanese Society of Irrigation, Drainage, and Reclamation Engineering (JSIDRE), takes place.
	In Tokyo, at the University of the United Nations, the International conference, "Central-Euroasian forum on water resources: Caspian, Aral and Dead seas - water crisis and perspectives," takes place (Caspian, Aral and Dead Seas: Central Eurasian Water Crisis).
	In Otzu, Japan, at the Research Institute of Lake Biwa, the international meeting, "Forum on the Caspian, Aral and Dead seas: perspectives of water resources management and policy" convenes.
	In Japan (in Japanese), the book, "Light and Shadow of Global Water Environments and International Disputes: Caspian, Aral, and Dead sea and the 21st Century of Eurasia and Central Asia" is published.
	UNDP publishes the brochure, "Crisis of the Aral," in Tashkent.
	In Vagening, Netherlands, the seminar on "Interaction of irrigation, a drainage, and environment in the Aral Sea Basin" takes place.
	Kluwer Academic Publishers releases the international magazine, "Geojournal" (vol. 35, no. 1), devoted to the Aral Sea Basin and ecological anthropogenous catastrophe.
	The "Glavgidromet" of the Uzbek Republic publishes the book, "Climate Variability in Central Asia."
	SOPS of the USSR State Committee for Planning publishes the scientific report, "Forecast of the Socioeconomic Development of the Circum-Aral Area until the 2020s."
	The Aral Sea level is +36.1 m.
1996	Government of Kazakhstan signs the preliminary agreement with the Japanese oil company, "Sekiu Codan," about the development of large oil fields at the Kazakhstan site of the Aral (near Aralsk).
	The third international symposium on management of ground and water resources for steady development in the Aral Sea Basin is held in Tokyo, Japan.
	By an expert estimation of the Republic of Uzbekistan's Glavgidromet (Meteorological Service), the water level of the Aral Sea reaches 36 m.
	In Bishkek, presidents of Republics of Kazakhstan, Kyrgyzstan, and Uzbekistan discuss problems of the use of water-power resources.
	In Tashkent, the international conference, "The Scientific substantiation and practical use of managing information systems by water and ground resources," is held.

Years	General Events
	In Ljubljana, Slovenia, a special technical session of International Commission on Irrigation and Drainage (ICID) on the Aral Sea problem convenes.
	In NATO ASI Series "Ecology" (V.12), the proceedings of the seminar of NATO, "The Aral Sea Basin," edited by Ph. Micklin and W.D. Williams is published.
	The University of the United Nations (Tokyo) publishes the book, "Regions of risk: Comparisons of threatened environments," edited by J.X. Kasperson, R.E. Kasperson, and B.L. Turner, in which the paper by N.F. Glazovskiy "Aral Sea" is included.
	On the bottom of the former Berg Strait, the first dam between the Large and Small Aral is rebuilt.
	N.V. Aladin and W.D. Willams publish the book, "Aral Sea," in Adelaide, Australia.
	The work of E.I. Pankova, I.P. Aidarov, I.A. Yamnov, A.F. Novikova, and N.S. Blagovolin, "Natural and Anthropogenic Soil Salinization in the Aral Sea Basin (Geography, genesis, Evolution)," is published.
	The Aral Sea level is +35.5 m.
1996–1997	SIC ICWC develops "Substantive provisions of Regional water strategy of the Aral basin" and presents them to the World Bank.
1997	At the meeting of the heads of the states of Central Asia on problems of the Aral Sea Basin, the Almaty Declaration is signed.
	During restructuring of existing interstate organizations, the International Fund for saving the Aral Sea (IFAS) is transformed. The President of the Republic of Uzbekistan, I.A. Karimov, is elected president of the Fund.
	The permanent executive and administrative body, The Executive Committee of the International Fund for saving the Aral Sea (IFAS Executive Committee), is created in Tashkent with branches in all states of Central Asia.
	The international technical meeting, "Programs of the Aral Sea Basin," of countries and donors convenes in Tashkent.
	The second phase of the UNESCO and Federal Ministry of Education and Research of Germany project begins (has finished in 2000).
	In "The World Atlas of Desertification," the paper "Estimation of desertification and mapping of the region of the Aral Sea" by Novikova N.M., Kust G.S., and Ptichnikov A.V. is published.
	M. Glantz and R. Figueroa publish an article, "Does the Aral Sea Merit Heritage Status," in the journal "Global Environmental Change," which is subsequently translated and issued in Russian.
	Governments of the Central Asian republics apply to the UNESCO Secretariat to consider a plan of action on the conditions of the Aral crisis.
	The Interstate Council on the Aral Sea is abolished and its functions are transferred to the Board of the International Fund for saving the Aral Sea.
	President of Uzbekistan I.A. Karimov is elected the President of the International Fund for saving the Aral Sea.
	UNESCO establishes the Scientific Advisory Board on the Aral Sea Basin (SABAS).
	Expert from the Colorado University, USA, Ted Shannon, carries out investigations on "Irrigation Development in the Aral Basin."
	In Mashad, Iran, the ICID Iranian National Committee organizes a workshop, "The Aral and Other Problems of the Aral Sea."

Years	General Events
	The Center for Research in Water Resources at the University of Texas, USA, publishes the work of S. Akmansoy, "Aral Sea Water Rights."
	In St.-Petersburg the book of M.I. Krivoshei, "The Aral and the Caspian: The Causes of Disaster," is published.
	A temporary dam separating the Large and Small Aral is replaced with a dam 20 km long and 26 m wide.
	The Aral Sea level was +35.1 m.
1998	S. Bruk, D. Keyser, J. Kutscher, and V. Moustafaev publish the book, "Ecological Research and Monitoring of the Aral Sea: A Basis for Restoration. Book 1. UNESCO Aral Sea Project 1992–1995 Final Scientific Report."
	With international donor support, development of the regional project, "Water resources management and the environment of the Aral Sea Basin," begins.
	NGO "Aral Tenizi" is established in Aralsk (Kazakhstan).
	"Uzbekneftegas" drills a search hole 1207 m deep on the island of Vozrozhdenia. No traces of oil or gas are found.
	Realization of the UNDP project, "Development of potential of basin of the Aral Sea," begins.
	University of the United Nations in Tokyo, Japan publishes the book, "Central Eurasian Water Crisis: Caspian, Aral, and Dead Seas," edited by I. Kobori and M. Glantz.
	Within the framework of the Global Environmental Facilities, the project, "Water Resources and Environmental Management in the Aral Sea Basin," is approved and launched.
	Tashkent holds the regional scientific workshop, "Eolian Salt Transfer in the Aral Sea Basin: Problems of Salt Migration."
	UNESCO, at its session of executive managers, launches the "Aral Sea Initiative" with a view on outlining future activities in the region.
	The ICID Spanish National Committee organizes a technical trip for experts of the Aral basin and for organizations connected with irrigation.
	In Paris, UNESCO publishes the collection of works, "Ecological Research and Monitoring of the Aral Sea Deltas."
	In Stockholm, Sweden, the International Conference on the Aral Sea, "Women, Children, Health and Environment," is held.
	The journal, "Science and Technology in Russia" (No. 3), publishes the article by Ye.G. Maev, "History of the Aral Sea."
	In Nukus, Karakalpakstan, the book, "The Aral, My Hope," a collection of sketches, poems, and songs, is published.
	The first book, "Environmental Studies and Monitoring of the Aral Sea: Basis for Restoration. Concluding Scientific Report of the UNESCO Project on the Aral Sea 1992–1995," edited by S. Brook, D. Kaizer, J. Katsher, and V. Mustafaev, is published.
	The Aral Sea level is +34.8 m.
1999	The dam at the bottom of the former Berg Strait totally collapses because the level of the Small Aral increases to about +43 m.
	The KaraKum canal is renamed by the decision of the president of Turkmenistan as the Karakum River.
	The meeting of the heads of the states of Central Asia convenes in Ashkhabad, at which S.A. Nijazov, the President of Turkmenistan, is elected President of the IFAS.
	In Almaty, the international conference on problems of the use of trans-boundary water resources of the Aral Sea Basin is held.

Years	General Events
	Cambridge University Press (UK) publishes the book, "Creeping Environmental Problems and Sustainable Development in the Aral Sea," edited by Michael H. Glantz.
	In the journal, "Problems in the Development of Deserts," the special heading, "Aral and its problems," in which original scientific articles and practical recommendations on the problem of the Aral are published, commences.
	The International Children Conference, "The Aral Crisis Viewed by Children," is held in Nukus.
	President of Turkmenistan, S.A. Niyazov, is elected President of the International Fund for saving the Aral Sea."
	The summit meeting of the Central Asian countries in Ashgabad adopts the Declaration on the Problems of the Aral Sea Basin.
	The Workshop, "Water and Peace in Central Asia," organized by UNESCO in cooperation with the International Fund for saving the Aral Sea, is held.
	In Moscow, the scientific reference book by B.S. Zalogin and A.N. Kosarev, "The Seas," in the series, "Nature of the World," is published. It contains a comprehensive geographical description of the Aral Sea.
	The ICID Indian National Committee organizes the Seminar, "Drainage and water resource management, including utilization of saline waters in agriculture and agroforestry on saline soils," intended for experts on the Aral basin.
	In Moscow, the monograph by G.S. Koust, "Desertification: The Principles of Ecological and Genetic Assessment and Mapping," is published, the greater part of which is devoted to the Circum-Aral Area.
	The Aral Sea level is +34.0 m.
2000	The book, "Ecological Research and Monitoring of the Aral Sea Deltas: A basis for restoration. Book 2. UNESCO Aral Sea Project. 1997–1999 Final Scientific Reports," edited by D. Keyser et al., is published.
	By the decision of the ICWC, the coordination metrological centre is established (CMC) in Bishkek (the Kyrgyz Republic).
	Construction of the "Lake of the Golden Age" begins in Turkmenistan in the Karakum desert to gather drainage waters from some regions of Turkmenistan and the Khorezm region of Uzbekistan.
	The book "Water resources and problems of the Aral and environment" is published in Tashkent.
	INCO-Copernicus-2 Project "Aral-Kum" (2000–2003), coordinated by Liege University, Belgium, begins.
	UNESCO publishes the work, "Water Related Vision for the Aral Sea Basin," as the first phase of the "Aral Initiative." The work is presented in The Hague, Netherlands, at the World Water Forum.
	SIC ICWC publishes the work, "Assessment of the Socioeconomic Consequences of the Environmental Disaster: Drying of the Aral Sea" (INTAS Project).
	The International Conference, "Reality and Perspectives of Sustainable Development of the Ecosystems in the Aral Region," is held in Almaty.
	The work of V.E. Chub, "Climate Variations and Its Influence on the Natural Resource Potential of the Republic of Uzbekistan," is published in Tashkent. One of its chapters is devoted to the Aral Sea, "Climate Variations in the Circum-Aral Area and the Modern State of the Aral Sea."

Years	General Events
	The second book, “Environmental Studies and Monitoring of the Aral Sea: Basis for Restoration. Concluding Scientific Report of the UNESCO Project on the Aral Sea 1992–1995,” edited by D. Kaizer et al., is published.
	The Aral Sea level is +33.3 m.
	21th century
2001	Large-scale exploration for oil on the Aral Sea begins. Drilling activity in the northern part of the Aral is carried out by the joint-stock company, “Kazakhkaspyiskshelf,” and on the western coast by the joint venture, “Kulandi-Energy Corporation.” Prospect drilling is to be carried out on Barsakelmes and Vozrozhdenia Islands, and also in the area of the Kokaral and Kaskakulan settlements.
	The Central Asian Regional forum of the University for Peace of the United Nations begins its activities.
	NATO CLG Project “Climate Change in the Aral Region: Past Variability, Present Challenges, and Future Scenarios” (2001-2003), headed by Liege University (Belgium) and the P.P. Shirshov Institute of Oceanology (Russia), begins.
	Springer Publishers (Germany) releases the book, “Sustainable land use in deserts,” devoted to land development in the Circum-Aral Area.
	In Britain, the book by T. Saiko, “Environmental Crisis: Geographical Case Studies in Post-Socialist Eurasia,” is published. One of the chapters is devoted to the Aral Sea: “Desiccation of the Aral Sea: The Hidden Costs of Irrigation.”
	“Minmakroeconomstat” of Uzbekistan approves the Project, “Creation of Local Water Bodies on the Sea Coastline in the Amudarya Delta.”
	The Aral Sea level is +32.1 m.
2002	In the Russian journal, “Land improvement and water management,” a section devoted to the decade of ICWC is issued.
	J. Nihoul (Belgium), A. Kosarev, A. Kostianoy, and I. Zonn (Russia) publish the book, “The Aral Sea: Selected Bibliography.”
	In Tashkent, the International conference, “The Problems of Aral and Priaral’e, an imperative for international cooperation,” is held. The organizers of the conference are the International Fund “Ecosun,” Tashkent club “Ecosun” (branch of the Roman club), representative office of “Roszarubezhtsentr” in the Republic of Uzbekistan, and Zhukorgi Kenges in the Republic of Karakalpakstan.
	E.S. Rakhmonov, the President of the Republic of Tajikistan, by the decision of the Central Asian states, is elected as the President of IFAS in February.
	The Major of Moscow, Yu. M. Luzhkov, contacts President of the Russian Federation, V.V. Putin, with concerns about the question of the use of superfluous and freshet waters from Siberian rivers for irrigation of the lands of Russia and Central Asia.
	On October 6, the meeting of the heads of the Central Asian states convens in Dushanbe on problems of the Aral Sea Basin. The Dushanbe Declaration is signed and 14 basic directions are approved: “Programs of concrete actions on the problems of the Aral Sea Basin for 2003–2010.”
	In Philadelphia, USA, an informal international meeting, “The Climate, water resources, and development of the river basin of the Amudarya,” convenes.

Years	General Events
	INTAS CLIMAN Project expedition takes coring material and environmental data.
	The Russian-Uzbek sea expedition of the P.P. Shirshov Institute of Oceanology (Moscow) and the Institute of Geology and Geophysics (Tashkent) commences in the Aral Sea.
	The Dushanbe Summit of the Central Asian states is held and approves the main directions of the Program of Actions on Improvement of the Ecological and Socioeconomic Situation in the Aral Sea Basin for 2003–2010.
	SIC ICWC publishes a collection of works for the project, "Dialog about Water and Climate: A Case Study of the Aral Sea Basin."
	The International Crisis Group (ICG) publishes the report, "Central Asia: Water and Conflict."
	The Aral Sea level is +30.5 m.
2003	In Kyoto, Japan, in the framework for carrying out the 3rd World Forum on Water Resources, a special session, "Regional cooperation on the divided water resources in Central Asia" takes place.
	In May, in Liege, Belgium, the 35th International Liege Colloquium on Ocean Dynamics and NATO Advanced Research Workshop on the "Dying and Dead Seas" (including the Aral Sea) is held.
	In Almaty, Kazakhstan, the conference, "Ecological stability and advanced approaches to water resources management in the basin of the Aral sea" is held.
	The international conference is held in April in Moscow, "Russia and the Central Asia: problems of water and strategies of cooperation."
	The Russian-Uzbek expedition (Zoological Institute of Russian Academy of Sciences and IFAS) to the Aral Sea continues.
	G.V. Voropaev, G.H. Ismailov, and V.M. Fedorov publish the book, "The problems of water resources management of the Aral-Caspian region," in Moscow.
	In the book, "Anthropogenic Impacts on Water Resources of Russia and Neighboring Countries in the Late 20th Century," prepared by the RAS Institute of Geography, the section of S.I. Shaporenko, "Irrigated Farming in the Aral Sea Basin in Different Historical Conditions," is included.
	The journal, "The Earth and the Universe" (N 2) publishes the paper of A.N. Kosarev, A.G. Kostianoy, and V.N. Mikhailov, "The Aral that we are losing."
	E. Rakhmonov, President of Tajikistan and Chairman of the International Fund for saving the Aral Sea (from 2002 to 2005) addresses the World Water Forum in Kyoto with "Pure Water – Long Life."
	A report, "Southern Circum-Aral Area: New Prospects," edited by V.A. Dukhovny and Yu. De Shutter, on the project of the NATO Scientific Council of the Program, "Science for Peace," is published in Tashkent.
	The International Fund for saving the Aral Sea publishes in Dushanbe the book by U.A. Ashirbekov and I.S. Zonn, "The Aral: The History of the Extinguishing Sea."
	In Moscow, academician of the Russian Agricultural Academy, B.S. Maslov, publishes the work, "There, Behind a Turn," which provides information about the origin, justification, and development of the projects on territorial redistribution of the river flow that were called in the Soviet times, "River Turning." The book also discusses the fate of the Aral.

Years	General Events
	The international conference on the Aral Sea basin is held in Bukhara, Uzbekistan.
	In Canada, the works of Rob Ferguson, "The Devil and the Disappearing Sea: A True Story about the Aral Sea Catastrophe" and "How I Tried to Stop the World's Worst Ecological Catastrophe," are published.
	The NATO Seminar on perspective investigations, "Desertification Problems in Central Asia and Their Regional Strategic Importance," is held in Samarkand.
	The Tashkent InfoCenter, "Habitat," publishes the first ecological almanac, "Under the Aral Sign," with the financial support of the US Environmental Protection Agency.
	In October P.P. Shirshov Institute of Oceanology (Moscow, Russia) in cooperation with Uzbek research institutes carries out a complex sea expedition in the Aral Sea.
	The Aral Sea level is +30.5 m.
2004	Construction of a new dam in the Berg Strait begins under financial support of the World Bank. The Aral Sea draft of the framework of ILEC-LakeNet Project is published on the Internet.
	Establishment of the International Association of Aral Sea Veterans.
	In April and August P.P. Shirshov Institute of Oceanology (Moscow, Russia) in cooperation with Uzbek research institutes carries out complex sea expeditions in the Aral Sea.
	Meeting of the NATO CLG Project, "Physical and Chemical Fluxes in Dying Aral Sea", is held in November in GeoForschungsZentrum, Potsdam, Germany.
	J.C.J. Nihoul, P.O. Zavialov, and Ph.P. Micklin edit the book, "Dying and Dead Seas: Climatic versus Anthropic Causes," published by Kluwer Academic Publishers (Dordrecht) in the NATO ARW/ASI Series.
	In Tashkent, the collection of materials, "Drainage in the Aral Sea Basin and the Strategy of Sustainable Development," prepared in cooperation with SIC ICWC, FAO, IPTRID, INCO-Copernicus and edited by V.A. Dukhovny, is published.
	The journal, "National Geographic Russia" (May), is first published in Moscow in the Russian language. It contains the article, "The sea recedes, but does not surrender," devoted to the Aral.
	The Elsevier Publishers release a special issue of the Journal of Marine Systems (V.47, N1-4), "The Dying Aral Sea," edited by A.G. Kostianoy (Russia) and W. Wiseman (USA).
	The Aral Sea level is +30.1 m.
2005	In Geneva, the World Meteorological Organization (WMO) publishes in English and Russian the booklet of M. Glantz (USA) and I.S. Zonn (Russia), "The Aral Sea: Water Problems, Climate, and Environmental Changes in Central Asia," containing 40 color maps of the Aral (from 1960 to 2004) showing its level fluctuations.
	In Tashkent, the collection, "Irrigation Management for Desertification Control in the Aral Sea Basin: Assessments and Instruments," edited by Professor L.S. Pereira (Portugal), Professor V.A. Dukhovny, and engineer P.G. Horst (Uzbekistan), is published.
	In Germany, Springer Publishers publish the monograph by P.O. Zavialov (IO RAS), "Physical Oceanography of the Dying Aral Sea."

Years	General Events
	In October P.P. Shirshov Institute of Oceanology (Moscow, Russia) in cooperation with Uzbek research institutes carries out a complex sea expedition in the Aral Sea.
	The Aral Sea level is +30.4 m.
2006	Transnational Consortium comprising LUKOIL, Chinese CNPC, Malaysian Petronas, South-Korean KNOC, and "Uzbekneftegaz" sign with the Uzbekistan government the product sharing agreement for the project on prospecting and development of perspective hydrocarbon fields in the Aral Sea sector.
	The Presidents of Kazakhstan, Kyrghyzstan, Tajikistan and Uzbekistan meet in Astana within the framework of the organization "Central Asian Cooperation" to discuss the issue of Aral revival. The participants of this meeting agree that restoration of the desiccated sea is not a regional problem, but "concerns Europe and many other countries."
	In the USA, the book by Fred Pearce, "When the Rivers Run Dry: Water, the Defining Crisis of the Twenty-First Century," is published. One of its chapters is devoted to the Aral Sea – "Aral Sea: The End of World."
	In Stockholm, the Swedish Academy of Sciences convenes a conference devoted to the Aral Sea at which M. Glantz presents the report, "Aral Sea Disaster: What Can We Do? What Can Be Done?"
	The M.V. Lomonosov Moscow State University and the Russian Academy of Natural Sciences publish the fundamental work, "Modern Global Changes in Nature," in 2 volumes. The second volume includes the article by V.N. Mikhailov, V.I. Kravtsova, and F.N. Gurov, "Hydrological and Coastline Variations of the Aral Sea."
	The Journal "Priroda" (N 8) uses as its cover page a photo from space of the desiccating Aral Sea.
	At the International Conference on Reduction of Risks of Natural Disasters in Davos, Switzerland, the Executive Board of the International Fund for saving the Aral Sea together with the Swiss Federal Institute for Snow and Avalanche Research (SLF) and with financial support of the Swiss Office on Cooperation and Development within the framework of the International Conference on Alleviation of the Disaster Impacts organize the invitational Session, "Harmonization of the Integrated Management of Water Resources in Central Asia."
	Within the framework of the ADB technical assistance is held the Training Seminar, "International and National Water Law," with participation of Professor Patricia Waters (Director of the UNESCO Center on Water Law, Policy, and Science, Dandi University, Scotland). During this seminar, the participants discuss existing agreements and drafts of agreements in preparation in terms of the basic provisions of the international water law and possibilities of their application in the Draft Agreement "On Utilization of Water and Power Resources of the Syrdarya River Basin."
	A Consortium of investors, including "Uzbekneftegaz", LUKOIL Overseas, Petronas Carigadi, CNPC International Ltd., and KNOC Aral Ltd. sign a Joint Operating Agreement and Single Operator Agreement to implement the Production Sharing Agreement (PSA) concerning the Uzbek sector of the Aral Sea.
	In Moscow, Nauka Publishers release the book by V.I. Lymarev, "Alexey Ivanovich Butakov, 1816-1869," in the series, "Scientific and Biographical Literature."

Years	General Events
	In March and September P.P. Shirshov Institute of Oceanology (Moscow, Russia) in cooperation with Uzbek research institutes carries out complex sea expeditions in the Aral Sea.
	The Aral Sea level is +29.6 m.
2007	In November P.P. Shirshov Institute of Oceanology (Moscow, Russia) in cooperation with Uzbek research institutes carries out a complex sea expedition in the Aral Sea.
	The Aral Sea level is +29.1 m.
2008	V.A. Dukhovny, P. Navratil, I. Ruziev, G. Stulina, and E. Roschenko edit the book "Complex remote sensing and ground investigations of the Aral Sea dried bottom" published by ICWC in Tashkent.
	An international conference "Problems of the Aral, their impact on the gene pool of population, fauna and flora, and measures of international cooperation on the mitigation of their consequences" was held in Tashkent on 11–12 March.
	In May-June P.P. Shirshov Institute of Oceanology (Moscow, Russia) in cooperation with Uzbek research institutes carries out a complex sea expedition in the Aral Sea.
	I.S. Zonn and M.H. Glantz publish the book "The Aral Sea Encyclopedia" in Russian edition (Mezhdunarodnye Otnosheniya).
	The Aral Sea level is +29.2 m in April 2008.
2009	I.S. Zonn, M.H. Glantz, A.G. Kostianoy, and A.N. Kosarev publish the book "The Aral Sea Encyclopedia" in Springer.
	A.G. Kostianoy and A.N. Kosarev edit the book "The Aral Sea Environment" published in Springer.

References

Akramov Z.M., Rafikov A.A., 1990. The Past, Present and Future of the Aral Sea. Tashkent: Mekhnat, 144 pp. (in Russian).
Aladin N.V., Williams W.D., 1996. The Aral Sea. Adelaide: Gleneagles Publishing, 350 pp.
Aral Crisis (Historical–Geographical Retrospective). Research and Coordination Center "Aral". Moscow, 1991, 309 pp. (in Russian).
Aral Yesterday, Today, Tomorrow, 1995. Nukus, 32 pp. (in Russian).
Berg L.S., 1908. The Aral Sea. An Experience of Physico-Geographical Monograph. Izv. Turkestanskogo Otd. Russk. Geogr. Obshchestva, St.-Petersburg, V.5, N.9, 580 pp. (in Russian).
Blinov L.K., 1956. Hydrochemistry of the Aral Sea. Leningrad: Gidrometeoizdat, 152 pp. (in Russian).
Center for International Projects, Goskompriroda SSSR, 1991. The Modern Condition of Nature, Population, and Economy of the Aral Sea Basin: Diagnostic study. Moscow: Center for International Projects (in English and Russian).
Dukhovny V.A., 1994. Aral Sea Problems: Review and Decisions. Interstate Coordination Water Commission. SANIIRI, 30 pp. (in Russian).
Fish Reserves of the Aral Sea and the Ways of Their Rational Use. Tashkent: Nauka, 1964, 194 pp. (in Russian).
Geller S.Yu. (Ed.), 1969. The Aral Sea Problem. Moscow: Nauka, 1969, 174 pp. (in Russian).
Glantz M.H. (Ed.), 1995. Creeping Environmental Problems and Sustainable Development in the Aral Sea Basin. Cambridge University Press, 291 pp.
Glantz M.H., Rubinstein A.L., Zonn I.S., 1994. Tragedy in the Aral Sea Basin: Looking back to plan ahead? In: Central Asia. Hafeez Malik (Ed.), St. Martin's Press, N.Y., pp. 159–194.
Glazovsky N.F., 1990. The Aral Crisis. Causative Factors and Means of Solution. Moscow: Nauka, 136 pp (in Russian).
Kiessling K.L., 1999. The Aral in crisis. In: Alleviating the Consequences of an Ecological Catastrophe. K.L. Kiessling (Ed.), Stockholm, 166 pp.
Kosarev A.N., 1975. Hydrology of the Caspian and Aral Seas. Moscow: Mosk. Univ., 272 pp. (in Russian).
Kuksa V.I., 1994. The Southern Seas (Aral, Caspian, Azov and Black) Under Conditions of the Anthropogenic Stress. St.-Petersburg: Gidrometeoizdat, 319 pp. (in Russian).
Letolle R., Mainguet M., 1993. Aral. France: Springer-Verlag, 357 pp.
Lymarev V.I., 1959. The Aral Sea. Moscow: Geografizdat, 63 pp. (in Russian).
Micklin Ph.P., Williams W.D. (Eds.), 1996. The Aral Sea Basin. NATO ASI Series. Partnership Sub-Series. Environment. 12. Springer-Verlag, 186 pp. (Proc. NATO Advanced Research Workshop, Tashkent, Uzbekistan, 1994).

Nihoul J.C.J., Kosarev A.N., Kostianoy A.G., Zonn I.S. "The Aral Sea: Selected Bibliography", Noosphere, Moscow (2002), 232 pp.
Nikanova S. (Ed.), 1990. The Aral Today and Tomorrow. Alma-Ata: Kaynar, 277 pp. (in Russian).
Oreshkin D.B., 1990. The Aral Catastrophe. Moscow: Znaniye, 46 pp. (in Russian).
Rafikov A.A., Tetyukhin G.F., 1981. Lowering of the Aral Sea Level and Changes of Natural Conditions of the Amudarya Lower. Tashkent: FAN, 199 pp. (in Russian).
Rogov M.M., 1957. Hydrology of the Amudarya Delta. Leningrad: Gidrometeoizdat, 255 pp. (in Russian).
Sergiev V.P., Beer S.A., El'piner L.I., Vinogradov V.G., 1993. Medical and Ecological Aspects of the Aral Sea Crisis. Moscow: UNEPCOM, 102 pp. (in Russian).
Socio-Economic Problems of the Aral and the Aral Sea Region, 1990. Academy of Sciences of Uzbekistan. Council for the Research of Productive Forces. Tashkent: FAN, 143 pp. (in Russian).
The Aral in Crisis. UNDP. Tashkent, 1995, 16 pp. (in Russian).
The Aral Sea. Diagnostic Study for the Development of the Action Plan for the Conservation of the Aral Sea. Published by UNEP, 1993, 127 pp.
The Aral Sea Crisis. Proposed Framework of Activities. The World Bank, 1993, March, 39 pp.
Tsinzerling V.V., 1927. Irrigation on the Amudarya. Moscow: Upr. Vodnogo Khozyaistva Sredney Azii, 808 pp. (in Russian).
Tum P.P., 1960. The Aral Sea. Navigation-Geographic and Hydrometeorological Essay. Moscow: Mortranskom, 44 pp. (in Russian).
Umarov U., Karimov A.Kh. (Eds.), 2000. Water Resources, Problems of the Aral and Environment. Tashkent: Universitet, 398 pp. (in Russian).
UNESCO, 2000. Water Related Vision for the Aral Sea Basin for the Year 2025. Division of Water Sciences, UNESCO, 238 pp. (in English and in Russian).
Witword P., 1992. Environmental Issues in the Aral Sea Basin. World Bank New York, 150 pp.
World Bank, EC of ICAS, and ICWC, 1996. Developing a regional water management strategy: Issues and work plan. Aral Sea Basin Program technical paper series, Washington, April 1996.
World Bank, 1998. Aral Sea Basin Program (Kazakhstan, Kyrgyz Republic, Tajikistan, Turkmenistan and Uzbekistan): Water and environmental management project. Washington, D.C., May 1998.
Zalogin B.S., Kosarev A.N., 1999. The Aral Sea. In: Morya (The Seas). Moscow: Mysl', pp. 391–397 (in Russian).
Zavialov P.O. Physical Oceanography of the Dying Aral Sea, Praxis Publishing, Chichester, UK, 2005, 146 pp.
Zonn I.S., Glantz M.H. The Aral Sea Encyclopedia, Moscow, Mezhdunarodnye Otnosheniya, 2008, 252 pp. (in Russian).

To find more information about the Aral Sea problems, we can recommend the first bibliography on the Aral Sea research, which was published by J.C.J. Nihoul, A.N. Kosarev, A.G. Kostianoy, and I.S. Zonn "The Aral Sea: Selected Bibliography" (2002).

Fig. 55 The sunrise at the shore of the Sudochie Lake in November 2007. Photo by Pavel Kosenko, http://pavel-kosenko.livejournal.com

List of Abbreviations

abs.	Absolute height over the World Ocean level (in meters)
A.D.	Anno Domini
ADB	Asian Development Bank
AS	Academy of Sciences
A.S.	Aral Sea
ASSR	Autonomous Soviet Socialist Republic
B.C.	Before Christ
BVO	Basin Water Management Associations
C.A.	Central Asia
CC (CPSU)	Central Committee of the Communist Party of the Soviet Union
CIS	Commonwealth of Independent States
CMC	Coordination Metrological Center
CNRS	Centre National de la Recherche Scientifique (France)
CPSU	Communist Party of the Soviet Union
cu.	cubic
DADIS	Department of the Amudarya Delta Irrigation Systems
DAIC	Department of the Amudarya Irrigation Canals
DFG	Deutsche Forschungsgemeinschaft (Germany)
EBRD	European Bank for Reconstruction and Development
ECO	Economic Cooperation Organization
ESCAP	The United Nations Economic and Social Commission for Asia and the Pacific
EU	European Union
FAO	Food and Agriculture Organization of the United Nations
GEF	Global Environment Facility
GFC	Greater Ferghana Canal

GGI	State Hydrological Institute (Russia)
GIS	Geographic Information System
GKNT (SCST)	The USSR State Committee for Science and Techniques
GNP	Gross National Product
GOIN	State Oceanographic Institute (Russia)
ha	hectare
HMS	Hydro-meteo station
HPS	Hydroelectric power station
IBRD	International Bank for Reconstruction and Development
ICAS	Interstate Council for the Aral Sea
ICID	International Commission on Irrigation and Drainage
ICSD	Interstate Commission for Sustainable Development
ICWC	Interstate Coordination Water (management) Commission
IFAS	International Fund for saving the Aral Sea
ILEC	International Lake Environment Committee
INCO-Copernicus	The INCO-COPERNICUS programme promotes scientific and Technological cooperation with the Countries of Central Europe and the New Independent States of the former Soviet Union
INTAS	International Association for the promotion of co-operation with scientists from the New Independent States of the former Soviet Union
IO RAS	P.P. Shirshov Institute of Oceanology, Russian Academy of Sciences
IPTRID	International Programme for Technology and Research in Irrigation and Drainage, FAO of the UN
ISEE	International Society for Ecological Economics
JSC	Joint Stock Company
KDS	Collector-drainage systems
KFAED	The Kuwait Fund for Arab Economic Development
KKA SSR	Karakalpak Autonomous Soviet Socialist Republic
KUOS	Committee for Sudochie Lake Management
m abs. elev.	Absolute height over the World Ocean level (in meters)
MSU	Moscow State University (Russia)
NATO CLG	NATO Cooperative Linkage Grant
NGO	A non-governmental organization

NIC MKVK or SIC ICWC	Scientific-Information Centre of the Interstate Coordination Water Commission
NOVIB	The Netherlands Organisation for International Development Co-operation
OECD	Organisation for Economic Co-operation and Development
OIC	Organisation of the Islamic Conference
OSCE	Organization for Security and Co-operation in Europe
RAS	Russian Academy of Sciences
RF	Russian Federation
RFBR	Russian Foundation for Basic Research
RGS	Russian Geographical Society
RSFSR	Russian Soviet Federal Socialist Republic
SANIIRI	Central Asian Research Institute for Irrigation (Uzbekistan)
SARNIGMI	Central Asian Regional Research Institute for Hydrometeorology (Uzbekistan)
SCO	The Shanghai Cooperation Organization
SCP	State Committee for Planning (USSR)
SCST (GKNT)	The USSR State Committee for Science and Technology
SNK	Council of People's Commissars (USSR, 1923–1946)
SOPS	Council for Investigation of Production Forces
sq.	square
SSR	Soviet Socialist Republic
TACIS	Technical Assistance for the Commonwealth of Independent States
TEMPUS	The Trans-European mobility scheme for university studies
thou	thousand
UGMS	Directorate for Hydrometeorology Service
UN	United Nations
UNDP	The United Nations Development Programme
UNEP	The United Nations Environment Programme
UNESCO	United Nations Educational, Scientific and Cultural Organization
UNIDO	United Nations Industrial Development Organization
UPRADIK	The Authority for Amudarya Inter-Republican Canals
USAID	United States Agency for International Development

USSR AS	The USSR Academy of Sciences
VASKhNIL	All-Union Academy of Agricultural Sciences of the Soviet Union (1929–1992)
VINITI	All-Union Institute for Scientific and Technical Information (USSR)
VKP(b)	All-Union Communist Party of bolsheviks (USSR, 1925–1952)
VNIIMORGEO	All-Union Scientific Research Institute for Marine Geology and Geophysics (Russia)
VNIRO	All-Russia Scientific Research Institute for Fishery and Oceanography
VSNKh	Supreme Council of the Russian (1917–1932) and the Soviet Union (1963–1965) National Economy
WEMP	Water and Environmental Management Project
WMO	World Meteorological Organization

The manufacturer's authorised representative in the EU is Springer Nature Customer Service Centre GmbH, Europaplatz 3, 69115 Heidelberg, Germany. If you have any concerns regarding our products, please contact ProductSafety@springernature.com

Printed and bound by CPI Group (UK) Ltd, Croydon, CR0 4YY

30/09/2026

02225599-0006